MINES AND MINERAL DEPOSITS OF MA: COUNTY, CALIFORNIA

By Oliver E. Bowen, Jr.,[*] and Cliffton H. Gray, Jr.[*]

OUTLINE OF REPORT

Illustrations

[*] Mining Geologist, California State Division of Mines. Manuscript submitted for publication November, 1956.

ABSTRACT

Mariposa County, embracing 1453 square miles of sparsely populated territory, lies almost wholly within the Sierra Nevada. From the southwestern boundary, where the terrain has moderate relief bordering the San Joaquin Valley, the county narrows gradually eastward, becomes increasingly rugged and terminates at the crest of the Cathedral Range—part of the glaciated wonderland of the High Sierra. Famous as a gold-mining county from earliest gold-rush days, it is even more noted for its vacationlands of Yosemite Valley, Tenaya Lake, Wawona and Mariposa Grove of giant sequoias.

Approximately three-fifths of the county, from El Portal westward, is underlain predominantly by complexly folded metasedimentary and metavolcanic rocks of Paleozoic and Upper Jurassic age. East of El Portal it is underlain almost wholly by late Jurassic or early Cretaceous granitic rocks. A thrust fault system formed in Upper Jurassic time passes across the county from Mormon Bar northwest to Coulterville. In general this fault system divides areas of Paleozoic and Upper Jurassic rocks, Jurassic rocks lying chiefly west of the fault system and Paleozoic rocks to the east. The fault system has been extensively mineralized by gold-bearing quartz veins which form the famous Mother Lode. Important belts of vein-filled fissures extend for 10 miles on both sides of the Mother Lode and certain fissure systems paralleling the Mother Lode have been mineralized by base metals.

Although primarily a gold-mining country, gold is not the only mineral resource. Substantial quantities of copper, lead, zinc, silver and tungsten have been found in the county and production of base metals is particularly important in wartime. The principal nonmetallic minerals are barite, limestone, dolomite, mica schist, slate, granite, silica, and sand and gravel. More than 20 mineral commodities have been produced in the county, recorded mineral production having reached a peak in 1939 when $1,759,286 was realized from mineral materials. Current mineral production is at the rate of between $200,000 and $300,000 annually.

Gold production, carried on almost continuously since 1848, has totaled more than $24,000,000 since 1880. Total production, including that prior to 1880, although imperfectly recorded, has been estimated at about $48,000,000. The most productive mines have been the Hite, Princeton, Mount Gaines, Pine Tree, and Josephine, Mariposa, Hasloe, Washington, Bandarita, Mary Harrison, Red Cloud and Clearinghouse.

Silver, almost invariably associated with gold in the quartz veins, has added materially to the mineral wealth of the county. Mesothermal quartz veins carrying silver sulpho-salts as the chief primary minerals are found in the Silver Bar and Silver Lane mines near Bootjack.

Platinum, recovered in small quantities in the county from placer gravels, was once the object of extensive search because it was reported to be present in quartz veins near Bagby. No platinum was ever produced from vein sources and the occurrence has not been verified.

Tungsten is widely distributed in the eastern part of the county where it occurs chiefly in limestone and allied carbonate rocks near their borders with intrusive granitic rock. Tungsten deposits were being developed and exploited in the El Portal district in 1954.

Copper mining, although currently inactive, has been important in several past periods and considerable copper reserves remain in the county. Peaks in copper production were reached in 1913 when 416,031 pounds was produced and 1942-1945 when more than 400,000 pounds was recovered.

Production of lead and zinc, which occur together, has been carried on principally in wartime. The Blue-Moon mine produced 50,000 tons of ore averaging 14 percent zinc in 1944-45. There has also been a small wartime production of chromium and manganese ore.

Limestone, quartz, and barite are the three principal non-metallic minerals utilized thus far. Nearly 2½ million tons of limestone were mined from deposits near Incline by the former Yosemite Portland Cement Company for its plant at Merced; these deposits were not depleted; the LeGrand quarry produced 147,964 tons of quartz between 1942 and 1952 for use in the manufacture of ferrosilicon. The El Portal barite mine was the source of 198,613 tons of barite between 1910 and 1948.

INTRODUCTION

Geographic and Cultural Features. Mariposa County was one of the original 27 counties created by the State Legislature early in 1850. It originally included most of Merced, Madera, Fresno, Kings, Tulare, and Kern Counties as well as parts of Mono, Inyo, and San Bernardino Counties (Laizure, 1928, p. 72). At present Mariposa County has an area of 1463 square miles, is roughly triangular, and lies almost wholly within the Sierra Nevada. The long direction of the triangle, oriented northeast, measures about 65 miles, with the southwestern boundary or base of the triangle roughly 35 miles in length.

Along the western edge of the county, which borders San Joaquin Valley, the terrain consists of rolling hills of moderate relief. Toward the northeast the surface becomes increasingly dissected and rugged and relief becomes extreme in the ice-sculptured canyons of the High Sierra east of Yosemite Valley. However, remnants of ancient, low-rolling surfaces still appear in the central part of the county where they stand far above the bottoms of the actively eroding river canyons. Much of the arable land and upland pasture is found on these old surfaces.

The county is drained chiefly by the Merced River and its forks, although tributaries to the Chowchilla River drain the southwestern part. Where the Merced River Canyon debouches onto the plain of San Joaquin Valley, elevations are only 350 feet, but the canyon floor rises to elevations of 4000 feet in Yosemite Valley, and the main stream has its headwaters on Mt. Lyell in Madera County at elevations above 13,000 feet. Parsons Peak, the highest in the county, rises to 12,120 feet.

The population of Mariposa County, numbering slightly over 5100 by the 1950 census, is concentrated chiefly in Mariposa (the county seat), in the Yosemite Valley resorts and in smaller towns scattered

FIGURE 1.   The Merced River Canyon as seen from Highway 49 two miles north of Bear Valley, observed from northwest. The youthful, rugged terrain is typified by the v-shaped profile of the canyon. Part of one of the nearly flat, early Tertiary surfaces into which the canyons were incised may be seen along the right skyline. Most of the downcutting has taken place since Miocene time. *Photo by Mary H. Rice.*

along the two chief state highways (140 and 49). From 1907 to 1945 the county was served by the Yosemite Valley Railroad which followed the course of the Merced River from Merced Falls to El Portal. Discontinuance of this line left the county without rail facilities. State Highway 140, the all-year, all-weather route to Yosemite, is the only road through the county adequately engineered for sustained heavy trucking.

Prospecting and mineral exploitation is prohibited in Yosemite National Park. Parts of the county east of Yosemite Valley have proven too rugged and inaccessible to attract mining operators and are largely devoid of mineral deposits. Mining is confined largely to the western two-thirds of the county.

Historical Summary of Mining and Mineral Utilization.   The early mining life of the county was identified first with gold and second with copper. Placer gold was discovered sometime prior to 1849, probably by Spanish Californians. Agua Fria and Mariposa Creeks were among the first to be worked. Before the end of the gold rush hardly a gulch within the gold belt had been left unworked by placer miners, and the piles of tailings they left can still be seen in most canyon bottoms. What gold-bearing placer ground was left by these miners was later worked by mechanized dredges. A small amount of platinum was produced as a by-product of gold-dredging operations. Discovery of lode gold in Mariposa County is generally credited to Kit Carson and two

Table 1. Mineral production of Mariposa County 1880-1954.

Year	Total value	Gold, value	Silver, value	Copper, amount and value		Lead, amount and value		Zinc, amount and value	Miscellaneous and unapportioned	
				amount	value	amount	value		Value	Substances
1880	$151,317	$150,017	$1,300							
1881	201,200	200,000	1,200							
1882	254,000	250,000	4,000							
1883	223,000	220,000	3,000							
1884	180,000	180,000								
1885	149,277	149,177	100							
1886	197,600	197,600								
1887	185,261	187,165	96							
1888	175,250	175,000	250							
1889	146,021	145,819	210							
1890	124,287	124,265	22							
1891	84,414	84,414								
1892	81,078	81,011	67							
1893	164,423	164,116	307							
1894	153,747	153,708	39							
1895	216,629	216,622	7							
1896	335,817	335,637	180							
1897	452,087	451,427	660							
1898	337,411	336,418	993							
1899	565,636	562,829	2,207			70,000 lbs.	$3,080		$600	Slate
1900	171,516	157,663	13,853	191,622 lbs.	$30,180					
1901	542,975	504,928	4,787	104,700	11,940					
1902	647,298	631,478	3,880	61,627	6,808					
1903	552,516	542,355	3,353	11,500	1,466					
1904	434,066	429,771	2,829	12,541	1,956					
1905	393,592	386,380	5,231						25	Platinum
1906	369,771	366,394	3,377							
1907	410,058	405,498	4,500							
1908	484,112	439,862	4,732	29,124	2,958	1,142	60		36,560	Miscellaneous stone
1909	470,035	396,465	2,729						62,430	Miscellaneous stone
1910	346,245	317,580	2,364						26,301	Miscellaneous stone, barite and unapportioned 1900-1909
1911	175,752	172,532	1,390	14,641	1,830				3,130	Miscellaneous minerals
1912	214,294	160,541	6,790	284,587	46,957					
1913	246,079	171,034	7,430	416,031	64,485					
1914	187,505	131,458	677	277,472	36,904				18,466	Miscellaneous stone, barite, marble
1915	412,326	385,577	2,175	38,630	6,760				17,814	Miscellaneous minerals
1916	487,971	401,718	2,680	162,280	39,930	1,857	128		43,515	Miscellaneous minerals

Year								Misc. minerals value	Minerals listed
1917	353,237	313,296	3,221	53,381	14,583	1,075	92	21,046	Miscellaneous minerals
1918	352,504	337,682	5,083	30,294	7,483	—	—	2,256	Chromite, lead, misc. stone
1919	269,566	253,392	4,139	24,879	4,627	—	—	408	Miscellaneous minerals
1920	271,031	261,830	4,705	—	—	—	—	4,496	Barite, copper, lead, misc. stone
1921	342,601	331,295	5,251	—	—	—	—	6,055	Barite, pyrite, misc. stone
1922	226,832	218,571	3,301	—	—	—	—	4,960	Barite, pyrite, misc. stone
1923	170,911	141,883	1,735	—	—	—	—	27,293	Barite, pyrite, misc. stone
1924	234,707	182,099	1,608	—	—	—	—	51,000	Miscellaneous minerals
1925	634,862	192,810	1,758	—	—	—	—	440,294	Miscellaneous minerals
1926	319,724	182,313	1,518	—	—	—	—	135,883	Barite, copper, pyrite, misc. stone
1927	499,878	183,805	1,376	—	—	—	—	314,697	Barite, pyrite, slate, granite, misc. stone
1928	282,165	120,568	2,199	—	—	—	—	159,398	Granite, silica, barite, copper, misc. stone
1929	244,017	91,052	651	6,302	1,109	—	—	151,205	Barite, silica, misc. stone
1930	143,465	58,985	318	3,690	472	—	—	83,690	Barite, granite, lead, misc. stone
1931	193,641	88,600	551	—	—	—	—	104,490	Barite, copper, granite, lead, silica, misc. stone
1932	379,254	169,627	636	—	—	—	—	208,991	Barite, copper, granite, lead, misc. stone
1933	575,118	254,663	1,112	—	—	—	—	319,343	Barite, copper, granite, ite, misc. stone
1934	807,908	517,443	3,214	1,771	142	1,438	—	287,109	Barite, granite, lead, misc. stone
1935	875,512	514,544	4,913	2,252	187	—	57	353,541	Barite, granite, misc. stone
1936	1,130,018	863,485	4,756	2,350	216	—	—	261,561	Barite, granite, lead, misc. stone
1937	1,270,774	1,025,010	6,084	11,927	1,443	—	—	238,237	Barite, granite, lead, mica, pumice, misc. stone
1938	1,588,861	1,081,815	5,154	4,328	424	—	—	501,468	Barite, granite, misc. stone
1939	1,755,776	1,296,155	13,181	3,810	396	50,357	2,367	443,677	Barite, granite, misc. stone
1940	1,224,286	949,640	6,615	7,611	861	27,725	1,386	265,784	Miscellaneous minerals

Table 1. Mineral production of Mariposa County 1880-1954.—Continued.

Year	Total value	Gold, value	Silver, value	Copper, amount and value	Lead, amount and value	Zinc, amount and value	Miscellaneous and unapportioned Value	Substances
1941	1,327,594	1,141,070	7,183	5,908 / 697	7,183 / 416	—	178,228	Barite, mica, misc. stone
1942	1,321,238	1,025,220	6,840	26,973 / 3,264	15,782 / 1,057	—	284,857	Barite, silica, manganese, misc. stone
1943	443,693	227,115	1,231	—	—	—	215,347	Barite, copper, lead, manganese, silica, tungsten, misc. stone
1944	1,306,411	192,115	82,210	— / —	255,657 / 20,453	6,688,655 lbs. / $762,507	224,448	Barite, silica, misc. stone
1945	1,171,094	114,450	68,639	179,848 / 24,280	236,315 / 20,323	6,621,320 / 761,452	181,950	Barite, lead, misc. stone
1946	613,477	122,150	898	— / —	— / —	571,137 / 69,679	420,750	Barite, copper, lead, silica, misc. stone
1947	301,938	222,705	1,749	4,000 / 840	—	—	76,644	Barite, silica, misc. stone
1948	280,739	263,375	1,960	—	—	—	15,404	Barite, platinum, slate, misc. stone
1949	325,732	233,835	1,558	200 / 39	—	—	90,300	Barite, silica, misc. stone
1950	392,500	236,075	1,801	—	2,700 / 365	—	154,259	Granite, platinum, silica, misc. stone
1951	—	—	—	—	— / —	—	—	Gold, platinum, slate, silver, lead, misc. stone
1952	323,980	23,310	333	—	4,300 / 692	—	299,645	Silica, soapstone, misc. stone
1953	223,006	24,185	117	—	—	—	198,704	Miscellaneous minerals
1954	104,366	14,420	205	—	800 / 110	600 / 65	89,566	Miscellaneous minerals
Totals	$33,200,032	$23,837,047	$345,224	1,974,279 lbs. / $313,237	676,331 lbs. / $50,586	13,881,712 lbs. / $1,593,703	$7,025,839	

¹ See under "Miscellaneous and unapportioned."

associates who located the Mariposa mine in the spring of 1849. However, presence of early day arrastras suggest that the Spanish Californians may have milled the first lode gold. By July of 1849 a stamp mill had been erected and was processing ore from the Mariposa mine (Logan, 1934, p. 184). A more elaborate stamp mill was built at the Jenny Lind mine near Hornitos in 1851 (Bowen and Crippen, 1948, p. 38). Although more elaborate crushing devices, such as the Huntington mill and various ball mills, were later developed and used extensively in Mariposa County, use of stamp mills has persisted to the present day.

Perhaps the greatest single influence on early-day gold mining in Mariposa County was John C. Fremont's Spanish land grant, Las Mariposas, containing 44,387 acres. Included in Las Mariposas was 12 miles of Mother Lode veins (extending from Merced River to Mariposa) as well as numerous other East Belt and West Belt properties. Gold was discovered on the grant before Fremont had established clear title to it and mines within its confines were involved in court actions and even in bloody battles over a period of ten years after discovery of gold. Largely unsurveyed and, because of grant title, not subject to mining laws of the day, the mining properties had indefinite boundaries and are difficult to locate on maps to this day. The Mariposa Mining and Commercial Company, organized by San Francisco financial giants Hayward, Flood, Mackay, Jones, and others in 1887 (Bradley, 1954, p. 32) took over the management of Las Mariposas and the Company was not wholly liquidated until after World War II. Internal dissension and inefficient management of the grant at some times during its existence probably retarded development of its gold properties and lowered the total gold production of the county through the years (Bradley, 1954, p. 32). Nevertheless, about 15 million dollars was produced from mines of Las Mariposas, nearly one-third of the estimated gold production of the county.

Another large property which figured prominently in Mariposa County gold mining was the 20,000-acre Cook Estate astride the Mother Lode in the Coulterville area. Operations of the Merced Gold Mining Company which took over holdings of the Cook Estate in 1895 lasted until near the close of World War I. The Cook Estate included most of the prominent mines in the Coulterville area.

Copper was discovered in Mariposa County sometime during early gold rush days but little or no mining of copper went on until 1863. By 1868 (Browne, 1868, p. 218) there were three copper smelters in Mariposa County, one each at James Ranch, Bear Valley and Hunter Valley. A drop in the price of copper in 1868 ended the first period of copper-mining activity. Other periods of activity in copper were 1901-1919 and 1929-1946. There have been few shipments of copper ore in recent years.

Mining of other base-metals in Mariposa County has been carried on largely during war years, although small amounts are produced intermittently as by-products of gold mining. Government price supports and the stockpiling program have stimulated some interest in strategic minerals such as tungsten, chromium and manganese. Tungsten is the most actively sought-for metal in the county with the possible exception of uranium. Commercial deposits of uranium have not been found.

Mariposa County has contributed to the construction industry since 1850. Soapstone, quarried at Greeley Hill (Heizer and Fenenga, 1948, p. 103) and used for building facings in Coulterville, and granite blocks quarried at Mormon Bar and used in the Mariposa jail, are among the earliest materials processed in the county. Many of the buildings erected during gold rush days were built of slabs of field stone, available in most parts of the county. Production of roofing slate dates back at least to 1897 (Aubury, 1906, p. 152). There are large reserves of black slate of good quality in the county but throughout most of California's history roofing slate has been unable to compete with lower-cost roofing materials. A material increase in population of the Central Valley towns might revive slate quarrying to supply flagstone and allied materials. There is a small production for such purposes. Colored materials for terrazzo chips and roofing granules are in considerable demand. There is a small production of green serpentine for terrazzo chips from a quarry north of Bagby.

Production of limestone began in Mariposa County in 1927 with the opening of the Yosemite Portland Cement Company plant at Merced (Merced County), and the establishment of quarries north of the Merced River between Jenkins Hill and Incline. These quarries remained active until 1944 when the Yosemite Valley Railroad ceased to function. Limestone reserves were not materially depleted, although more than 2,400,000 tons of limestone was removed.

Barite mines near El Portal produced throughout most of the period from 1910 through 1948. The El Portal barite mine produced 398,613 tons of barite during this period, about 70 percent of the total California production. Substantial reserves remain in the El Portal mine; the mill is still functional and mining may well be resumed.

The existence of large numbers of veins of nearly pure quartz in Mariposa County have attracted manufacturers of refractories and ferrosilicon in recent years. From 1942 to 1952 147,964 tons of silica was produced from the LeGrand (White Rock) silica quarry for use in manufacture of ferrosilicon. There had been a small intermittent production of silica prior to 1942 and a little has been mined since 1952.

Acknowledgments. The authors are particularly indebted to Walter D. McLean and John P. Fulham who supplied much valuable information on mines in the Coulterville and Colorado-Sherlock Creek-Whitlock Creek districts, and who kindly criticized parts of the manuscript. Much valuable information on the Hornitos district, particularly on the Mount Gaines mine, was supplied by Francis H. Frederick.

Other Mariposa County property owners who have cooperated with the authors in many ways are P. R. Bradley, Jr., George Matlock, Clyde Call, Thomas M. Bains III, Alfred Stickney, H. H. Odgers, Earle Williams, John Moss, Charles Owen, Russell Wilson, John Paul Jones, Peter Mulas, Lee Speaker, Tom Perrin, Peter Jericoff, J. W. Radil, Chris Miller, and Harry Barrett.

Kenneth L. Arndke and staff of the Mariposa County Assessors office made available public records dealing with mine locations and ownerships and R. V. Maurer and Robert Wallace of the San Francisco office of the U. S. Bureau of Mines assisted greatly in compilation of the mining histories and production statistics.

Denton W. Carlson, William B. Clark and Philip Lydon of the Sacramento office of the Division of Mines contributed field notes which were indispensable in preparing the manuscript and most of the photographs used in the report are the work of Mary H. Rice and Elisabeth L. Egenhoff of the Division of Mines editorial staff. R. A. Crippen, Jr., assisted materially in compilation of several of the mine maps.

GEOLOGY

General Geologic Features

The western two-thirds of Mariposa County is underlain chiefly by metasedimentary and metavolcanic rocks of Palezoic and Upper Jurassic age. Among the metasedimentary rocks, slate, quartz-biotite hornfels and quartz-biotite-graphite schists predominate, but there are some large masses of limestone, chert, dolomite and quartzite. Metavolcanic rocks are predominantly greenstones and green schists derived from submarine-laid pyroxene andesite and basalt. All of these metamorphic rocks have been intruded in late Mesozoic time by a variety of granitic and peridotitic igneous rocks, among which biotite-hornblende granodiorite is the predominant type.

The eastern third of the county is underlain mainly by intrusive granitic rocks of late Jurassic or early Cretaceous age varying in composition from hornblende gabbro to biotite granite. Hornblende biotite granodiorite, biotite quartz monzonite and biotite granite are most common. The granitic suite of rocks is exceptionally well-exposed in the walls of Yosemite Valley.

The metamorphic rocks in general form broad northwest-trending belts. They are so distributed because they have been thrown into a series of acute, commonly isoclinal, northwest-trending folds by compressional forces acting essentially from the northeast and southwest. The Mother Lode thrust-fault system which trends northwest across the county from the vicinity of Mariposa through Coulterville, nearly everywhere divides Paleozoic and Upper Jurassic rocks. Those east of the fault system, although not studied in detail thus far, are largely of Paleozoic age whereas those to the west are Upper Jurassic. This would indicate that the entire Upper Jurassic stratigraphic section, which once aggregated at least 20,000 feet, has been uplifted and eroded from the block lying on the east side of the fault. The Mother Lode fault system in Mariposa County is not a single vein-filled fissure but rather is a series of essentially parallel to en-echelon breaks. For several miles in the Peñon Blanco district a single, vein-filled fissure stands out from among all other veins in the area. In other parts of the county the Mother Lode is hard to define. Many of the most productive veins in the county are not directly related to the Mother Lode fault system although they formed under similar conditions and at about the same time.

Rock Units

Thus far the Paleozoic rocks have not been subdivided into units. Turner and Ransome (1897, pp. 4, 5) described all of the metasedimentary rocks of probable Paleozoic age in Sonora quadrangle under the name Calaveras formation. More recently these rocks have been referred to as the Calaveras group (Taliaferro, 1943, pp. 281-282). The

FIGURE 2. Glaciated walls of Yosemite Valley in the High Sierra over which Upper and Lower Yosemite Falls tumble. The gently inclined surfaces covered with trees have formed along joint systems in the granite. *Photo by Mary H. Rice.*

Paleozoic rocks in the very fine section exposed along the Merced River east of Bagby probably could be divided into several slate and one or more carbonate, chert and greenstone units. Carbonate rocks were the only units separately mapped under the Calaveras formation by Turner and Ransome.

Taliaferro (1943, pp. 283-285; 1951, p. 120) divides the Upper Jurassic rocks along the Merced River west of Bagby into an upper formation, the Mariposa slate, at least 6000 feet thick, and a lower, predominantly volcanic series below it, the Amador group, totaling nearly 15,000 feet. He subdivides the Amador group into five units, oldest to youngest: the lower volcanics of unknown thickness, the pillow basalts (1200 feet thick), the Hunter Valley chert (lenticular, 950 feet thick, maximum), the Peñon Blanco volcanics (chiefly greenstone, 9000+ feet thick) and the Agua Fria volcanics (3500 feet thick). The Hunter Valley chert and the pillow basalt units thin out south of the Merced River and neither are present in the section exposed along the Mt. Bullion-Merced Falls road.

Calkins (1930, pp. 120-129) has subdivided the granitic rocks of the Yosemite region into 13 units: the Johnson granite porphyry, Cathedral Peak granite, Half Dome quartz monzonite, Sentinel granodiorite, Bridalveil granite, Leaning Tower quartz monzonite, Taft granite, El Capitan granite, granodiorite of the Gateway, biotite granite of Arch Rock, diorite and gabbro, porphyry of Red and Gray Peaks, and the Mount Clark granite. Rocks similar to these units occur elsewhere in the county but are as yet unnamed and uncorrelated. In addition to the rock types just listed there are small masses of gabbro, diorite, quartz diorite, and olivine norite as well as lamprophyres and hornblende porphyries. All of these belong to the Jura-Cretaceous suite of granitic rocks. Also belonging to this group is a rare dike-rock, albitite, composed almost entirely of albite feldspar. In the Flyaway Gulch district albitite is associated with quartz veins and in a few mines is the host rock for gold-bearing ore-shoots.

Another important suite of intrusive igneous rocks, emplaced somewhat earlier than the granitic series but also of Upper Jurassic age, is the peridotite group—most bodies of which are thoroughly serpentinized. Saxonite and dunite were the principal types of peridotite that formed prior to serpentinization, but within the large masses of serpentinized peridotite are small segregations of gabbro, norite, pyroxenite, hornblendite and chromite. Two large, lenticular masses of the serpentine rocks are found adjacent to the Mother Lode from Bagby to Coulterville. The more southerly of the two extends north from Bagby a distance of 6 miles and reaches a maximum width of about 2 miles.

Two groups of contact metamorphic rocks are worthy of special consideration inasmuch as they are potential sources of economic minerals. These are the contact-altered slates and carbonate rocks. Where large masses of granitic rock have intruded slate, well-defined alteration zones commonly develop. These are particularly conspicuous in western Mariposa County. Approaching a granitic contact there is a regular progression from uniform slate as follows:

Stage 1. Spotted slate with small dots of biotite mica, cordierite or quartz, singly or together.

Stage 2. Knotted, glossy slate or phyllite with larger clots of the above minerals and (or) indistinctly formed chiastolite or well-formed andalusite crystals in a groundmass characteristically containing sericite mica and graphite.

Stage 3. Spotted to knotted schist with well-crystallized, micaceous groundmass material plus large clots or large crystals of andalusite, cordierite, quartz, etc.

Stage 4. (Nearest the granite) Dense, horn-like rock, called hornfels made up chiefly of small, irregular-shaped crystals of quartz and biotite with minor sillimanite or andalusite.

Some andalusite-bearing slates and schists in western Mariposa County are possible sources of refractory aluminum silicate.

FIGURE 3. Contorted chert and slate of the Paleozoic Calaveras group in Merced River Canyon between Briceburg and Clearinghouse. The cherts and limestones of the Calaveras group although perhaps the most conspicuous, form only a minor part of the predominantly metasedimentary sequence. Slate derived from shale and schist derived from shale, siltstone, and sandstone are the most common rock types in the Calaveras group. *Photo by Mary H. Rice.*

FIGURE 4. Tombstone rocks of andesite greenstone protruding through soil cover on a highland meadow in the Buckhorn Peak district 7 miles southeast of Coulterville. The greenstone sequence in this vicinity has been long considered to be part of the Paleozoic Calaveras group but may possibly be similar in age to the greenstones of the Amador group. *Photo by Mary H. Rice.*

FIGURE 5. A typical outcrop of weathered, laminated slate of the Upper Jurassic Mariposa formation as seen in a roadcut along Highway 49 in Hell Hollow 1 mile south and slightly west of Bagby. The clay shale from which it is formed has been recrystallized by heat and pressure into a glossy, black, cleavable rock made up chiefly of micas, graphite, and quartz. *Photo by Mary H. Rice.*

Silicated zones have formed along contacts where granitic rocks have invaded carbonate rocks; they commonly contain suites of calcium and magnesium-bearing minerals such as garnet, diopside, epidote, tremolite and wollastonite. It is in such mineral assemblages, called tactite rock or skarn rock, that tungsten minerals most frequently occur. Most of the promising tungsten deposits in the county are in rocks of this sort.

Tertiary rocks are found in Mariposa County in isolated patches only. Erosion remnants of the middle Eocene Ione formation are found along the western border of the county in the vicinity of Merced Falls where they do not exceed 100 feet in thickness. Coarse, red sands and gravels, coarse white- to cream-colored andalusite-anauxite sands and a little fine, white quartz sand are the conspicuous rock types in the Ione formation in Mariposa County. Three isolated patches of Eocene river-channel gravels have been found in the county, two astride the

FIGURE 6.   A typical pasture scene in western Mariposa County along the Hornitos-Merced Falls road. In this place the tombstone-like rocks are tuffaceous slate of Upper Jurassic age, probably part of the Mariposa formation. *Photo by Mary H. Rice.*

FIGURE 7.   Bouldery outcrops of massive pyroxene andesite greenstone as seen along the Bear Valley-Hornitos road 3 miles west of Bear Valley. Massive outcrops of this kind are generally agglomerate or, less commonly, intrusive masses of coarse-crystalline rock. In this instance they are parts of the Peñon Blanco member of the Upper Jurassic Amador group. *Photo by Mary H. Rice.*

north county line. One is a mile northwest of Buck Meadows Lodge on Highway 120, another is west of Lake McClure two miles northwest of Fortynine Gap, and the third is northwest of Bower Cave in the Jordan Creek vicinity. A small patch of Pliocene andesite tuff overlies the Eocene channel gravel near Buck Meadows Lodge.

The most important Quaternary deposits are the glacial moraines scattered through the eastern part of the county and the lakebeds on the floor of Yosemite Valley. Isolated patches of terrace gravels are found along the Merced River and its tributaries.

Geologic History

The earliest record upon which to base the geologic history of this part of the Sierra Nevada lies in the Paleozoic metamorphic rocks. These are marine sediments and possibly some submarine volcanic rocks, indicating that marine waters covered the land for a long period of time. A thick sequence of shale, limestone, chert (in some places manganiferous), sandstone and andesitic volcanic rocks accumulated which later was transformed by heat and pressure into the slate, crystalline limestone, metachert, quartzite and greenstone seen today. Inasmuch as the degree of metamorphism is greater among the

FIGURE 8. Closeup of one of the coarse-grained, intrusive facies of the Peñon Blanco pyroxene andesite greenstone. The speckled appearance is caused by black crystals of altered pyroxene. Pick handle is 12 inches long. *Photo by Mary H. Rice.*

FIGURE 9. When the greenstones similar to those shown in figures 7 and 8 have been strongly sheared during metamorphism they crop out in the tombstone-like forms pictured above. Hornitos district. *Photo by Mary H. Rice.*

Paleozoic rocks than among the younger Jurassic rocks, two metamorphic episodes may have taken place, and the crustal motion involved in the first episode may have raised the Sierran area above sea level.

Triassic rocks are unknown in the Mariposa County portion of the Sierra Nevada. The base of the Amador group rests unconformably on Paleozoic rocks, showing that a period of erosion preceded deposition of the Jurassic rocks. Consequently, much of the area of the Sierra Nevada must have been above sea level and actively eroding during Triassic and Lower Jurassic time. By Middle Jurassic time much of the Sierran area had again fallen below sea level, and 15,000 feet of predominantly submarine volcanic material was piled onto the old erosion surface. The manganiferous cherts originated as sea floor chemical precipitates from a volcanic source and became interbedded with the volcanic sequence. Part of the Mariposa formation, which is essentially slate, was laid down during the waning period of volcanism, and locally the lower part of the formation contains interbeds of volcanic material. Most of the 6000-foot thickness of the Mariposa formation is, however, a monotonous sequence of graphitic slate derived from organic clay shale, and its homogeneity is broken only locally by lenticular sandstone and conglomeratic layers.

Following deposition of the Mariposa formation, the series of crustal disturbances began which resulted in rise of the ancestral Sierra Nevada. Prior to warping of the rocks above sea level, the marine sedimentary and volcanic deposits were invaded by dikes, sills, and irregularly shaped masses of peridotite. These were either serpentinized as they were emplaced or were altered soon after. Chromite deposits were formed as segregations in the original peridotite. The ancestral Sierra Nevada, less lofty than the current mountain chain, nevertheless covered a much broader area, extending at least half way across what is now our Central Valley. During folding, large masses of granitic rock were emplaced in the core of the range. The crustal deformations involved in the folding resulted in deep-seated fracturing and faulting of the metamorphic roof rocks as well as the outer granitic core. Thus part of the vapors and solutions given off at the time of invasion of the granitic rocks could rise upward and be deposited in the fractures. Such is the origin of the gold-, silver-, copper-, lead- and zinc-bearing veins in Mariposa County.

A long period of erosion followed the rise of the ancestral Sierra Nevada. During this time the mountains were greatly reduced in elevation and in area, the metamorphic roof-rocks being deeply eroded, and the veins and ultimately the granitic core of the range laid bare. Gold and the other heavy minerals of the veins and roof rocks became concentrated in stream deposits or in deltaic deposits at the edge of the western sea. Erosion of the rocks was aided greatly by the chemical decomposition characteristic of tropical climates, tropical conditions

FIGURE 10.   Platy, laminated metasediments of the Hunter Valley chert member of the Upper Jurassic Amador group in the bottom of Temperance Creek just west of the Hunter Valley road. *Photo by Mary H. Rice.*

FIGURE 11. Serpentine terrain along the Bagby Grade of Highway 49, about 3 miles northwest of Bagby. The strongly fractured rock, on which there is virtually no soil cover, supports a sparse growth of digger pines and scattered toyon or California holly. Fires have reduced the already sparse growth to a degree suggestive of an arid region. Rainfall here, however, averages well over 20 inches a year. *Photo by Mary H. Rice.*

FIGURE 12. A typical slickensided serpentine outcrop on the Bagby Grade of Highway 49. Almost every piece of serpentine observed has a polished surface. *Photo by Mary H. Rice.*

FIGURE 13.   Low, flat topped buttes of sandstone resting on slate bedrock, Hornitos-Merced Falls road near the county line. The red and buff sandstones, which belong to the middle Eocene Ione formation, lie almost horizontally upon the upturned edges of the much older Jurassic slates. *Photo by Mary H. Rice.*

having slowly developed at the end of the Cretaceous period as elevations were reduced. Middle Eocene deposits are, therefore, of the sort produced by chemical weathering—quartz-andalusite-anauxite sands, kaolinite-type clays and the like. These are very sparsely represented in Mariposa County.

Still more sparsely represented are remnants of the Mio-Pliocene period of volcanism, which literally overwhelmed the Sierran province with volcanic debris. In Miocene time the range was again raised by warping and faulting, and numerous volcanoes broke out along the crest. These erupted rhyolitic ash first and then enormous volumes of andesitic fragmental material and flow-rock. Old stream systems were obliterated under piles of volcanic debris and new systems had to evolve. Eocene gold-rich stream deposits were buried under the volcanic mantle and were thus protected from later erosion—to survive to the present day. Volcanism continued intermittently into the Pleistocene with extrusion of flows of basaltic aspect and latitic composition.

Late in the Pliocene great faults developed east of the former crest of the range and highlands bordering the present Sierra began to drop down, probably accompanied by some re-elevation of the range-crest by tilting and by rise of fault blocks (Hudson, F. S., 1955, pp. 835-869). This faulting continued through the Quaternary along the east side of the Sierra Nevada and is going on today.

The tropical climate of Eocene time had gradually cooled to a temperate one by Mio-Pliocene time. This cooler trend climaxed in the Pleistocene with development of the alpine type of valley glaciers. These sculptured the High Sierra into the form we see it today. Throughout the Pliocene and Pleistocene epochs, great river systems carved the deep, V-shaped canyons so characteristic of the western three-fourths of the Mariposa County landscape. Although erosion has largely eaten away such early gold-placer deposits as once may have existed in the county, active erosion and temporary stream deposition are responsible for accumulation of the placer deposits so important to the early life of the county. Exhumed remnants of early Tertiary land surfaces yield most of the arable land and gently rolling pasture land in the county.

MINES AND MINERAL DEPOSITS

Statistical Summary

The recorded value of minerals produced in Mariposa County from 1880 through 1954 has amounted to $33,516,967. Gold production has amounted to approximately 71 percent of this total—$23,837,047, exclusive of 1951. Inasmuch as the estimated total gold production to date has been about $48,000,000 (Bradley, 1954, p. 32), it follows that approximately $24,000,000 in gold was produced prior to 1880. The estimated total mineral production, therefore, is more than $57,548,000, gold accounting for 83 percent of the estimated total.

The table on pages 40-42 lists the mineral production of the county by year and commodity from 1880 through 1954. Items listed under unapportioned include those wherein producers numbered less than three. In addition to the commodities listed in the table, 398,613 tons of barite valued at $2,760,493 was produced in the county between 1910 and 1948 and 2,430,842.5 captive tons of limestone was quarried in the period 1927 to 1945. The value of the limestone and barite is included in the unapportioned list.

FIGURE 14. A typical exposure of Mother Lode quartz as seen along the west side of Highway 49 just south of Coulterville. The vein, which strikes away from the observer and dips steeply toward the right (northeast), is part of a multiple vein system which locally reaches widths of several hundred feet. The edge of the quartz sheet shown here is 15 to 20 feet thick. *Photo by Mary H. Rice.*

Table 2A. Summary of economic deposits of metallic minerals, Mariposa County.

Ore or source material	Origin and mode of occurrence	Time of formation	Associated rocks	Associated minerals	Utilization
Chromium	Magmatic segregations in serpentinized peridotite.	Upper Jurassic—prior to intrusion of the granitic rocks.	Serpentine, gabbro, pyroxenite, peridotite, dunite.	Chromite, antigorite, chrysotile, olivine, pyroxene.	Mainly in wartime as a source of chromium for ferro-alloys and refractories.
Copper	Mesothermal veins. Replacement and cavity fillings in shear zones.	Upper Jurassic—related to the intrusion of granitic rocks.	Quartz-rich vein matter, sheared, hydrothermally altered igneous and metamorphic rocks.	Oxidized zone—azurite, malachite, chalcocite, limonite. Below oxidized zone—chalcopyrite, bornite, pyrite, galena, sphalerite.	Mainly in wartime as a source of copper and copper chemicals.
Gold (lode)	Mesothermal veins and wallrock replacements adjacent to veins.	Upper Jurassic—related to intrusion of granitic rocks.	Quartz and quartz-ankerite-mariposite vein matter; albitite, granitic and greenstone dikes cutting numerous types of igneous and metamorphic rocks.	Native gold-silver alloys, pyrite, galena, arsenopyrite, sphalerite, chalcopyrite, tetrahedrite, quartz, ankerite, calcite, mariposite in numerous combinations.	Gold ore commonly yields silver, copper, and lead as well as gold.
Gold (placer)	River gravels. Soil mantle.	Eocene channels (rare). Pleistocene benches. Recent channels and benches.	Quartz-rich sand, gravel boulders. Soil mantle.	Native gold-silver alloys, ilmenite, chromite, magnetite, platinum, garnet, zircon, quartz, rock fragments.	California placer gravels commonly yield gold, silver, platinum and zircon (for refractories).
Lead	Mesothermal veins. Replacements and cavity fillings along shear zones.	Upper Jurassic—related to the intrusion of granitic rocks.	Quartz, quartz-carbonate, quartz-barite vein matter; sheared, hydrothermally altered metamorphic and igneous rocks.	Galena, sphalerite, pyrite, chalcopyrite, native gold, quartz, ankerite, calcite, limonite, in numerous combinations.	In addition to lead, zinc, copper, gold and silver are commonly recovered from lead ore.
Manganese	Marine sediment, probably of volcanic origin.	Paleozoic and Upper Jurassic.	Chert and jasper, greenstone.	Black manganese oxides, rhodonite, rhodochrosite, spessartite, chalcedony and jasperoid chalcedony.	Principally wartime—for steel alloys and as a deoxidizer in steel manufacturing.
Nickel	Mesothermal veins.	Upper Jurassic.	Quartz-mariposite-ankerite vein matter.	Niccolite, millerite, pyrite, ankerite, mariposite, quartz gold.	Not recovered thus far in Mariposa County. Used in alloys and as protective coating on other metals.
	Weathering of serpentine.	Early Tertiary to Recent.	Weathered serpentine, clay, chalcedony and opal.	Garnierite, serpentine minerals, clay, chalcedony, opal.	

Table 2A. Summary of economic deposits of metallic minerals, Mariposa County—continued.

Ore or source material	Origin and mode of occurrence	Time of formation	Associated rocks	Associated minerals	Utilization
Platinum	Probably magmatic segregations disseminated in serpentinized peridotite. Erosion of serpentines.	Upper Jurassic—prior to intrusion of granitic rocks. Placers—Eocene to Recent.	Alluvial gravels.	Native gold-silver alloys, platinum, magnetite, ilmenite, chromite, garnet, zircon.	Recovered in small amounts occasionally as a minor constituent in placer concentrates. Used for jewelry and chemical utensils.
Silver	Mesothermal veins.	Upper Jurassic—related to intrusion of granitic rocks.	Quartz, quartz carbonate vein matter cutting many metamorphic and igneous rocks.	Native gold-silver alloys, pyrite, chalcopyrite, proustite, pyrargyrite, etc.	Ores commonly yield gold, copper, and lead in addition to silver. In rare cases zinc is also saved.
	Gold placers.	River gravels—Eocene to Recent.	Alluvial gravels.	Same as for gold and platinum.	See gold.
Tungsten	Contact metamorphism of ancient marine sediments. Mesothermal veins.	Upper Jurassic—related to the intrusion of granitic rocks.	Limestone, tactite, limy slate.	Scheelite, calcite, garnet, epidote, quartz. Occasionally wolframite.	Machine-tool alloys, electric light filaments. Predominantly wartime.
Zinc	Replacements and cavity fillings along shear zones. Mesothermal veins.	Upper Jurassic—related to intrusion of granitic rocks.	Sheared, hydrothermally altered metamorphic and igneous rocks. Quartz-rich vein matter.	Sphalerite, galena, pyrite, chalcopyrite, quartz, barite, limonite.	Coverings for ferrous alloys and constituent of nonferrous alloys, pigments, chemicals.

Table 2B. Summary of economic deposits of nonmetallic minerals, Mariposa County

Ore or source material	Origin and mode of occurrence	Time of formation	Associated rocks	Associated minerals	Utilization
Andalusite	Contact metamorphism of shale.	Upper Jurassic.	Slate near contacts with granitic rocks.	Sericite, graphite.	Not utilized thus far in Mariposa County. Porcelains and refractories.
Asbestos	Deuteric (?) alteration of peridotite.	Upper Jurassic—prior to intrusion of granitic rocks.	Serpentine rock.	Antigorite, bastite.	Insulation.

Commodity	Origin	Age	Associated rocks	Associated minerals	Uses
Barite	Hydrothermal replacement of limestone and related crystalline rocks.	Upper Jurassic (?).	Limestone, quartzite, slate, schist, zinc ores.	Barite, witherite, quartz, lead-zinc-copper ore minerals.	Weighting of oil-well drilling mud. Barium chemicals and pigments.
Clay	Weathering of a wide variety of rocks; alluvial concentration and sorting.	Eocene to Recent.	Quartz sand and white clay. Alluvium and mantle with dark clays.	Various clay minerals and detritus.	Brick, tile, other ceramic articles.
Limestone	Marine sediment.	Paleozoic and Upper Jurassic.	Limestone, slate, greenstone.		Portland cement, lime.
Magnesite	Hydrothermal veinlets.	Upper Jurassic, probably prior to intrusion of granitic rocks.	Serpentine rock.	Serpentine-family minerals.	Not utilized in Mariposa County thus far. Magnesia refractories, oxychloride cement.
Mica schist	Metamorphism of rhyolite tuff.	Upper Jurassic.	Slate, quartzite, meta-sandstone.		Lubricant in rubber goods. Mineral filler.
Roofing granules and terrazzo chips	Numerous igneous and metamorphic rocks—serpentine, slate, etc.	Paleozoic to Upper Jurassic.			Protective covering on asphalt-base roofs. Aggregate in polished concrete floors and walls.
Sand and gravel	Alluvial deposits. / Mine dumps. Mill tailings.	Tertiary to Recent.	Alluvial debris.		Road metal, railroad ballast, road base, concrete aggregate, plaster sand.
Silica	Vein quartz.	Upper Jurassic.			Manufacture of ferro-silicon, chemicals.
Stone miscellaneous — Greenstone	Submarine volcanism.	Paleozoic, Upper Jurassic.			Building stone, road base.
Granite	Igneous intrusion.	Upper Jurassic.			Building stone, aggregate.
Chert	Marine volcanic sediment.	Paleozoic and Upper Jurassic.			Road base, railroad ballast.
Soapstone	Hydrothermal alteration of serpentine.	Upper Jurassic.			Building stone, insecticides.
Slate	Contact metamorphism of shale.	Upper Jurassic.			Roofing slabs, granules.

Table 3.   *Mariposa County Mines having an estimated production greater than $100,000 up to 1954.*

Name of mine	Value of products	Type of mine
1. Princeton	$5,000,000	Gold
2. Pine Tree and Josephine	4,000,000	Gold
3. Mount Gaines	3,590,000	Gold
4. Clearinghouse	3,350,000	Gold
5. Hite	3,000,000	Gold
6. Hasloe	3,000,000	Gold
7. Jenkins Hill	3,000,000	Limestone
8. El Portal Barite	2,760,493	Barite
9. Mariposa	2,395,405	Gold
10. Washington	2,377,000	Gold
11. Blue Moon	2,054,000	Zinc-lead-copper
12. Bandarita	1,520,000	Gold
13. Mary Harrison group	1,500,000	Gold
14. Red Cloud	1,500,000	Gold
15. Malvina group	1,000,000	Gold
16. Mountain King	1,000,000	Gold
17. Virginia	824,000	Gold
18. Diltz	750,000	Gold
19. Ruth Pierce	600,000	Gold
20. Whitlock group	500,000	Gold
21. Oakes and Reese	500,000	Gold
22. Spread Eagle group	425,000	Gold
23. Quail	400,000	Gold
24. Bondurant	390,000	Gold
25. Sierra Rica	300,000	Gold
26. Mt. Ophir	250,000	Gold
27. White Rock Silica	200,000	Silica
28. Pyramid	200,000	Gold
29. Marble Springs	200,000	Gold
30. Feliciana	159,500	Gold
31. Schroeder	158,400	Gold
32. Golden Key group	154,000	Gold
33. Champion I	150,000	Gold
34. Greens Gulch	119,140	Gold
35. French	116,750	Gold
36. Pocohontas	113,800	Copper
37. Tyro	110,000	Gold
38. Long Mary	108,928	Gold
39. Red Bank	100,000	Gold
40. Doss	100,000	Gold

The principal deposits for which the authors were able to collect considerable data are listed alphabetically by commodity and property name. Such data as were available for the other mines and prospects of the county will be found in the tabulation which follows the text. All mines known to the authors are covered in the tabulation.

METALLIC MINERALS

Chromium

Chromium most commonly occurs in the mineral chromite ($FeCr_2O_4$), and in California is found in economic accumulations only in peridotite or its alteration products. Chromite may occur as a massive segregation in peridotite or it may be disseminated in grains or pebble-shaped clots in a matrix of magnesian silicate minerals. Ore is generally classed as massive, disseminated or leopard, leopard ore being clots of chromite in a serpentine matrix. All of the known occurrences in Mariposa County are of massive type, but masses of ore seldom exceed a hundred tons. According to Cater (1948, p. 8), large concentrations of chromite are found only in dunite, a variety of peridotite made up chiefly of olivine. Most of the peridotite in Mariposa County is the variety saxonite, containing abundant enstatite as well as olivine. Serpentinization is well advanced in all of the Mariposa County peridotite masses.

The two main masses of serpentine, elongated in a northwesterly direction to the Mother Lode, stretch almost continuously from Mt. Bullion north through Coulterville and across the northern county boundary. A maximum thickness of about 2 miles is attained by the more southerly of the two main masses just north of Bagby. A narrow, northwest-trending belt of serpentine is also present adjacent to the north bank of the Merced River a few miles northwest of Bagby.

Chromite has been mined on three properties in Mariposa County, all active during World War I. The following excerpts are from Cater (1948, pp. 14-15) who abstracted the material from unpublished wartime reports by N. L. Taliaferro.

Fossow Property (1). "The deposits on the Fossow property are in the NE¼, sec. 29, T.2S., R.16E., immediately at the crossroads known as Peñon Blanco, about 3 miles by road northwest of Coulterville. The only ore found occurred as float plowed out of the deep soil covering a small flat near the crossroads, and was on the west side of a small draw near the west contact of a highly sheared serpentine mass. At the time of Taliaferro's visit in 1918, about 600 pounds of such ore had been recovered, but none was ever shipped." (The property apparently is now part of patented claims owned by Oro Rico Mines Co., c/o A. D. Vencile, Room 14, 1584 Washington Boulevard, Los Angeles).

Reed Property (2). "The deposit on the Reed property is on Blacks Creek, 1 mile southeast of the Fossow deposits in SW¼, sec. 27, T.2S., R.16W. Workings consist of a few trenches and a small open cut about 4 feet deep. The ore occurred as very small lenses striking N.40° W., parallel to the schistosity of highly sheared serpentine country rock. This deposit, like the Fossow deposit, is near the west contact of serpentine with Mariposa slates. At the time of Taliaferro's visit in 1918, about 600 pounds of medium-grade chromite had been mined and a small amount of ore was in sight in the cut." The property now is part of the holdings of Walter J. Lautenschlager, 626 So. Catalina Street, Los Angeles 5, California.

Riverside Chrome Mine (Purcell-Griffin) (3). "The Riverside chrome mine is about 3½ miles south of Coulterville in the NE¼, sec. 22, T. 3 S., R. 16 E. The deposit was located and worked by T. C. Purcell who shipped 32 long tons of ore from it in 1918. Workings consist of an east-trending open-cut 15 feet long and 8 feet deep, at the end of which is a pit 4 feet by 6 feet [in plan] and 10 feet deep. The ore consisted of an irregular lens of chromite striking about N. 80° W., near the north end of a narrow serpentine mass. Near the workings the serpentine is soft and highly sheared, but a short distance west of the pit it is more compact and contains fine streaks and disseminated grains of chromite. The schistosity of the serpentine in the vicinity of the workings strikes N. 25°–30° W., at an angle to the attitude of the ore body. No ore remained in sight in the workings at the time of Taliaferro's visit, and it is assumed the deposit was exhausted." The property is on or adjacent to the Adelaide gold mine, a patented property owned by Alfred W. Stickney, 435 Hillcrest Road, San Mateo, California.

Copper

Mining of copper in Mariposa County is confined at present to small-scale, intermittent activity with occasional small shipments of oxidized ore. Some shipments of gold ore contain recoverable copper and there is a small supplementary production from this source. The recorded production of copper since 1880 is about 2,205,000 pounds valued at $344,-500. Although there was some unrecorded production in the 1860s the total copper mined in the county prior to 1880 was probably small. As the total production of copper in California has amounted to about 600,000 tons (O'Brien and Crosby, 1950, p. 308) the amount of that total contributed by Mariposa County has been less than 1 percent.

The primary copper deposits in Mariposa County are all of meso-thermal origin and may be divided roughly into three classes: those where copper is a by-product of gold mining, those where copper, zinc and lead minerals are associated, and those in which copper ore-minerals predominate. However, most copper ores contain some gold and silver and many gold ores contain some copper so that the classes are grada-tional. Copper ore is generally concentrated in shoots in shear zones which cut a wide variety of wall rocks. Quartz or quartz and barite and pulverized wall rocks are the common gangue materials. The principal copper mines occur in northwest-trending belts that roughly parallel the Mother Lode vein system. Although the copper deposits are probably not directly related to the more extensive gold-quartz veins—some can be demonstrated to have formed earlier than some of the gold deposits— the structures which allowed egress to the mineralizing solutions of both gold and copper formed during the same orogenic period, and both metals are believed to have stemmed from solutions given off during late-stage cooling of the intrusive granitic rocks. The various theories of origin of the copper in the Sierran foothills are covered by Heyl (1948, pp. 25-28).

Most of the copper and zinc-copper mines are concentrated in three narrow belts, each trending roughly N. 35° W. The most southerly be-gins in the vicinity of Green Mountain School and extends somewhat northwest of White Rock ranger station. Principal mines in this group are the Green Mountain group, Lone Tree, Cavan-San Jose Great North-ern and Pocahontas. A second belt, chiefly zinc-copper-lead mines, ex-tends from somewhat northeast of Hornitos to Exchequer Dam. The Blue Moon and American Eagle mines are the principal examples of this group. A third belt of copper mines is in Hunter Valley between Cotton Creek and Jasper Point on the Merced River. The La Victoria and Bar-rett mines are the principal copper mines in this district. The Blue Moon and Pocahontas mines are by far the largest copper producers in the county but the Blue Moon mine is primarily a zinc mine. Gold mines which have produced notable amounts of copper are the Mount Gaines, Washington and Pine Tree and Josephine.

Barrett (Baretta, Barette, Beaudry) Mine. Location: Secs. 29. 30. 32, T. 3 S., R. 16 E., M.D., in hill land at the northern end of Hunter Valley. Accessible from the paved Bear Valley-Hornitos road by 3½ miles of paved road and 4½ miles of graded dirt road. Ownership: An-gelo Beaudry and Inez Bouvier, 1431 Moraga Street, San Francisco. California, own three patented claims, the Wildcat, Mountaineer, and

New Year, and a rectangular piece of land known as Berettas Enclosure which includes most of the old Barrett and Beaudry claims.

The Barrett mine is a very old property discovered during the 1860s (Aubury, 1905, p. 215). It was owned in 1902 by Joseph Baretta (ibid.) and at that time had a shaft 200 feet deep. More than $90,000 had been realized from the mine up to 1907 (Lang, 1907, p. 966) most of this amount having come from a pocket of coarse gold. In 1920 the property was under lease to C. H. Burt of Mountain View (Castello, 1920, p. 102) and by that date a tunnel had been driven northeast to intersect the northwest-trending ore zone. For a short period during 1931-32 it was mined by D. Jacobs of Coulterville and produced 100 tons of high-grade gold ore from which 70.7 oz. of gold and 16 oz. of silver were recovered (U. S. Bureau of Mines records). In recent years the mine has been idle and there is no equipment on the property.

The Barrett mine is among a large group of mines and prospects that include the Castagnetto, Bruschi, Ferrari, Oaks and Reese and Iron Duke mines. Many of these mines and prospects contain both gold and copper, probably as the result of separate mineralization. The Barrett has produced pocket gold ore and gold ore of milling grade as well as high-grade, oxidized and sulfide copper ores. Upper parts of the present, partly caved workings, apparently on the Wildcat claim, expose a copper-stained quartz vein 1 to 2 feet wide, striking N. 60° to 65° E. and dipping 65° southeast. Lang (1907, p. 966) describes banded vein matter composed of alternating bands of calcite or calcite and barite and quartz. Pyrite and chalcopyrite more or less impregnate the vein matter in unoxidized parts of the veins. Some massive sulfide ore showing some malachite was on the dumps on the Wildcat claim in September 1954. Lang (1907, p. 966) mentions vein structures trending N. 15° W. and there are probably two sets of mineralized structures. Soil cover is heavy and the area is brushy. Wall rocks are chiefly meta-augite andesite greenstone although chert has been described in the workings. No chert was present on the dump. The greenstone is part of the Peñon Blanco member of the Upper Jurassic Amador group of rocks (Taliaferro, 1943, pp. 282-284). Rhythmically banded chert and associated sedimentary rocks of the Hunter Valley chert member of the Amador group, exposed along the road south of the Barrett mine, strike N. 40° W. and dip 63° northeast and it is assumed the enclosing greenstones in general follow a similar trend in this area.

Blue Moon (Blue Cloud, Red Cloud, Porcupine) Mine. Location: Secs. 19, 30, T. 4 S., R. 16 E., M.D., 6 miles north of Hornitos. Ownership: Not determined. Last operator was Hecla Mining Company, Wallace, Idaho (1945).

The Blue Moon mine is primarily a zinc mine but produced a notable amount of copper. 55,656 tons of ore milled in 1944-45 yielded 406,038 pounds of copper, 13,687,920 pounds of zinc, 553,753 pounds of lead, 3,446 ounces of gold and 208,965 ounces of silver (Eric and Cox, 1948, p. 145). The mine is more fully discussed under zinc.

Good View (Cavan, Cavan-San Jose). Location: 3 miles west of Green Mountain School in secs. 5, 8, 9, T. 8 S., R. 18 E., and sec. 32, T. 7 S., R. 18 E., M.D. Ownership: Mine-Metal Properties, Inc., c/o C. C. Kellogg, 50 Twenty-sixth Street, Merced, California, owns 161.3

acres of patented land including the following claims: Copper King, Crown Point, Good View, Little Giant, Rothschild, San Jose, Stonewall Jackson and Sunset. Accessible from Green Mountain School by 2½ miles of graded dirt road and 1¼ miles of truck trail.

The Good View mine is an old property discovered in the early 1860s (Lowell, 1916, p. 573). Little is known of its history until the period of operation by Cavan Mining and Milling Company of Stockton from about 1900 to 1910. It was then known as the Cavan Copper Mine. According to old U. S. Bureau of Mines records, 22 tons of ore was shipped from the Good View claim in 1902 which yielded 8,673 pounds of copper, 8 ounces of silver and .87 ounces of gold. In 1907, 45 tons of ore yielded 4,624 pounds of copper and 1.21 ounces of gold. By 1913 the property had passed into the hands of S. L. Pearce who shipped 7 carloads of ore to the Selby, California, smelter which averaged $1000 per car (Lowell, 1916, p. 573), the ore carrying $2.50 in gold per ton (at the price of gold then). By 1927 the mine had passed to the ownership of Samuel L. Thrift, et al., of Stockton and patent had been acquired by the property (Laizure, 1928, p. 125). Since then the mine has been known as the Good View. The property has been idle for a considerable period and all equipment had been removed (1954).

Workings are driven chiefly in black slate, mica-andalusite schist and blue-black quartz-biotite hornfels, all derived from clay shale, and schistose metadiorite (?) intrusive into the metasediments. Some thin, lenticular beds of metavolcanic material, probably tuff, are interbedded with the metasediments in minor amounts and some metasandstone is included in the slate. The metamorphic sequence is part of the Mariposa formation of Upper Jurassic age. Regional trend of the folds in the metasediments is northwest and the fracture systems along which ore has developed also trend northwest.

As most of the workings have no means of access or are watered, and as individual workings and claim boundaries are difficult to identify, the workings can only be described in a general way. In 1902 a tunnel, then 240 feet long, was being driven to intersect the parallel Rothschild and Good View veins. This tunnel was estimated to strike the Good View vein at a depth of 375 to 400 feet and the Good View at 250 feet (Aubury, 1905, pp. 207-208). A 20-foot shaft at the south end of Good View vein is also mentioned which followed a 6-8 inch streak of ore running 15-25 percent copper. This streak was traceable on the surface for 1400 feet. The main shaft, apparently on the Good View claim, was described as 140 feet deep, exposing a vein 6 inches to 6 feet wide. Ore was described as oxidized to a depth of 75 feet with massive sulfides below 75 feet. (Azurite, malachite and chalcopyrite are the chief ore minerals in the district.) Aubury (1905, p. 208) also mentions a 58-foot shaft on the Sunset claim, exposing ore containing 17 percent copper from 10 feet below the surface to the bottom of the shaft in widths of 6 to 30 inches; a 20-foot and a 60-foot shaft and a 65-foot crosscut tunnel on the Copper King claim (no drifts); and a cross-cut tunnel 170 feet long on the San Jose claim from which three winzes 40, 80, and 115 feet deep had been sunk. The San Jose workings developed an ore body 65 feet long and 4 to 4½ feet wide and at least 115 feet deep. Surface work only had been done on the Crown Point, Stonewall Jackson, and Little Giant claims.

Green Mountain (Copper Mountain, Legioneer, Copper Queen) Group. Location: Secs. 3, 10, T. 8 S., R. 18 E., M.D., 2 miles northwest of Green Mountain School. Ownership: Dr. Felix A. Smith, 507 Medical Building, Oakland 12, California, owns 11 unpatented claims, the Amador, Copper Chief, Copper King, Copper Mountain, Copper Peak, Discovery, Francisco (Frisco), Green Mountain, Juliet, and Last Chance. The Copper Queen and Buena Vista patented claims are owned by Marie D. Kellogg, et al. (½), 50 Twenty-sixth Street, Merced, California, and Dr. Felix A. Smith (½).

Croppings of copper ore were found in the Green Mountain area in 1861 (Lowell, 1916, p. 572) and the claims were worked intermittently between 1863 and 1900. From the early 1900s to 1923 the property was owned and intermittently worked by the Legioneer Gold Mining Company, O. R. Sydney, et al., of LeGrand or their authorized lessees. The Legioneer company also worked the Lone Tree mine in the early 1900s. During 1903, 170 tons of ore was shipped which yielded 14,245 pounds of copper. In 1904, twenty tons of ore were shipped from which 9,500 pounds of copper was extracted (U. S. Bureau of Mines records). These figures are believed to be representative of the amount and type of ore shipped at various times during the early 1900s but are not the only shipments made. In 1919 the United Chemicals Company of San Francisco completed construction of a leaching and precipitation plant at Raymond, Madera County, to process ore from the Green Mountain mines. Copper oxide, copper sulfate and iron sulfate were produced in this plant, the average ore treated running 6 percent copper (Castello, 1921, p. 103). The Legioneer company continued to operate the mine while the Raymond plant was in operation, O. R. Sydney acting as superintendent. From 1923 to about 1929 the properties were operated by the Floraferro Company of San Francisco. This company sold pulverized, oxidized ore as a soil additive and insecticide. The Copper Producers Trust of Boston, Massachusetts, held title to the properties for a while in the late 1920s.

Ore bodies occur as replacements in schist along irregular-shaped shear zones, a majority being oriented roughly northeast with steep dips toward the southwest. The schist is cut by dikes and irregular intrusions of granitic rocks. Ore in the upper or oxidized parts of the shoots contains various combinations of the minerals, malachite, azurite, chalcocite, cuprite and native copper. Primary ores in the lower parts of the shoots are made up chiefly of chalcopyrite and pyrrhotite. Veins and irregular replacements adjacent to veins occur over a width of as much as 1200 feet and some masses of sulfide ore reach a width of 60 feet (Aubury, 1902, p. 253). Much of the ore proved difficult to handle because it crumbles and heats upon exposure to air (Aubury, 1905, p. 254).

There are more than 4,000 feet of underground workings on the property (Castello, 1921, p. 103) including numerous openings to the surface. Aubury (1905, p. 204) states "... the most important (openings) are two tunnels several hundred feet in length, above which are the stopes and chambers from which the best ore has been taken. The lower or east tunnel is in 600 feet (1902). At about 400 feet it has crosscut a vein 60 feet in width. ... The main body of ore lies back of this, the openings or workings of which are about 60 feet above the tunnel level, to which they are all connected with an upraise. . . . About 900 feet west from the above tunnel and 50 feet above it, another tunnel has been run several hundred feet, and from it a large quantity of carbonate and sulfide ore has been extracted . . .''.

Johnnie Green (Green, Johnny Green, Johnny Green Jr.) Mine. Location: NE¼, sec. 31, T. 7 S., R. 18 E., M.D., accessible by 4½ miles of truck trail from the graded LeGrand-White Rock (Ganns Creek) road or by 3½ miles of truck trail from the graded Green Mountain School road. Ownership: undetermined.

There is little record of the early history of this mine. In October 1866, fifty men were employed at the mine and about 50 tons of copper were smelted and shipped per week (Min. and Sci. Press, vol. 17, no. 17, p. 262, 1866). In 1907 high-grade copper ore was still being mined which was described as massive chalcopyrite replacing schist (Lang. 1907, p. 964).

The mine workings are in a belt of quartz-mica and quartz-mica-chiastolite schist, the schist being derived by contact metamorphism of clay shale of the Mariposa formation of Upper Jurassic age. The strike of the schist is N. 30°-40° W. and the dip is roughly 60° northeast. Primary ore consists of chalcopyrite and pyrite or pyrrhotite impregnating schist in replacement masses bordering narrow quartz veins. There is some carbonate ore in the upper parts of the shoots.

A shaft of undetermined depth inclined 60° northeast is the principal working but there is a second shaft opening 75 feet northeast of the main shaft, also inclined 50°-60° E., but partly caved and of unknown depth. North of these workings, several hundred yards, is a second group consisting of 2 inclined shafts and a caved tunnel driven about N. 20° W. The workings have no means of access and there was no equipment on the property in September 1954. A small shipment of oxidized ore was shipped from this area in 1954, but the authors were unable to determine which property the ore was from.

La Victoria (Victoire, La Victoire, Tandem) Mine. Location: Secs. 4, 9, 10, T. 4 S., R. 16 E., M.D., in Hunter Valley, 11 miles by graded road north of Hornitos. Ownership: Ralph E. and Libbie N. Dailey, 1165 West 22nd Street, Merced, California.

The La Victoria Copper Mine was thoroughly investigated by Manning W. Cox and Donald G. Wyant of the U. S. Geological Survey as part of the wartime research on the copper-zinc belt of the Sierran foothills. The results of this investigation were published in 1948 by the California Division of Mines on pages 127-132 of Bulletin 144, entitled "Copper in California." The following discussion is largely abstracted from this source.

The La Victoria deposits were discovered in April 1864 and explored by the La Victoria Mining Company. This company and its successors mined and sorted high-grade ore for direct shipment to smelters and also constructed a small furnace on the property to treat low-grade ore. By 1873, most of the readily accessible ore had been removed and the mine was shut down. Some ore was shipped during the 1890's and in 1917. From May 1943 to March 1944 the mine was under lease from the owner, Herbert Lang, by R. B. Lamb. During this period the mine workings were partly rehabilitated and the property was sampled with the aid of a R.F.C. loan. Since that time the mine has been inactive.

Mine workings are driven in meta-augite andesite greenstone which has prominent pillow structure. The pillows range in diameter from a few inches to 4 feet. Many are surrounded by a thin shell of mudstone or by masses of epidote and in some places the greenstone pillows occur in a matrix of brecciated greenstone and red jasper. These metavolcanic rocks are part of the Upper Jurassic Amador group and the mine workings are near the crest of a large northwest-trending anticline (Taliaferro, 1943, p. 125).

The La Victoria ore zone, exposed for 125 feet along the main level, is found along the main shear zone of a system which strikes N. 50° W. and dips 10°-45° northeast. A second mineralized shear zone near the Cavagnaro shaft trends N. 45° W. and dips 30°-60° northeast. The two mineralized zones are truncated by the Poupion shear zone which strikes N. 60°-80° W. and dips 65°-87° north. This zone is not mineralized. In the La Victoria ore zone the copper-bearing part of the shoot is restricted to the part of the shear zone that strikes N. 15° W. The more west-trending part of the zone is principally quartz-pyrite rock with very little copper. Ore occurs in pods a few inches to 15 feet long and from 1 to 3.2 feet wide, although masses of oxidized ore removed during the early history of the mine may have been larger. Primary ore consists of fine-grained chalcopyrite, pyrite and quartz. Secondary minerals are chiefly azurite, malachite and chalcocite. Browne (1868, p. 213) mentions a considerable quantity of black oxide in ore mined in the 1860's. The average copper content of pillar samples in the upper workings of the mine is 5.3 percent and 4.1 percent on the bottom level. The ore was partly oxidized for the full depth mined and completely oxidized for a depth of 30 feet below the outcrop of the mineralized zone. Copper had been almost completely leached from the uppermost 20 feet of vein matter.

According to Cox and Wyant (1948, p. 131) the enriched 5 percent ore of the mine is nearly exhausted, but a small amount remains in pillars above the bottom level. A 2-foot-wide ore shoot in the north-trending part of the La Victoria ore zone continues below the present mine workings and may have a strike length of 110 feet. The average grade of primary ore on the bottom level, which runs 4 percent copper, 2 percent zinc and less than 0.1 ounce of gold per ton, may represent the grade of ore to be expected below. The only other zone believed to carry copper ore is the one just east of the Cavagnaro shaft.

Pocahontas Mine. Location: NW ¼ sec. 12, T. 7 S., R. 17 E., M.D., one mile north of White Rock School by good dirt road. Ownership: Estate of Eben N. Briggs, c/o L. M. Olds, 57 Post Street, Room 402, San Francisco, California.

The Pocahontas mine has been the most productive copper mine in Mariposa County, having yielded more than 700,000 pounds of copper as well as considerable gold and silver. The gross amount realized from the mine has been well over $100,000. The Pocahontas deposits were discovered in the early 1860s but apparently were not extensively mined until the early 1900s. The property was owned in 1903 by J. E. Waller of LeGrand, and was under lease to C. and S. Wilcox of White Rock. These lessees mined and shipped over $30,000 worth of copper ore in a three-year period (Aubury, 1905, p. 210). The mine then passed into the ownership of Mrs. Abbie Waller and was under lease for a short time by William McIntosh and W. M. Darling of San Francisco. By June 1907 the mine had been taken over by the Pocahontas Copper Mining Company, was operated under the direction of J. B. Roberts for a short time, and then under direction of David Ross. By August 1907, the shaft had been sunk to a depth of 145 feet, plans had been made to sink 500 feet deeper and a smelter was planned (Eng. and Min. Jour., vol. 84, no. 12, p. 563, 1907). Late in September 1907 the property was leased to the Michigan Steamship Company of San Francisco (Eng. and Min. Jour., vol. 84, no. 19, p. 896, 1907) but this lease apparently was short lived. The Pocahontas Copper Mining Company or its lessees worked the mine during World War I. According to U. S. Bureau of Mines records, the Pocahontas mine produced 6,985 tons of ore from 1902 to 1913, inclusive, that yielded 621,273 pounds of copper, 383.92 ounces of gold, and 12,618 ounces of silver. A small amount of ore was shipped by lessees in 1914 and in 1916 1,200 tons of ore was shipped that yielded 63,220 pounds of copper. The mine has been idle since World War I but is still owned by heirs of officers in the Pocahontas Company.

The Pocahontas mine workings are driven in contact-altered rocks of the lower (?) member of the Upper Jurassic Mariposa group. These wall rocks consist principally of quartz-biotite and quartz-biotite-andalusite schist and quartz-biotite hornfels derived from tuffaceous sediments. Along the mineralized zone a sheet of hornblende-rich granitic rock has been intruded parallel to the schistosity and other thin sheets of biotite quartz diorite have penetrated the schists near the mine openings. A stock of biotite quartz diorite crops out less than 400 feet southeast of the main shaft. The hornblende-rich dike has been hydrothermally altered and mineralized and the original nature of the rock is, in many places, unrecognizable (Forstner, 1908, p. 748). Some hornblendite is on the dump but this rock does not crop out conspicuously because of masking limonite-quartz gossan.

The copper mineralization is concentrated in a shear zone which strikes N. 35°-45° W. and dips northeast at angles between 65° and 80°. The primary sulfide ore occurs in lenticular masses in a gangue of brecciated wall rock and introduced quartz. One of the sulfide ore shoots mined in the early 1900's was 50 feet long, 4 feet wide and more than 100 feet deep (Aubury, 1905, p. 210). Much of the primary ore is a dark bluish, massive, fine-grained rock composed predominantly of chalcopyrite, pyrite, sphalerite and pyrrhotite. According to Aubury (1905, p. 210) this type of ore ran 6 to 12 percent copper and $2.50 per ton in gold (at the 1902 price). The ore-bearing zone has been oxidized and secondarily enriched to a depth of about 100 feet but most of this ore along the main ore zone has been already mined. Much of this oxidized ore, consisting principally of malachite, chrysocolla and azurite, ran from 30 to 35 percent copper. Minable oxidized ore reached a width of 20 feet (Forstner, 1908, p. 747-748) in some shoots.

Workings (September 1954) consist of 7 shafts from 15 to 75 feet deep and a tunnel driven northwest, opening four or five hundred feet southeast of the main shaft. Most of the shafts are inclined steeply east and follow the ore zone. The main shaft is surmounted by a wooden

FIGURE 15.   A typical Coulterville scene, observer facing southeast from the edge of Highway 49. To the right of the large quartz vein standing in the middle ground is the Louisa mine, one of several mines of the productive Mary Harrison group. *Photo by Mary H. Rice.*

headframe and ore bin, both in poor condition. According to Lowell (1916, p. 573) this shaft was 300 feet deep in 1913 and there was a total of 900 feet of drifts from the various shafts. Julihn and Horton (1940, p. 166) place the depth of the main shaft at only 200 feet. In 1954 the shaft was caved at a depth of 75 feet. The tunnel, described as 200 feet long in 1913 (Lowell, 1916, p. 573) is driven N. 33° W. on mineralized material. It is caved a few feet from the portal, and is inaccessible.

Estimation of reserves is not possible because the condition of the workings makes entry unfeasible. There are numerous gossan outcroppings which do not appear to have been explored and several shear zones parallel to the principal ore zone have not been thoroughly prospected. Showings of copper minerals may be seen over a width of more than 1000 feet. A number of small masses of high-grade copper ore were found south of the present workings at the contact of the metasedimentary series and coarse-grained quartz diorite (Forstner, 1908, pp. 747-748). None of the old accounts indicates that there is a lowering in the grade of the sulfide ore at depth.

Gold

Although gold mining currently is at a low ebb because of high operating costs in relation to the average grade of ore obtainable at

FIGURE 16.   An outcrop of green mariposite-ankerite-quartz vein rock exposed
along Highway 49 between Mary Harrison mine and Coulterville. This pretty, durable
rock is finding its way into many California gardens as an ornamental stone. It is
quarried for this purpose farther south along Highway 49. Large, lenticular masses
of the rock are found in the Mother Lode vein system alternating with sheets of
milky quartz. *Photo by Mary H. Rice.*

most mines, gold mining has always been vital to the economy of Mariposa County. The total gold production of Mariposa County through
1954, which amounts to about $48,000,000, is only part of the wealth
brought into the county through businesses connected directly or indi-
rectly with mining. A variety of factors, particularly a rise in the price
of gold, could quickly change the economic structure of gold mining
and result in rejuvenation of the gold mining industry. A majority
of the mines has not been exhausted of ore and their reactivation
awaits some significant change in the economics of precious metal
mining.

Placer-gold was discovered in Mariposa County about 1848, although
precise details are lacking. Among the earliest placer miners were
Mexicans or Californians of Spanish descent. Much of the geographic
nomenclature still in use in western Mariposa County is drawn from
the Spanish place names such as Mariposa (butterfly), Agua Fria
(cold water) and Hornitos (little ovens). Discovery of lode-gold is
commonly credited to Kit Carson and two associates at the Mariposa
mine in 1849, but again the more secretive Mexicans may already have
been working their arrastras in the Hornitos district. John C. Fremont's Las Mariposas Spanish land grant was interpreted as including
the southern end of the soon-to-be-famous Mother Lode, and after
lengthy litigation a very large area was set aside for private adminis-
tration and not developed in the same fashion as claims located on
public lands. As the Las Mariposas Grant mines were not located and
surveyed according to early public land survey practice, land plats

within the grant to this day are difficult to locate with respect to survey lines outside of the grant.

In addition to Las Mariposas grant, a large block of land north of the Merced River was held under single management by the Cook Estate and these two great properties greatly affected the course of gold mining in Mariposa County. Under these large enterprises there were notable periods of efficient and profitable management of the mines. At other times the mines suffered from mismanagement and lack of incentive toward vigorous development. It is a moot question whether the mines of the two properties as a whole would have proven more productive in the course of normal development under public land regulations than they have under the collective management of the two large interests.

As in other Sierran foothill counties, the lode-gold mines of Mariposa County have been conveniently grouped, for the sake of discussion, into the Mother Lode, East Belt and West Belt groups. The Mother Lode is a system of mesothermal quartz veins occupying a major thrust-fault zone. It is not a single great, continuous vein but rather is a series of discontinuous veins which in places are parallel to subparallel, en echelon, and in other places anastomosing or braided. In general, the veins of the Mother Lode have a northwest strike and a steep northeast dip but in the Bear Valley-Bagby segment of the lode the strike swings more to the north. In the vicinity of Coulterville the Mother Lode consists of two well defined, roughly parallel veins the more easterly of which diverges south of the Virginia Belmont mine and veers more to the southeast. Along the main length of this branch are the Louisa and Mary Harrison group of mines and the Virginia-Belmont group. A series of pocket mines is located astride the southeast-veering segment, called the Flyaway group. The more westerly vein in the Coulterville area, considered by some to be the main Mother Lode, is almost continuously traceable from the Malvina group south through the Midas and Adelaide mines and southeast toward the Merced River at Bagby. In addition to the predominating northwest-trending veins of the Mother Lode system there are structurally related minor cutter veins that join the northwest examples roughly at right angles. In some of the mines of the Peñon Blanco segment of the Mother Lode the richest ore occurred in the cutter veins rather than in the larger veins of northwest trend. In the vicinity of Mormon Bar, a few miles south of Mariposa, the Mother Lode terminates in granitic rocks.

The West Belt is made up arbitrarily of all mines located west of the Mother Lode system and the East Belt of all mines located east of it. There is no well-defined, parallel gold-quartz vein system either east or west of the Mother Lode in Mariposa County that has any comparable magnitude, although a number of East and West veins do roughly parallel the Mother Lode. Within the East and West Belts are veins of almost all attitudes, and east-northeast strikes prevail in some parts of the county instead of the more usual north to northwest trends found in the Mother Lode system.

Gold in paying concentrations is generally found localized in shoots adjacent to a large volume of valueless vein matter. The conspicuous masses of milky quartz cropping out along the Peñon Blanco are largely devoid of gold and paying ore is found principally along the upper (hanging) wall or lower (foot) wall of the vein. In other places the high-grade ore is found in the middle of the vein flanked on both sides by barren vein matter. Masses of ore (shoots) commonly have one long and two short dimensions and in Mother Lode veins the longest dimension commonly dips downward in the vein at a steep angle. For example, most of the ore taken from the Princeton mine was from a single, almost vertical, chimney-like shoot more than 600 feet deep (Knopf, 1929, p. 84). Knopf (ibid. p. 26) states that the average slope length of most ore shoots on the Mother Lode as measured along the strike of the vein is 200 to 300 feet, although the shoots tend to be much more persistent in depth. Ore commonly occurs where veins bulge, where there is an abrupt change in strike or dip of the vein, or where a cutter vein joins the main vein. Conditions of ore formation in one mine do not, however, necessarily persist in other mines of a district or even in adjacent mines.

Lode Mines

A-J (Burkhart) Mine. Location: Sec. 26, T. 4 S., R. 16 E., M.D., near the head of Burns Creek 3½ airline miles west of Bear Valley. Owner: B. F. and Ruth L. Burkhart, Bear Valley, California own one unpatented claim.

The A-J mine was first opened in the 1920s at which time a 220-foot adit was driven. The property had been worked intermittently up to the fall of 1954, the most recent operation by the Richardson Brothers of Bear Valley terminated in May 1954. Recent production has been small with ore averaging about 0.4 oz. of gold per ton.

Gold occurs free and with sulfides in a quartz vein 1 foot to 4 feet wide, the vein striking north and dipping steeply east. The wall rock is meta-augite andesite (greenstone) of Upper Jurassic age belonging to the Peñon Blanco member of the Amador group (see Taliaferro, 1943, pp. 282-284). A 150-foot shaft and three levels 35 feet apart are the principal active workings. Vein minerals include pyrite, chalcopyrite and free gold with minor amounts of green copper carbonates. Milling is done in a small Gibson mill which remains on the property.

Adelaide and Anderson Mines. Location: NE¼ sec. 22, NW¼ sec. 23, T. 3 S., R. 16 E., M.D., in the southern end of the Big Bend Mountains 4 miles south of Coulterville and 1 mile north of the Merced River. Ownership: Alfred W. Stickney, 435 Hillcrest Road, San Mateo, California owns two patented claims, the Adelaide and Anderson, and two unpatented millsites on the Merced River.

The Adelaide and Anderson claims were located during the late 1860s or early 1870s. By 1874 the claims had become part of a group that included the Midas, Crown Lead and Crown Peak claims. These were surveyed for patent purposes in 1874 and patent was established by Fred and Mary MacCrellish about 1875. The Adelaide and Anderson claims were worked by the MacCrellishes in the late 1870s and early 1880s first by open cuts and then underground. During this period most of the workings were driven and most of the stoping on the main ore shoot in the Adelaide mine was accomplished. A mine accident in 1885 resulted in the death of Fred MacCrellish and injury to his wife, and mining was discontinued. By 1895 ownership of the mines had passed to Robert E. McSherry of Coulterville and the Mary C.

MacCrellish estate. By 1914 the properties were under management of W. P. Edwards of Alameda. No material amount of work was done on either claim between 1885 and 1933. In 1934 leasers drove 440 feet of tunnels and drifts about 2000 feet south of the main workings on the Adelaide claim. A satisfactory ore body was not encountered there and the mines again became idle. In 1943 the properties were acquired by the present owner and some cleaning out and sampling was done on the Adelaide claim, particularly, near the main ore shoot. Since that time both claims have been idle.

The quartz vein at the *Adelaide* mine, part of the Mother Lode system, strikes N. 27° W. to N. 35° W. and dips 50° northeast (average). Its course is slightly sinuous and vein matter thickens and thins between the limits of 6 and 10 feet. The principal known ore shoot is 25 feet wide, measured along the strike of the vein, and has been stoped to a depth of 215 feet as measured in the plane of the vein. The shoot pitches northwest at about 68°. The vein has developed in a fissure that follows the east contact between a narrow mass of serpentine, on the footwall side, and a belt of black slate of the Mariposa formation on the northeast or hanging wall side. Both slate and serpentine are believed to be of Upper Jurassic age. The slate occupies a faulted synclinal trough between a large serpentine mass on the east and a thick section of meta-augite andesite greenstone of the Upper Jurassic Peñon Blanco formation (Taliaferro, 1943, pp. 282-284) on the west. The narrow mass of serpentine which forms the footwall of the main Adelaide vein has been intruded between much broader masses of slate and greenstone. Farther north the greenstone is in direct contact with slate.

In addition to the massive milky quartz, so prominently exposed on the surface of the Adelaide claim, vein matter in numerous places consists of a banded or ribbon-structure of alternating quartz and slate or quartz and talcose serpentine layers. In other places vein matter consists predominantly of quartz, ankerite and mariposite, both massive and banded. Pyrite heavily impregnates the vein material at numerous points. Gold occurs partly in the free state and partly in intimate mixture with pyrite. Associated ore minerals are galena and chalcopyrite. Presence of tellurides mentioned in an old account by Fairbanks (1890, p. 57) has not been confirmed. According to Fairbanks the best ore is found in a 2-foot-thick tabular mass on the footwall side of the vein.

The most important workings are at the north end of the claim in the vicinity of the principal ore shoot. These originally consisted of the 100-foot-deep Cobb inclined shaft; the 350-foot-long Cobb tunnel, driven northwest on the vein and intersecting the Cobb shaft 230 feet from the tunnel portal; a winze or interior shaft sunk about 100 feet from the tunnel portal; and two shallow inclined shafts with minor drifts. One of the two auxiliary shafts is 200 feet north of and the other 140 feet south of the Cobb shaft.

About 1,000 feet south of the Cobb tunnel portal is the entrance to the Boarding House tunnel driven south to southeast along a secondary vein and having three segments 60, 150 and 30 feet long, respectively. About 210 feet from the portal of this tunnel is a winze 60 feet deep sunk on the vein. This is believed to have been connected with a lower west-trending tunnel, now inaccessible. Two shafts are located south of the Boarding House tunnel, one 100 feet deep and the other 60 feet,deep.

About 2,000 feet southeast of the Cobb shaft is a 290-foot crosscut tunnel and connecting 150-foot drift driven west and northwest. The drift appears to be on a different vein than the main workings, a narrower one about 2 feet wide. These workings are largely southeast of the Adelaide claim boundaries.

In addition to the aforementioned workings there are several superficial open cuts, pits and short tunnels. The total length of development workings is more than 2,500 feet. The stoped part of the ore shoot totals roughly 130,000 cubic feet.

Little is known concerning the tenor of the ore in the Adelaide mine and the present owner was unable to reach the ground below the caved stopes in the vicinity of the main ore shoot. Samples taken by the owner at various places in the workings assayed from 1 dollar to 7 dollars per ton at the present price of gold. Future development work should explore the ground below the stoped area by cleaning out and deepening the winze from the Cobb tunnel.

The *Anderson* mine is a quarter of a mile east and slightly south of the Adelaide mine. The main vein strikes about N. 15° W., dips northeast at an average of 42° (Turner and Ransome, 1897, economic map), and averages 4 to 12 feet wide. Little or no work has been done on it since 1885. The workings, all caved and inaccessible, consisted of two inclined shafts, three tunnels and two short open cuts. The amount of stoping done is not known. Wall rocks are slate but there is a small intrusion of granodiorite on the southwestern part of the claim. The more northerly shaft, 75 feet deep, is sunk at the geographic center of the claim and once connected with a 175-foot tunnel to the surface bearing N. 75° E. The other shaft was originally 120 feet deep and was connected to the surface by a 350-foot-long tunnel trending N. 60° W. A third tunnel 120 feet long was driven N. 15° W. along the vein. There is no record of production for the Anderson mine and no data as to the tenor of the ore.

Annabelle Prospect. Location: Sec. 19, T. 35, R. 17 E., M.D., 4 airline miles southeast of Coulterville and 3½ airline miles northwest of Bagby on the northwest side of Scotch Gulch at its intersection with Highway 49. Ownership: Mr. and Mrs. Sherman S. Pickard and Bryan A. Miller, P. O. Box 36, Coulterville, California, own one unpatented claim.

The Annabelle prospect is a small pocket mine opened in 1952. The mine workings lie along a line of small, discontinuous quartz veins which strike slightly east of north and dip 40° east. The vein system lies at or close to the contact of serpentinized peridotite lying to the west, and slightly older metavolcanic greenstones lying to the east. A quartz diorite dike intruded along serpentine greenstone contact forms the footwall of the veins and the hanging wall is meta-andesite greenstone. Next to the quartz veins is a narrow zone of fractured greenstone and talc schist commonly cut by quartz stringers. The fractured greenstone and in some places the adjacent greenstone wallrock is mineralized and carries pyrite and auriferous pyrite. The quartz veins and stringers carry free-milling gold, often in pockets. Serpentinized gabbro forms the hillslope to the west across Highway 49.

Workings consist of a shallow shaft inclined 40° east with a 16-foot drift driven northeast along the vein from the bottom of a short crosscut. About 30 feet northwest of the shaft a horizontal tunnel has been driven northwest.

Some good specimen material has recently been obtained from this mine and there has been a small but fairly steady production since 1952. Work is done entirely by hand methods and there is no mill connected with the property.

Argo (Pioneer) Mine.   Location: Secs. 15, 16, T. 2 S., R. 17 E., M.D., 7 airline miles northeast of Coulterville in the Greeley Hill district. Accessible by three-quarters of a mile of dirt road north from the Coulterville-Kinsley road. Ownership: Walter D. McLean, Coulterville, California, owns two unpatented claims, the Argo and the Pioneer.

The Argo property has been owned by the McLean family since the early 1920s. It was leased from Walter McLean in 1931 by C. A. Gillis of Tuttletown and J. H. Hollbrook, Jr., of San Francisco. U. S. Bureau of Mines records show an intermittent production from the mine between 1923 and 1935. During the winter of 1937 two men recovered $76.00 worth of gold in 3 days from soil mantle in a small gulch south of the Argo shaft (Julihn and Horton, 1940, p. 135). Milling of 24 tons of dump material in 1949 yielded 18 oz. of gold

FIGURE 17.   Headframe, ore bin and water tanks at the Argo gold mine in the Greeley Hill district 8 airline miles northeast of Coulterville. The mine is noted for pockets of high-grade ore.

or an average of $26.25 per ton. Except for assessment work the claims have been idle since 1949.

The Argo vein occurs in a shear zone striking N. 10° W. to N. 15° W. and dipping 50°-55° E. It has an average width of 2 feet and a maximum width of 5 feet. In some places brecciated wall rocks have received a stockwork of quartz stringers. Wall rocks are blue-black quartz-biotite hornfels, black quartzite and a medium-grained, altered intrusive rock, either diabase or diorite. According to the owner the high-grade ore occurs in small shoots and pockets in which specimen gold is associated with quartz and pyrite. Two ore shoots were discovered at or near the surface and about 100 feet apart (Julihn and Horton, 1940, p. 135). One about 40 feet long and 75 feet deep yielded 150 tons of ore from which 120 oz. of gold was extracted. The second shoot has been worked from the surface to a depth of 200 feet and has yielded 408 oz. of gold from 800 tons of ore or an average of slightly better than 0.5 oz. per ton ($17.85 per ton). This ore shoot was 30 feet long on the 50-foot level, but lengthened to 120 on the 100-foot level and tapered to 40 feet long on the 200-foot level. Ore remaining in the shoot on the 200-foot level is said to average about 1 oz. of gold per ton (Julihn and Horton, 1940, p. 135)—calculated to the present price of gold—from slightly more than 1000 tons of ore. The total recorded production of the mine is about $18,500.

Workings consist of several open cuts and a 200-foot inclined shaft with several levels and stopes. Equipment on the property includes a 4-inch jaw crusher, 3 stamps, amalgamating plates, 2 concentrating tables, a dismantled steam hoist and skip. The wooden headframe and shaft are in a fair state of repair but the mine is partly flooded. There are several serviceable buildings as well as the mill. Sufficient water for milling purposes is obtained from the mine.

Badger (Prescott) Mine. Location: Sec. 2, T. 5 S., R. 16 E., M.D., on the Eldorado Creek road between the Mount Gaines and Number Nine mines or about 6½ miles by road from Hornitos. Adjoins the Number Five mine on the north. Ownership: Mrs. Charles B. Cavagnaro, Hornitos, California owns the Prescott lode claim and mill site.

The Badger mine was discovered in the 1850s and is reported to have yielded $80,000 from surface workings during the early days of its history. Most of the later activity at the mine has not been recorded, but ownership has been in the Cavagnaro family for more than 30 years. The last work done on the mine was by C. S. Shafer and son of Hornitos in 1934. A small production was recorded at that time.

The Prescott vein is a rather sinuous, north-trending vein 1 foot to 3 feet wide dipping 30-45° east. Judging from the numerous shallow pits and other surface workings there must be several other parallel stringers or small blanket veins. Near the end of the line of main workings on the Prescott vein is a west-striking, nearly vertical cutter vein which has been stoped for a distance of over 500 feet. It is 1 foot to 3 feet wide and consists principally of milky quartz with abundant pyrite and some chalcopyrite. The quartz contains many vugs lined with crystals. The Prescott vein is of similar character. The authors were unable to determine whether the west-trending line of workings is part of the Badger mine or not. Wall rocks are blue-black, massive, spotted slate, schistose greenstone, black hornblende schist and horn-felsic slate, probably of Upper Jurassic age.

The Prescott vein is developed by several west and northwest-trending crosscut adits driven from the east slope of the ridge and a few short ones driven from the west side of the ridge. There are extensive drifts and stopes connecting with these workings and the whole hillside is covered with pits. Several of the crosscuts are accessible but in need of cleaning out, and workings below the lower adit levels are flooded (January, 1956).

Bandarita (Bandaretta, Eclipse, Goodwin). Location: NW¼, sec. 7, T. 3 S., R. 18 E., M.D., on the North Fork of the Merced River at its confluence with Gentry Gulch 11 airline miles east of Coulterville. Ownership: W. Lee Brown, c/o W. J. Beatty, Coulterville, California, owns 4 patented claims, the Bandarita Nos. 1 and 2, Little Fort Knox and Fort Knox No. 2. Nelson M. Leoni, c/o Dolores Pharmacy, Carmel-by-the-Sea, California, owns three patented claims, the Kinsley Nos. 1, 2, 3. In September 1954, the mine was accessible only by trail because of a washout on the Gentry Gulch road.

The Bandarita mine was discovered in 1856 (Browne, 68, p. 33) by Thomas E. Palmer and profitably worked almost continuously for more than 30 years. During the first 20 years the Goodwin Brothers and Peter Wynant were the principal operators. In 1873 operation of the mine was taken over from the Goodwins and Wynant by a San Francisco company under management of a man named Hanagan (Eng. and Min. Jour., 1873, vol. 16, no. 26, p. 412). By 1874 management had reverted to the Goodwin Brothers (Min. Sci. Press, 1874, vol. 29, no. 9, p. 133) who in a single mill cleanup took 25 pounds of gold from 100 tons of ore! By August 1876 management of the mine had passed to Levi Keyes, presumably still under ownership of the Goodwin Brothers (Min. Sci. Press, 1876, vol. 33, no. 10, p. 157). More than $160,000 was spent for mine machinery between 1872 and 1876 entirely from proceeds of the mining operations. About Jan. 17, 1880 a new 10-stamp mill run by water power was put into operation (Min. Sci. Press, 1880, vol. 40, no. 3, p. 37). Ownership passed to P. P. Mast of Coulterville about this time and C. L. Mast took over as superintendent. The Masts began to drive a tunnel from the level of the mill to crosscut the main vein 1200 feet from the surface outcrop (as measured along the vein). This ultimately reached a length of 1364 feet and encountered ore averaging $25.00 per ton. According to Julihn and Horton (1940, p. 141), the period 1881 to 1887 was the most productive in the mine's history. By 1889 the most of the ore above the lowest tunnel level had been stoped out and the mine became idle (Eng. and Min. Jour., vol. 42, no. 22, p. 407). Intermittent work was done from 1890 to 1898. Early in 1898 Lafayette Gold Mining Company of San Francisco leased the mine with the idea of cleaning out the lowest tunnel and sinking lower from the tunnel level. (Eng. and Min. Jour., vol. 65, no. 8, p. 229.) Little is known of the results of this operation, but by 1904 ownership had passed to Rodgers and Loomis of Springfield, Ohio. There is no record of work of any extent on the mine from 1904 to 1928 when ownership passed into the hands of Gentry Gulch Consolidated Mines Company of San Francisco (Laizure, 1928, p. 89). This operation was apparently short-lived and there is no known record of production from the Bandarita mine at that time. About 1937 the mine was leased from the owners, Nelson M. Leoni and J. E. Brown, by four partners, William Beatty, W. M. Boyer, Carl Crouse and G. A. Ogden (Julihn and Horton, 1940, p. 141) who operated the mine until 1943. Since 1943 the mine has been idle.

The Bandarita mine is credited with a total production of $1,520,000 much of which was mined between 1881 and 1887, according to Julihn and Horton. During the last period of activity, from 1937 to 1943, approximately $40,000 in gold and silver was obtained from 7362 tons of ore (U. S. Bur. of Mines records), the tenor of the ore being $5.42 per ton or considerably lower than the ore milled in the 1880s.

The principal vein at the Bandarita mine strikes N. 84° W. and dips about 47° S. Its width varies from 2 to 10 feet. Wall rocks are slate, quartz-biotite hornfels and granitic dike rocks. The metasedimentary series has been assigned to the Paleozoic Calaveras formation (group) by Turner and Ransome (1897).

In addition to native gold the ore contains pyrite, galena, sphalerite and tetrahedrite, the best ore containing considerable tetrahedrite. Presence of tellurides, mentioned by S. Rudolph (Min. and Sci. Press, 1880, vol. 40, no. 11, p. 162) has not been confirmed by later writers. Julihn and Horton (1940, p. 141) visited the property in 1938 and the following data is drawn largely from their report. There are four granodiorite dikes which strike roughly north and intersect the vein approximately at right angles to it. The vein has been faulted at these intersections with throws of from 1 foot to 42 feet.

The mine is developed by a 1364-foot crosscut adit, the portal of which is 50 feet above river level, and by six adits driven along the strike of the vein, two on the east side of the hill and four on the west slope of the hill. The longest of these adits, 800 feet in length, is driven into the east slope and taps three ore shoots stoped from the surface to various depths between 75 and 650 feet. The crosscut adit taps the vein about 850 feet vertically from the surface or 1200 feet as measured along the vein.

In 1938, lessees stated that there was 275,000 tons of gob (broken ore) in the old stopes of the mine that would pay to mill at costs prevailing at the time and at the prevailing price of gold. The lessees were, however, mining unbroken ore from the Goodman (Goodwin (?)) shoot which was reputed to average 0.57 oz. of gold per ton.

Black Hill (Pumpkin) Mine.   Location: E½, sec. 33, T. 2 S., R. 16 E., M.D., half a mile northwest of Coulterville. Adjoins the Margaret claim on the north. Ownership: Louise A. Boyd, 265 California St., San Francisco 11, California owns 40 acres of patented land.

The Black Hill or Pumpkin mine is the most northerly mine in the Mary Harrison group, all of which are aligned on a branch of the Mother Lode vein system commonly called the east branch. The east branch is roughly en echelon to the west branch on which are aligned the Malvina group of mines. The two branches are approximately 4700 feet apart.

The mine probably was discovered and first worked at about the same time as the Mary Harrison, about 1867. There is little record of activity before the 1890s although considerable development work must have been done. There was a 600-foot tunnel on the property as early as 1895 (Storms, 1896, p. 216) at which time the owners were Clark and McLaymont of Coulterville. By 1900 ownership had passed to John Boyd and the mine had been reopened under the management of J. J. Dolan. A new shaft was sunk to a depth of about 100 feet (Eng. and Min. Jour., vol. 70, no. 22, p. 647) and some rich ore was discovered. Very little work has been done on the property since 1902.

At the Black Hill mine the east branch of the Mother Lode strikes N. 35° W. and dips about 45° northeast. The vein is close to the contact between a serpentine intrusion and slate of the Mariposa formation, but at the surface the wall rocks are entirely serpentine. Vein matter is chiefly quartz-ankerite-mariposite rock and the vein is more than 20 feet wide. Considerable low-grade ore is found as pyritic impregations in serpentine wall rock adjacent to the vein.

Workings consist of a main inclined shaft, surmounted by a wooden headframe and ore bin, two tunnels driven northeast and several auxiliary shafts and stopes. The main shaft is watered and partly caved and the other workings are all caved and inaccessible. There is a tool shed in good condition on the property (September 1954) but no mining or hoisting machinery.

Bondurant (Hathaway-Bondurant). Location: Sec. 25, T. 2 S., R. 17 E., M.D., 12 miles east of Coulterville near the North Fork of the Merced River. Ownership: A. E. Adams, et al., 367 Maude Ave., San Leandro, California own one patented claim.

The Bondurant mine is the second oldest patented mine in the county, patent having been issued in 1856 (Julihn and Horton, 1940, p. 138). It was mined by hand methods and ore was treated in arrastras during the early years of operation, and one of the early owners was Judge Bondurant for whom the mine is named. By the late 1870s ownership had passed into the hands of Joshua Hendy of San Francisco who in 1879 sold the Bondurant and Martin-Walling mines to an eastern company for $80,000 (Min. and Sci. Press, vol. 39, no. 1, p. 5).* At that time a 10-stamp mill was operating 24 hours a day on $30.00 per ton ore. After a period of inactivity in the early 1880s the Hathaway-Bondurant Gold Mining Company was organized in 1887 with a capitalization of $1,000,000 (Min. and Sci. Press, vol. 44, no. 10, p. 173) and the mine was placed under the management of T. T. Hathaway. This operation ceased in 1891. After 5 years of intermittent activity the mine was unwatered and sampled by the Bruner Brothers of St. Louis, Missouri, Charles Bruner, superintendent. About 1902 the mine was taken over by the Boston and Mariposa Mining Company, Fred Whitman, superintendent, but this operation was also short-lived. In July 1911 under the management of George H. Gerkin, specimen ore was discovered on the 300-level and another high-grade shoot was discovered soon thereafter between the 150 and 200 levels (Min. and Sci. Press, vol. 103, no. 3, p. 90, 148). In 1913 ownership passed to the Bondurant Gold Mining Company, A. L. Adams, Bridgeport, Connecticut, the principal owner. A crosscut adit started by previous owners near the level of the Merced River, calculated to strike the vein at a distance of about 1400 feet from the portal, apparently was lengthened

* Walter McLean, Coulterville, California, suggests that this transaction probably involved the Bandarita mine rather than the Bondurant.

under this management. The adit apparently was only driven 945 feet (Lowell, 1916, p. 577). In 1916 the mine was sold by Adams to George R. Stone, Bridgeport, Conn., but little or no work was done under this ownership (Min. and Sci. Press, vol. 102, no. 18, p. 106). From 1926 to 1928 or 1929 the mine was operated by the Bondurant Mining Trust of Modesto, a partnership of Walter and Arthur Ritz (Laizure, 1928, p. 81). The Ritz management extended the crosscut adit started by the Adams operation to 1400 feet. From about 1929 to October 12, 1940 the Bondurant Mining and Milling Company of San Francisco operated the property and it was during this period that the largest production was recorded. The known production up to October 12, 1940 is about $292,-000. From that date to October 15, 1942, when closed by War Production Board order, the Boston-California Mining Company mined between 12,000 and 14,000 tons of ore which yielded approximately $98,000. Consequently the total known production of the mine is $390,-000. Between 1911 and 1915 approximately 746 tons of ore mined yielded 431.95 oz. of gold, 259 oz. of silver and 3672 pounds of copper. From 1931 to 1940 a total of 16,596 tons of ore mined yielded 7879.92 oz. of gold and 5176 oz. of silver.

There are three major, roughly parallel veins on or closely adjacent to the Bondurant property known as the Bondurant, Reynolds, and Louisiana. No material amount of work has so far been done on the Reynolds or Louisiana veins on the Bondurant property. The Bondurant vein strikes generally N. 65° W. and dips 35° to 45° E. at the surface. It is about 3 feet wide where it crops out at the surface but averages 4 feet wide at depth (McLean, W. D., personal communication, 1956). Vein matter now exposed at the collar of the shaft is chiefly hydrothermally altered, micaceous dike-rock, but most of the gold occurs in quartz with pyrite, chalcopyrite, sphalerite and galena. According to the recorded production figures ore has average 0.48 oz. of gold and 0.31 oz. of silver per ton. Much of the dump material is pyrite-impregnated slate and other wall rock which must carry some gold.

Wall rocks enclosing the Bondurant vein are crumpled slates, slaty mica schists and coarser-grained gneisses, presumably part of the Paleozoic series generally known as the Calaveras group. The precise character of the altered dike-rock associated with quartz vein-matter cannot be determined from hand specimens, but it probably is a fine-grained, biotite-rich granodiorite.

In August, 1954, the principal working at the Bondurant mine was a 10x12-foot shaft inclined 45° toward the northeast. This was filled with debris below an inclined depth of about 100 feet. According to Julihn and Horton, the last writers to visit the mine while it was operating, the shaft was 412 feet deep and had a total of 2500 feet of drifts on 5 levels, the 100, 150, 250, 300 and 350. In 1938 the crosscut tunnel, begun near river level and driven 945 feet in 1913-14, had been lengthened to 1100 feet. At that time the company planned to continue the tunnel 600 feet farther until it intersected the vein. The projected point of intersection is approximately 1100 feet from the surface outcrop of the vein as measured along the dip (Julihn and Horton, 1940, p. 138). There is no record showing whether the projected 1700-foot tunnel was ever completed or not. The mine was shut down about 1942 by War Production Board Order L-208 and has been inactive since that time. All equipment has been removed from the property and all of the workings except the upper part of the shaft were inaccessible at the time of the authors' visit in August 1954.

Buena Vista Group (Busch, Washington-Buena Vista). Location: Secs. 20, 21, 28, T. 4 S., R. 18 E., M.D., in the Colorado district 5 airline miles north and slightly west of Mariposa via State Highway 49 and the Saxon Creek Road. Ownership: C. W. and Velma Worley, Mariposa, California own three patented claims, the Washington, Buena Vista and Phoenix. Once included the Talc, Lucky Lindy and Charles claims.

The Buena Vista group of mines was first opened in 1882 or 1883. In January 1884 a 10-stamp steam mill was put into operation by the O'Gorman Brothers (Min. and Sci. Press, vol. 48, no. 5, p. 88). By May 8, 1886 the main shaft had reached a depth of 100 feet with Lorenzo Alvord in charge of operations (Mariposa Gazette, May 8, 1886). Early in 1887 the mine was leased or bonded from a Dr. Turner to Emmanuel San Pedro who operated the property for several years. (Min. and Sci. Press, vol. 54, no. 4, p. 56) In 1892 an 8- or 9-inch wide ore shoot was struck in the vein at a depth of 15 feet which ran $62.00 per ton in gold and $1.67 in silver. (Eng. and Min. Jour., vol. 54, no. 17, p. 397) The mine was idle much of the time between 1893 and 1916. In 1917 it was reopened by W. M. Brice, probably under lease from L. F. W. Busch (Castello, 1920, p. 109). About 1919, Busch sold all of his holdings except a talc claim, the Washington claim going to Pacific States Syndicate of Los Angeles (Castello, 1921, p. 141). None of the claims appear to have been operated to any extent until 1927 when the Buena Vista group of claims was acquired by Consolidated Gold Fields of Mariposa, Belle McCord Roberts of Long Beach, President (Laizure, 1928, p. 120). This company did considerable cleaning out and sampling and several shoots containing milling-grade and high-grade ore were found (Laizure, 1928, p. 120-121). This operation apparently did not survive the crash of 1929 and depression of the 1930's and the claims have been idle during most of the period 1929-1955.

The principal vein on the Buena Vista claims is 2-7 feet wide, strikes N. 70°-75° W. and dips steeply northeast. There is at least one parallel stringer vein. Wall rocks are slate and metavolcanic porphyries. In 1914 the principal working was a 180-foot shaft with 140 feet of drifts, one winze of unknown length and a 67-foot crosscut tunnel (Lowell, 1916, p. 578). According to Laizure (1928, p. 1?0) most of the workings are on the Washington and Buena Vista claims. He mentions a tunnel driven on a stringer vein, a crosscut opening about 200 feet from the main shaft and drifts on the vein from the crosscut tunnel. Assays made in 1927-28 by Techow and Davis of Sacramento varied from $6.40 to $32.96 per ton (Laizure, 1928, p. 120). The mine is idle and the property was not visited during this investigation. The foregoing data were abstracted from the literature and from records of the U. S. Land Office at Sacramento.

Buffalo (San Domingo) Mine. Location: SW¼, sec. 13, T. 4 S., R. 18 E., M.D., about 4 airline miles north of Midpines on Trabucco Creek. Ownership: Vernon and Clara Tharp, 1850 Central Ave., Alameda, California own one claim and a mill site.

Little has been written of the early history of this mine. It was described as an old mine in 1910 at which time a large body of milling-grade ore had been developed under direction of R. L. Mann (Min. and Sci. Press, vol. 101, no. 15, p. 486). The most recent period of activity was from 1932 to 1940, mainly under the management of the Gabriel Mining Company of Midpines. There is no machinery on the property and the workings are covered by dense brush (September, 1954). Local prospectors report that entry to the mine may be made through the air shaft and that the tunnel is open for nearly 300 feet past the caved portal. During the last period of activity, a little more than $15,200 in gold and silver was taken from the mine from an unknown tonnage of ore.

Rocks exposed in the vicinity of the workings are slate, greenstone and various granitic rocks. According to Castello (1920, p. 137) early workings explored a quartz vein 2 feet wide carrying free milling gold, having a northwest strike and northeast dip of 50°. He described the footwall as slate and the hanging wall as granite. Workings at that time consisted of a 400-foot tunnel, a 180-foot air shaft and several winzes and raises. Laizure (1935, p. 29) states that the main quartz vein varies from a few inches to 7 feet wide with several feeder veins, all badly fractured. Ore shoots were described as short and irregular. According to his account the vein crops out at the surface, has a north strike and an east dip of 26°. Workings in 1935 were reported to consist of a 310-foot tunnel and connecting winze, reaching 400 feet below the outcrop and with 500 feet of drifts on the vein. Also, 1000 feet of upraises and intermediate drifts and one stope 100 x 150 feet.

Campo (Campodonica Italian) Mine. Location: NW¼, sec. 16, T. 5 S., R. 16 E., M.D., 1 mile east of Hornitos. Ownership: Not determined.

The Campo mine was discovered in early gold rush days and worked in a small way with an arrastra during the 1850s and 1860s. In 1874 the lower level of the Campo mine reached 110 feet below the surface and the shaft was 205 feet deep on a 45° incline (Min. and Sci. Press, vol. 29, no. 4, p. 53). At that time the vein was 8 feet wide, but only a small amount of the width constituted ore. By 1876, a 10-stamp mill had been erected which milled 10 tons of ore in a 24-hour period, ore running $15.00 per ton at the prevailing price of gold (Min. and Sci. Press, vol. 33, no. 12, p. 189). Chinese labor was employed. After a period of inactivity of unknown duration the mine was reopened in 1896 by the Campodonica Mining Company under the direction of James Kennedy. (Min. and Sci. Press, vol. 73, no. 11, p. 222.) Under this management the mine yielded 387 oz. of gold from unknown tonnage of ore (U. S. Bureau of Mines records) but the mine was again closed down in 1898 or 1899. During 1937-1939 the mine was operated by a group headed by N. L. Wagner, L. G. Corwin and Claude Shafer of Hornitos. In this period 356 tons of ore were mined which yielded 238 oz. of gold and 96 oz. of silver. In 1938, the old dump sampled $8.42 per ton (Julihn and Horton, 1940, p. 125.) The mine has not been operated since 1940.

According to Julihn and Horton, the last writers to visit the mine while it was operating, the quartz vein averages 4 to 5 feet wide between a schist footwall and a porphyry (greenstone) hanging wall. It strikes north and dips 40° to 45° east. The principal working is an inclined shaft 230 feet deep from which a large but unknown footage of drifts has been driven. In 1938, the shaft was being cleaned out and retimbered, a cave-in having occurred 40 or 50 feet below the collar of the shaft. The rocks in the vicinity of the mine are slates and schists of the Upper Jurassic Mariposa and Amador groups of rocks.

Carson (Bouvier) Mine.   Location: Secs. 3, 4, T. 4 S., R. 16 E., M.D., in Hunter Valley 8 airline miles north of Hornitos. Adjoins Iron Duke mine on the south. Ownership: Not determined, probably Inez Bouvier, et al., 1431 Moraga St., San Francisco, California.

The Carson mine consists of four adjoining unpatented claims, the Carson, Carson No. 1, Carson No. 2 and Bouvier. The Carson claim adjoins the Iron Duke claim on the south, the others extending southeast in the order named. All are on a well-known bluish quartz vein, locally called the Blue Lead, which strikes N. 30°-35° W. and dips about 60° east. The Blue Lead is crossed nearly at right angles by several cutter veins wherein some of the best ore is found. The Blue Lead varies from 2 to 20 feet wide and the cutter veins from 2 to 4 feet wide. The main vein is on or close to the contact between the Hunter Valley chert and meta-augite andesite greenstone of the Amador group. Both these units are of Upper Jurassic age. In general, chert forms the hanging wall and greenstone the footwall. The principal working, located on the Carson claim on a cutter vein about 150 feet east of the Blue Lead, is a 125-foot inclined shaft. There are several other prospect pits. The mine was once equipped with a 5-stamp mill, but this burned in 1868 (Castello, 1921, p. 108). The mine is idle and there is no equipment on the property.

Champion I Mine.   Location: Secs. 28, 33, T. 2 S., R. 16 E., M.D., on Blacks Creek 1¾ miles N.W. of Coulterville. Ownership: Car Da Mining Company, a trust estate (½) and Adelaide M. Ray (½), c/o C. P. Rose, 1259 North Fuller Ave., Hollywood, California, own 1 patented claim consisting of 10.37 acres.

The Champion I mine is a very old property probably located in the 1850s. Little is known of the early history of the mine except that it was known to be active in 1878. From 1903 to 1905 the mine was operated intermittently by C. E. Van Meter under the management of a Mr. Chandler of San Francisco. The main shaft was 100 feet deep in 1904 (Min. and Sci. Press, vol. 88, no. 11, p. 184) and 200 feet deep in 1905 (Min. and Sci. Press, vol. 90, no. 12, p. 197). By 1906 ownership of the mine was held by N. S. Ray, the Mentzer Brothers and Daniel Wagner. A strike of rich ore was reported in the mine in September 1906 (Min. and Sci. Press, vol. 82, no. 12, p. 561). In 1910 the mine was under lease to Bagby, Wilburn and McQuinn who had an operating mill and employed 15 men. In February 1910, a single pocket of gold yielded $74444 (Min. and Sci. Press, vol. 89, no. 7, p. 388). In the fall of 1910 one mill cleanup from 250 tons of ore milled yielded $14,-800 (Min. and Sci. Press, vol. 90, no. 19, p. 926). About 1914, the mine was leased to C. I. Mentzer, et al., of Coulterville who milled several thousand dollars' worth of ore averaging $45 per ton. In 1917, the Champion Mining Company of Los Angeles was permitted to issue 675,000 shares of stock to C. H. White and associates for reactivation and redevelopment of the mine. This company constructed a 10-stamp mill and milled a small tonnage of ore which ran about 5 oz. of gold per ton (U. S. Bur. Mines records). By 1920, ownership of the mine had passed to Mrs. D. Wagner, Mrs. N. C. Kay and Mrs. M. A. Mentzer (Castello, 1921, p. 110). By 1936, ownership of the mine had passed to the Car Da Mining Company, Frank A. Notterman of Sonora, president and general manager. In 1938, 16 men were employed. The mine has been idle since the period 1936-38. All equipment and buildings have been removed from the property except a tool shed and cabin. The workings are full of water (September, 1954).

The recorded production of the Champion I mine since 1900 is about $146,200.00 with the total production somewhere between $150,000.00 and $200,000.00. About $67,000.00 in gold and silver was produced during the last period of activity, 1936-1939.

The long direction of the main shaft is oriented N. 70° W. and it dips 55° northeast. The Champion claim is also oriented about N. 70 W. (U.S. Land Office records) and this is presumably parallel to the strike of the vein, which does not at present crop out strongly at the surface. Castello (1921, p. 110), Laizure (1928, p. 83) and Julihn and Horton (1940, p. 111) all state that the main vein strikes N. 17° E. and dips 45° SE. between slate walls, but these data are evidently in error. There is no slate on the dumps or near the workings and no evidence that the vein strikes northeast. Rock on the dumps is serpentine and greenstone and there is a serpentine-greenstone contact less than 100 feet east of the main shaft. Vein matter on the dumps consists of sulfide-impregnated quartz and quartz-ankerite-mariposite rock. Also in the dumps is a large quantity of sulfide-impregnated wall rock which probably constitutes low-grade ore. According to Julihn and Horton the ore contains free gold, pyrite, galena, tetrahedrite and sphalerite. They state that the main shaft is 440 feet deep with levels at 140, 240, 340, and 440 feet. They also mention a 200-foot exit shaft and 2000 feet of drifts, crosscuts and raises. In 1938 the mine made about 40,000 gallons of water per day in summer and 60,000 gallons per day in winter (Julihn and Horton, 1938, p. 111). A complete account of the milling practice in 1938 is described by Julihn and Horton. The mill has been removed from the property.

There are two shafts of unknown depth about 1,000 feet southwest of the main shaft which apparently are on the Champion claim. The relationship of these shafts with the other workings could not be determined. One of these may be the exit shaft mentioned by Julihn and Horton.

Champion II Mine. Location: Sec. 34, T. 3 S., R. 18 E., M.D., 1¼ miles south of Colorado School and east of the Colorado-Sherlock Creek road, or 5 airline miles northeast of Mariposa. Ownership: Belle McCord Roberts, 2625 East Tenth St., Long Beach, California, has retained mineral rights on one claim of about 20 acres, which is part of a parcel of patented agricultural land.

The Champion II mine was discovered sometime before the 1880s about the same time as many other mines in the Colorado district. In October 1889 the mine was being operated by a partnership of James Ridgway, Colonel Dunbar and Mr. Hay. At that time it was worked from a 4x10-foot nearly vertical shaft 90 feet deep with drift levels at 60 and 90 feet run both directions from the shaft (Min. and Sci. Press, vol. 59, pp. 320, 450, 1889; Goodyear; 1890, p. 303). Five hundred tons of ore taken out during 1889-90 was reported by Goodyear to average about $12 per ton. Ore was crushed for a time in a steam-powered arrastra but this proved impractical and a 5-stamp mill was planned. Management disputes apparently resulted in termination of this period of activity. In 1919 the mine was owned by G. E. Dunbar of Mariposa but only assessment work was being done (Castello, 1921, p. 110). Late in 1927 the Champion mine was acquired along with a large group of other mines in the Colorado district by Consolidated Gold Fields of Mariposa, Inc., in which the present owner was the principal partner (Laizure, 1928, p. 84). No sustained mining was done before the depression years starting in 1929 and little has been done with the mine in recent years.

The Champion vein strikes N. 65-70° W. and varies only a few degrees from vertical. The average width is perhaps 4 feet but it locally thickens to as much as 12 feet. Vein matter is in part massive milky quartz and in part a ribbon-structure of laminated white and glassy quartz including thin sheets of slate wall-rock. Ore minerals are pyrite, galena and native gold. The vein is on or close to the contact between slaty and schistose metasediments of the Paleozoic Calaveras group and hornblende granodiorite. The granodiorite is cut by several granitic porphyry dikes the largest of which parallels the strike of the vein and is about 300 feet northeast of it.

Workings in 1919 consisted of a 96-foot shaft with levels at 60 and 90 feet. On the 60-foot level the east drift was 78 feet and the west 50 feet. On the 90-foot level there is a 60-foot west drift only. In addition there are shallow workings strung out along the vein for over 1300 feet. In August 1955 the shaft was open to a depth of 40 feet.

Clearinghouse Mine (Original and Ferguson, Anderson). Location: Secs. 16, 21, T. 3 S., R. 19 E., on the Merced River and State Highway 140 at Clearinghouse, 7 miles west of El Portal. Ownership: Mrs. Frank E. Gallagher, 211 26th St., Merced, California, owns ten patented claims, the Original, Original No. 2, South Original Extension, El Portal, Anderson, North Extension of the Anderson, Golden Rule, Golden Rule No. 2, Moonstone and Moonstone No. 2, totaling about 190 acres.

History: The more westerly group of claims, best known as the Ferguson mine, but also called the Anderson mine, was discovered about 1860, probably by the Ferguson brothers. During the 1860s the mine was operated by the Fergusons and a large tonnage of high-grade ore was extracted. The *Sonora Herald*, February 22, 1868, states that 174 oz. of gold were recovered during a week of operation of the 8-stamp mill at the Ferguson mine. This mill processed 6 tons of ore in 24 hours so that the ore must have run several ounces of gold per ton. In 1870 the mine was sold by the Fergusons to a San Francisco company for $80,000 (Min. and Sci. Press, vol. 21, no. 15, p. 252) but ownership had reverted to E. Ferguson by 1871. In February 1871, the ore was averaging $44.00 per ton (Min. and Sci. Press, vol. 23, no. 16, p. 244) and the mill had been enlarged to 15 stamps. The mine was intermittently idle during the middle 1870s but was operating full scale in 1878 under the ownership of E. Ferguson, with Robert Frances

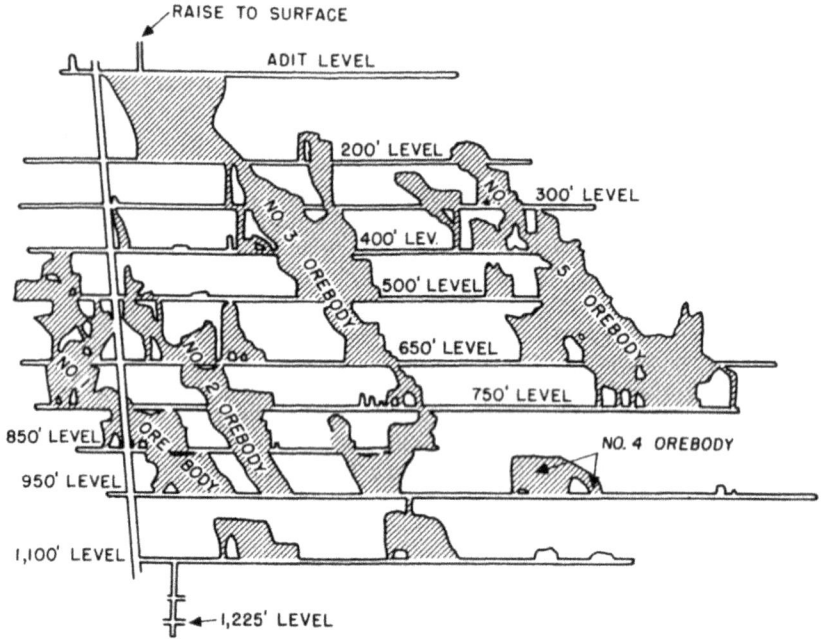

FIGURE 18. *A.* Longitudinal section (in the plane of the vein) of the Original vein workings, at the Clearinghouse mine just north of the Merced River in the El Portal district. The disposition of the ore-bodies is well shown. *Reproduced from fig. 1 of Engineering and Mining Journal, vol. 103, no. 16 (Young, 1929, pp. 45-48).*

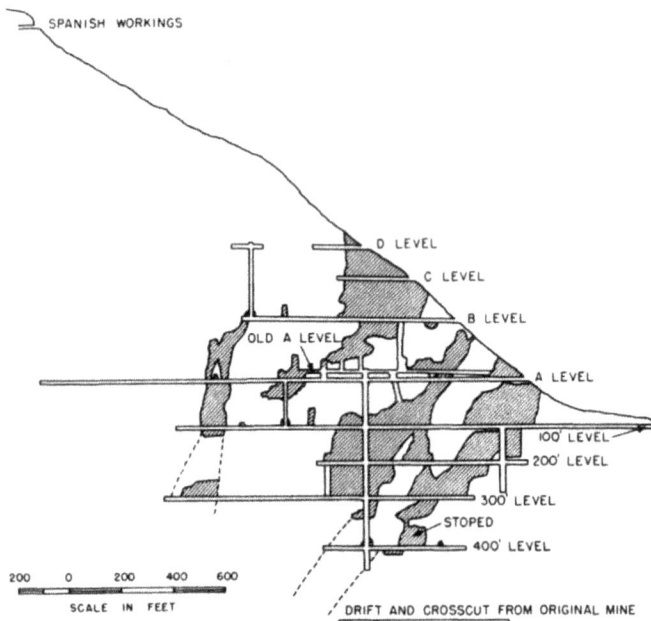

FIGURE 18. *B.* Longitudinal section of the principal workings on the Ferguson vein of the Clearinghouse mine in the El Portal district, showing disposition of the ore.

FIGURE 18.   C. A typical underground view of the Original vein as seen in the early days of mining on the number one adit about 280 feet from the portal. The width of the vein matter, attitude and ribbon structure are clearly shown. *Reproduced from an old photograph by courtesy of Elisabeth L. Egenhoff.*

as superintendent. Ore at that time ran from $10 to $40 per ton, but in 1879 some ore ran between $75 and $100 per ton (Min. and Sci. Press, vol. 37, no. 22, p. 34; vol. 38, no. 12, p. 181).

As early as 1870 the main adit had reached a length of 1100 feet. About 1871 an underground shaft or winze was sunk from the end of the 1100-foot tunnel. This was 100 feet deep in February 1871 (Min. Sci. Press, vol. 22, no. 8, p. 115). About 1878 a second underground shaft was sunk from the main adit at a distance of about 1000 feet from the portal. This No. 2 shaft was 200 feet deep by fall of 1878 (Min. and Sci. Press, vol. 37, no. 14, p. 341).

Through 1880 the Ferguson mine continued to be an active producer and in July 1880, fifty pounds of specimen gold-quartz was sold to a jeweler for $1200 (Min. and Sci. Press, vol. 41, no. 18, p. 273). From 1881 to 1934 the mine was largely inactive. It was bonded for a short time in 1899 to J. F. Joseph and associates of Sonora (Eng. and Min. Jour., vol. 68, no. 16, p. 465); to L. and E. Mason, Incline, in 1905; and to George S. Barber of Oakland in 1910. There was no recorded production under any of these operators. By 1920, ownership had passed to Mrs. S. A. Hall and Anna K. Sherwood under the name Ferguson Consolidated Gold Mines, but the mine remained idle.

The more easterly group of claims, called the Original mine, was first located in 1908 (Laizure, 1928, p. 107). It was operated almost continuously from 1911 to 1933 by the Original Mining and Milling Company (E. C. Kocher, pres., G. M. Egenhoff, mgr., J. W. Warford, supt.), from Jan. 1, 1933 to October, 1934 by a group of lessees headed by J. W. Warford (Laizure, 1935, p. 40), and from 1934 to 1937 under lease by San Juan Ramsey Company (Boston, Mass., W. C. Smith, pres., A. S. Wyner, Incline, mgr.). About 1925 the claims of the Ferguson mine were acquired by the Original Mining and Milling Company and all ten claims of the present property were patented during that year (Laizure, 1928, p. 107). In December 1937 a flood washed out both the Yosemite Valley Railroad and the state highway. The mine closed down and has since been idle (Julihn and Horton, 1940, p. 143).

The Clearinghouse mine has been among the ten highest-producing mines in Mariposa County, the total production being in excess of $3,350,000. Claims of the Ferguson mine have yielded an early estimated production of $1,250,000 (Julihn and Horton, 1940, p. 134) and a recorded production between 1935 and 1940 of $107,000 for a total of $1,357,000. Between 1911 and 1937 the claims of the Original mine yielded 79,636.79 oz. of gold, 25,364 oz. of silver, 340 pounds of copper and 134 pounds of lead from 175,156 tons of ore. This amounted to slightly more than $2,000,000 at the prevailing gold prices, but would amount to about $2,797,900 at the present price of gold.

Geology: The Original and Ferguson veins lie on opposite sides of a north-trending sill-like mass of coarse-grained granitic rock approximating biotite granodiorite. The granodiorite has been intruded into metasediments of the Paleozoic Calaveras group roughly parallel to the slaty cleavage and schistosity. The Original vein, lying on the east side of the mass strikes N. 7°-10° E. and dips 78° east. To a depth of about 400 feet the vein lies entirely in slaty metasediments, but below that depth the wall rocks are chiefly granodiorite (Young, 1929, p. 46).

The Ferguson vein, lying on the west side of the granitic mass, strikes about N. 21° E. and dips about 70° west. In general, it follows the contact between the granodiorite and the metasediments. Metasedimentary rocks on the dump include slate, quartz-biotite hornfels, graphite-mica schist and some quartz-mica gneiss. According to Young (1929, p. 46) the slaty cleavage dips 40° east near the apex of the Ferguson vein but steepens to 78° at adit level. Three lateral quartz veins join the Ferguson vein at various angles from a northwesterly direction, the Spanish, Golden Rule and Moonstone veins. These have been superficially explored and their form is not yet certain.

Five ore shoots have been discovered in the Ferguson vein north of the underground shaft. These are roughly parallel and pitch north at an average of 60°. The shoots average 4-5 feet thick and 100 to 150 feet wide. Two of the five shoots have been stoped to depths exceeding 1000 feet, as measured on the axis of pitch. Four ore shoots have been worked in the south end of the Ferguson mine. These pitch north at angles between 40° and 70°. The character of ore is similar in most ore-shoots, arsenopyrite, galena, sphalerite and native gold occurring in ribbon quartz. In general, the sulfide content of the ore is about twice as great in the shoots of the Ferguson vein than in shoots of the Original vein (Julihn and Horton, 1940, p. 145).

Mine Workings: Inasmuch as the workings were not accessible to the authors, the following mine descriptions have been drawn largely from Young (1929) and Julihn and Horton (1940). Julihn and Horton were the last authors to visit the Clearinghouse mine while it was still in operation.

There are extensive workings on both the Original and Ferguson veins. The Original vein workings are entered by a 175-foot, west-trending crosscut adit which intersects the vein 180 feet below the outcrop. The portal of this adit is on the west side of a ravine at an elevation of approximately 1600 feet. A 2-compartment underground shaft or winze, inclined 78° east has been sunk from near the end of the crosscut adit to an inclined depth of approximately 1170 feet. The shaft serves 9 levels (below the adit level), the 200, 300, 400, 500, 650, 750, 850, 950 and 1100-foot levels. A 250-foot vertical winze with short levels at 1125 and 1150 feet is sunk from the 1100-foot level at a point 150 feet north of the main shaft and there is a 150-foot raise to the surface from the adit level. The total vertical depth explored in the Original vein is, therefore, approximately 1530 feet. Development work north of the main shaft amounts to an average of 1300 feet of drifts per level, the longest drifts being on the 650 and 950 levels. South of the main shaft the maximum distance penetrated is about 265 feet with some stoping between the 400 and 850 levels. There are extensive stoped areas north of the shaft between the adit and the 1100 levels.

From the 650 level of the Original workings 510 feet north of the Ferguson shaft a 1200-foot crosscut connects with the Ferguson vein. Owing to a faulty survey, the 600-foot drift along the vein from this crosscut did not strike the number 1 ore body and the drift was never connected to the other workings above it on the Ferguson vein. In addition to the crosscut and drift from the Original workings, the Ferguson vein is developed by 5 adit levels which are top to bottom, the D-, C-, B-, A- and 100-levels. There is also an unused former A-level roughly 50 feet above the main level (A or haulage level). The A and 100-levels are slightly over 1200 feet long; the others average less than half as long. From the A-level, a 450-foot winze or underground shaft has been sunk above 400 feet from the portal. This shaft serves the 200-, 300- and 400-foot levels. A second winze or underground shaft connects the A- and 100-foot levels about 600 feet from the portal of A-level, and there is a 200-foot winze sunk about 375 feet north of the portal of the 100-level. The 100-foot winze sunk from the end of the 1100-foot A-level in 1871 and a 200-foot winze sunk from A-level 1000 feet from the portal in 1878 do not show on the latest available mine diagrams. The latter apparently was stoped out and the former may be filled with gob or debris. Ore has been stoped to a vertical depth of slightly more than 450 feet and over a vein length of about 1000 feet.

About 1800 feet north and 1350 above the portal of the A-level are a group of old Spanish workings consisting of several short drifts with superficial connecting workings. These apparently are on the North Extension of the Anderson claim.

Grade of Ore: The ore milled from the Original vein from 1911 to 1934 averaged 0.454 oz. of gold, and 0.14 oz. of silver (calculated from U. S. Bureau of Mines figures) and a little copper and lead. Zinc although present was not recovered. The Original vein ore from 1935 to 1940 average 0.274 oz. of gold and 0.083 oz. of silver. There is no indication of the quality of the ore from early operations except the historical notations made earlier in this account.

Colorado Mine. Location: Near the center of sec. 27, T. 4 S., R. 18 E., M.D. on the east side of Long Canyon Creek a quarter of a mile west of Colorado School or 5 airline miles north of Mariposa. Accessible by graded dirt road via either Saxon or Sherlock Creek roads. Ownership: Property is jointly owned by Corinne Kratzer ($\frac{1}{4}$), 555 Thirty Second Ave., Richmond, California, F. E. and P. W. Judkins ($\frac{1}{4}$), 2817 San Pablo Ave., San Pablo, California, Charles and Irma St. Johns ($\frac{1}{4}$), 619 Humboldt St., Richmond and Ellen R. Weston ($\frac{1}{4}$), 443 Ninth St., Richmond. Property consists of 1 patented fractional claim of 11 acres and 2 unpatented claims.

Little of the early history of the Colorado mine has been recorded. It already had extensive workings in 1914 when the mine was being developed by P. W. Judkins, C. H. Weston and I. L. Dearborn of Mariposa. Ownership of the mine still remains with these individuals or their heirs.

The Colorado vein averages $1\frac{1}{2}$ to 2 feet wide, strikes N. 50° W. and dips 60°-80° NE. Locally it reaches a width of 7 feet. Wall rocks are chiefly slate and sandy metasediments belonging to the Paleozoic Calaveras group. The metasediments have been intruded by greenstone dikes and sills 1 foot to 3 feet wide. The strike and dip of the slaty cleavage near the shaft is roughly parallel to the vein and to the bedding planes. Vein matter is chiefly banded quartz carrying free gold and a little pyrite. According to Julihn and Horton (1940, p. 153) the ore shoots are 30 to 75 feet long and pitch north at about 45°.

Workings are reported to consist of a 200-foot inclined shaft, a 500-foot working level that intersects the shaft 125 feet from the collar, and a 200-foot drift on the 95-foot level south of the shaft (Julihn and Horton, 1940, p. 153). According to Laizure (1928, p. 83) there is a stope above the 95-foot level 120 feet long and 35 feet high and Julihn and Horton state that most of the known ore has been stoped to the surface above the 95-foot level. The workings are caved and inaccessible, and there were no buildings or equipment on the property in August 1954.

Nothing is recorded on the early production of the Colorado mine. In the late 1920's, about 2500 tons of ore was mined which ran $10 to $11 per ton with some "development rock" running $8 (Laizure, 1928, p. 83). Between 1915 and 1938, 7129 tons of ore was milled which yielded 2011.91 oz. of gold and 351 oz. of silver (U. S. Bureau of Mines records). The average tenor of ore was, therefore, 0.282 oz. of gold per ton or about $15.87 at the present price of gold. The total production of the mine probably does not greatly exceed $50,000.

Cotton Creek (Hauser) Mine. Location: NW ¼, sec. 24, T. 4 S., R. 16 E., M.D., on Cotton Creek ¼ mile east of the Hunter Valley road or 7½ airline miles northeast of Hornitos. Ownership: not determined. Last known operator (1939) was Cotton Creek Mining Company, W. H. Hauser, Richmond, California, manager.

In July, 1937, the Cotton Creek Mining Company began development of a hitherto little-known prospect (Julihn and Horton, 1940, p. 120). This company operated the mine until 1939 since when it has been idle except for assessment work.

The vein, which averages about 2½ feet wide, strikes N. 40° W. and dips 20-25° S. In places it reaches a width of 5 feet. Vein matter consists of quartz, auriferous pyrite, chalcopyrite, galena, and native gold. Large areas of wall-rock adjacent to the vein have been impregnated with pyrite. Wall rocks are greenstone (meta-augite andesite) of the Upper Jurassic Peñon Blanco formation.

Workings consist of a shaft inclined 20° to the S.W. and an unknown number and footage of drifts. In August 1938, the shaft was 175 feet deep with 75-foot drifts north and south from the shaft on the 100-foot level (Julihn and Horton, 1940, p. 120).

In August 1954 the mill and hoisting equipment were in good condition, but the shaft was caved just below the collar and the headframe was in need of repair. Equipment includes a hoist, compressor, jaw crushers, Straub mill, rake classifier and other miscellaneous equipment. The shaking tables and flotation cells which once were part of the mill had been removed.

Between 1937 and 1939 the mine produced 4897 tons of ore which yielded 1304 oz. of gold, 725 oz. of silver, 1669 lbs. of copper and 1730 lbs. of lead (U. S. Bureau of Mines records). According to Julihn and Horton (1940, p. 120) the ore averaged $15 per ton at the present price of gold, but by U. S. Bureau of Mines figures the yield was nearer $11 per ton.

Diltz (Diltz and Mann, W.Y.O.D.) Mine. Location: E. ½ sec. 29 T. 4 S., R. 18 E., M.D., on the east side of Sherlock Creek 5 airline miles north and slightly west of Mariposa. Accessible by good, graded dirt road via Whitlock Creek. Ownership: Diltz Mines (Allen F. Grant, Earl E. Baker et al.), c/o John P. Fulham, Mariposa, California, owns two patented claims, the Diltz and Mann and W.Y.O.D., and 6 unpatented claims (3 full and 3 fractional).

FIGURE 19.   Mill building and water tank at the Diltz gold mine in the Sherlock Creek district. The Diltz mine, with a total production in excess of $750,000, is noted for its pockets of specimen ore found particularly in the feathered parts of the vein and in stringer veins. Much of the early day rich material was placered from the weathered, partly eroded vein system at or close to the surface. *Photo by Mary H. Rice.*

FIGURE 20.   Parallel stringer veins of Diltz vein system at the portal of one of the mine adits. The vein system has a general northerly strike and a gentle east dip near the surface, but steepens somewhat at depth. *Photo by Mary H. Rice.*

The Diltz mine was discovered during the 1860s either by Tom Early or by Captain John S. Diltz. The *Mariposa Gazette* of April 10, 1886, states that Early worked the mine before Captain Diltz became owner, taking out 232 oz. of free gold by sluicing and sending 250 tons of ore to the Whitlock mill which returned $32 per ton. Captain Diltz operated the mine under very adverse circumstances from the 1870s until the time of his death in 1895, chiefly by washing the surface soil and soft, brecciated parts of the vein. In 1884 a landslide buried the Rickard shaft (the most northerly of several shallow shafts) and landslides seriously hampered operations in January, 1886. In 1886 Diltz recovered a mass of gold weighing 10 lbs. 6 oz., avoirdupois, for which he received $2000 (Min. and Sci. Press, vol. 52, no. 16, p. 260). Throughout the period of Diltz's ownership the mine was not equipped with a mill and very little ore was custom-milled elsewhere. Early in 1887 the mine was leased for a short period by Joe Shantz and Sam Jacoby and in 1890 George Stewart worked the mine for a time in partnership with Captain Diltz (Min. and Sci. Press, vol. 16, no. 13, p. 216).

After the death of Captain Diltz in 1895, the mine was held for a time as part of the Diltz estate and then was acquired by Hugh Devaney. In 1914 it was purchased from the Devaney estate by Mr. and Mrs. S. J. Harris of Jerseydale and some work was done at the mine that summer by the Mariposa Mines and Development Company, a partnership of C. A. Schlageter and T. L. Diven (Castello, W. O., 1921, p. 113). By 1919 ownership had passed to Mrs. Jenny Diven of Oakland, California. In 1927, George Ahart of Mariposa leased the mine from Mrs. Diven and installed a 10-stamp mill (Laizure, C. M., 1928, p. 85). The present owners acquired the property in 1931 and operated it during the periods 1931-42 and 1943-50. In May 1932, a mass of specimen gold weighing 52 pounds, troy, was taken from the 100-level. It yielded 43 lbs. 2 oz. of gold worth $10,707 at the old price of $20.67. Another specimen was recovered weighing 20 lbs. 2 oz. worth $3,862. Pockets containing $100 to $500 in free gold are not unusual.

The total estimated production of the Diltz mine is between $750,000 and $1,000,000. From 1924 through 1954 the mine produced 81,657 tons of ore from which 21,135 oz. of gold and 3395 oz. of silver were extracted (compiled from records of the U. S. Bureau of Mines). These figures do not include the period during which Captain Diltz operated the mine nor the specimen gold which was marketed elsewhere than the mint. Most of the recorded production was made since the price of gold was raised to $35 an ounce. The average tenor of ore milled was about 0.24 oz. per ton. The *Mariposa Gazette* of April 25, 1885, states that ore and float from the surface workings averaged $15 per ton (at the former price of gold). A later entry for December 12, 1885, indicates that the surface material averaged $14 per ton by sluicing and that the remaining uncrushed ore ran $30 per ton.

The veins at the Diltz mine cut an ancient series of pyroxene andesite greenstones of probable Paleozoic age. Both intrusive and flow-rocks as well as pyroclastic masses are represented in the series. The principal vein strikes N. 10° E. and dips 35-40° east. At depth it has an average width of about 2½ feet, but in places reaches a width of 10 feet. Near the old open workings of the mine northeast of the main shaft, the vein flattens or rolls and feathers out into the wall rocks in places becoming ,15-20 feet thick. According to old accounts the vein never apexed at the surface. The greenstone of the hanging wall differs considerably in appearance from that of the footwall side which may indicate a considerable throw on the fault the vein occupies.

In addition to native gold, ore minerals include arsenopyrite, pyrite, chalcopyrite, tetrahedrite and galena. Gangue minerals include milky and ribbon quartz, calcite and oxides of iron and manganese. According to Julihn and Horton (p. 146, 1940) the chief sulfide mineral is arsenopyrite. Ore shoots have been found in the vein over a distance of more than 1400 feet of strike length. These average about 350 feet long (Laizure, 1928, p. 29).

The principal working is a 625-foot inclined underground shaft sunk from the compressor level 200 feet below the top of the vein and 300 feet from the portal. This shaft serves 6 levels aggregating 4700 feet of drifts (Julihn and Horton, 1940, p. 146). There are also several shallow inclined shafts and several tunnels at various places along the vein which are now largely caved and inaccessible. The vein is well exposed in a glory hole 300 feet north of the portal of the main tunnel. When active, the mine made about 30,000 gallons of water per day (Julihn and Horton, 1940, p. 146).

The mill is equipped with an 8" x 20" jaw crusher, ore bins, fifteen 1250-lb. stamps, screens, amalgamation plates, drag classifier, and dewatering cones. Eighty percent of the water is returned to the mill circuit when the mill is operating. Mine and mill have been largely idle since 1952.

Dolman (Oyler Lode, Hickman, Bear Valley Mountain) Mine. Location: SE¼, sec. 29, NW¼ sec. 33, T. 4 S., R. 17 E., M.D., 1½ miles south of Bear Valley. Ownership: Not determined; probably Ellen T. Simpson, et al.

According to Laizure (1935, p. 31), the Dolman mine was discovered in 1894. In 1902 Charles Adair and Henry Oyler leased the mine from the Mariposa Commercial and Mining Co. and a 50-foot shaft was sunk (Min. and Sci. Press, vol. 75, no. 1, p. 608). During the period of activity some ore was mined and shipped. Laizure (1935, p. 31) states that the mine ". . . produced about 800 tons of ore which had an average value of $13 per ton . . .". Official records of the Mariposa Commercial and Mining Company, however, indicate that only 59.1 tons of ore was shipped which returned $756.78 or an average of $12.80 per ton. In 1934 the property passed into the hands of a company in which R. L. and E. P. Hickman of San Francisco and Phil B. Dolman of Bear Valley were the chief active participants. During 1939 and 1940 some 393 tons of ore was shipped which yielded 24 oz. of gold and 7 oz. of silver, returning about $2.15 per ton (U. S. Bureau of Mines records). The mine has been idle since 1940.

The Oyler vein has a general strike of N. 35° W. and dips 60° southwest. This direction of dip is unusual in mines of this area which generally dip northeast. It occupies a well-defined shear zone which is traceable for 2 miles. Wall rocks are pyroxene andesite greenstones of probably Upper Jurassic age. The vein varies from a fraction of a foot to 5 feet wide. Ore shoots are short—commonly less than 30 feet

(Laizure, 1935, p. 32). Vein matter is chiefly milky quartz with pyrite, pyrrhotite, chalcopyrite, oxides of iron and native gold. About 60 percent of the gold is found in sulfide minerals.

The principal group of workings, developed mainly by Hickman and Dolman between 1934 and 1940, consist of a 265-foot inclined shaft and 1445 feet of drifts on four levels, the 100, 175, 200, and 250. The longest drift, driven 760 feet north of the shaft is on the 200-level. In this drift an ore body was encountered 625 feet from the shaft which assayed $8 per ton (Julihn and Horton, 1940, p. 129). There is a main ore shoot south of the main shaft penetrated on the 100- and 175-foot levels and two smaller ore shoots were found on the 175-foot level. Some ore in the main shoot on the 175-foot level ran $18 per ton, according to Julihn and Horton. A geologic map and sections of the mine together with some sampling data are found in Laizure (1935, p. 31).

Doss (Doss and Thorne, Ginaca, Brooks) Mine.   Location: Secs. 21, 28, T. 5 S., R. 16 E., M.D., 2½ airline miles southeast of Hornitos or 2 miles northwest of Indian Gulch. Accessible by 2 miles of ungraded dirt road from the paved Hornitos-Mt. Bullion road. Ownership: Frank Trabucco Jr., Hornitos, California, owns 2 patented fractional claims, the Doss and the Thorne Extension, totaling 18.27 acres.

The Doss mine was discovered sometime prior to 1870. In 1874 the shaft was down 110 feet and averaged 10 feet wide to that depth. At that time the owner was a Mr. Brooks of Hornitos. One lot (200 tons) of ore ran $17 per ton and another of 400 tons yielded between $10 and $11 per ton (Min. and Sci. Press, vol. 29, no. 9, p. 85) plus a sulfide concentrate assaying $24 per ton. Storms (1896, p. 218) visited the mine in 1896, finding the shaft still 110 feet deep. Gold to the amount of 677.25 oz. was recovered from an unknown tonnage of ore in 1897. By 1920 ownership had passed to L. A. Ginaca and C. Ginaca of San Francisco. The shaft had been deepened to 400 feet and an adit level driven 475 feet. The 200-level had 300 feet of drifts. In 1933 the mine was under option to A. D. Hadsel and then was acquired by the Doss Mining Company of San Francisco who operated it until July, 1934. Between 1934 and 1938 it was intermittently active under several operators, but has been idle since 1938.

The vein at the Doss Mine strikes N. 40-45° E. and dips 55° southeast. Wall rocks are chiefly slaty and schistose volcanics belonging to the Upper Jurassic Amador group described by Taliaferro (1943, pp. 282-284). The vein averages 8-10 feet wide, reaching a maximum width of about 22 feet (Laizure, 1935, p. 114). Ore shoots average 250 feet long as measured along the strike and ore shoots have been discovered over a vein length of 1000 feet and to a depth of 400 feet. Vein matter consists of milky quartz with pyrite, galena, sphalerite, chalcopyrite and gold.

Workings consist of a 250-foot (vertical depth) inclined shaft, the last to be worked, and an older shaft 400 feet deep. There is also a 540-foot adit. None of these are in working condition.

The early production of the Doss Mine was not recorded other than the figures listed above. In 1929, thirty-eight tons of ore was milled which yielded 39.95 oz. of gold and 76 oz. of silver. Between 1933 and 1938, ten thousand two hundred and ninety-five tons of ore was milled which yielded 2025.39 oz. of gold and 1713 oz. of silver or an average of about $7 per ton. The total recorded production of the mine is approximately $94,400 and the estimated production is in excess of $100,000.

Duncan (Lost Douglas) Mine.   Location: Secs. 16 and 21, T. 5 S., R. 16 E., M. D., just south of the Hornitos—Mt. Bullion road. Adjoins Martinez mine on the northwest. Ownership: George D. Turner, Rt. 1, Box 1049, Ceres, California, owns one patented claim of 44 acres.

The Duncan mine is a very old property dating back to before 1870. The first owner of record was Jerome B. Brown who acquired the property in 1873 or 1874 (Min. and Sci. Press, vol. 35, no. 18, p. 278). Brown operated the mine intermittently until 1895 or 1896 when ownership passed into the hands of William and George Turner of Hornitos. In addition to ore run through the mill Brown took out a $3000 pocket of specimen material in 1877 (Min. and Sci. Press, vol. 35, no. 8, p. 278) and a $1000 pocket in 1888 (Min. and Sci. Press, vol. 57, no. 24, p. 400). A 5-stamp mill was operating on the property as early as 1885 (Min. and Sci. Press, vol. 50, no. 26, p. 416). Except for a short period of activity in 1917 there is no record of activity at the Duncan mine until 1920. In March 1920, Onito and Tura took out a pocket of specimen ore yielding $1000 (Castello, 1921, p. 114). In 1926 the Duncan Mining Company was formed in which Bert Thurber and D. L. Onito were the principal active partners. Considerable ore was milled in 1927 by this company. The mine was operated under lease for a time in 1932 by Austin Davis of Hornitos, in 1933 by Chris Peterson and in 1934-35 by H. L. Berkey and his associates David Onito and J. A. Bodies (U. S. Bureau of Mines records; Laizure, 1935, p. 32). The mine has largely been idle since the late 1930's and there is no equipment on the property.

The vein system at the Duncan mine is well defined at the surface, strikes N. 50° W. and dips about 45° NE. At the surface it varies from 2 to 5 feet wide and consists chiefly of vuggy quartz and pyrite. Although several ore shoots of fair grade have been worked, the mine is noted for its pockety character and the last ore milled in the middle 1930's was of marginal grade. Wall rocks are schistose greenstone, quartz biotite schist and quartz biotite hornfels derived from volcanic tuffs and flows of the Upper Jurassic Amador group.

Workings consist of a series of open pits and caved shafts strung out along the vein for nearly half a mile. None were accessible in September 1954. The main shaft was down 325 feet along a 45° incline in 1934 (Laizure, 1935, p. 32).

Production records on the Duncan mine are too incomplete to give much indication of the total yield of the mine. In 1897-98, two hundred and forty-six oz. of gold was recovered from an unknown quantity of ore. In 1927, five hundred tons of ore yielded 219.82 oz. of gold and 70 oz. of silver (U. S. Bureau of Mines records). From 1932 to 1935, eleven hundred and forty-six tons of ore yielded 90 oz. of gold and 45 oz. of silver (U. S. Bureau of Mines records).

Early (Revel, Louisa, Felix, George Placer) Group.   Location: Secs. 20, 21, T. 4 S., R. 19 E., M. D., 1½ miles northwest of Jerseydale or 7½ airline miles northeast of Mariposa. Accessible from Mariposa by about 10 miles of paved road, via Triangle Road, Darrah and Jerseydale, and

2½ miles of graded dirt road from Jerseydale. Ownership: Robin H. Jackson and Edythe E. Jackson, Box 142, Mariposa, California own a patented property of 80 acres, recorded as the George Placer mine, which includes the former Early, Louisa, Revel and Revel No. 2 lode claims. Also the Felix lode claim embracing 19.16 acres.

Recorded history of the Early mine reaches back to 1884 when Rice and Ferguson operated the property. At that time there was a shaft 100 feet deep on the Early claim and a 260-foot tunnel and 8-stamp mill on the Mt. View claim (Min. and Sci. Press, vol. 49, no. 16, p. 2488, 1884). The name Mt. View has been dropped, but it may be the same as the Revel. By 1902 ownership had passed to August and Eli Revel who operated the mine for a short time that year with A. T. Mitchell as superintendent (Eng. and Min., Jour., vol. 74, no. 10, p. 320, 1902). Between 1902 and 1916 it was active for short periods of time under various lessees. In 1916 it was leased by Mrs. C. N. McIntyre who opened the old drift (presumably on Revel claim) and sunk a 50-foot shaft from the drift level (Min. and Sci. Press, vol. 112, p. 528, 1916). After a year of inactivity operations were again started under new management in June 1917 with George H. Hook as superintendent. A 10-stamp mill was put into operation employing amalgamation, but, owing to the high percentage of sulfides in the ore, returns were not very satisfactory (Eng. and Min. Jour., vol. 104, no. 4, p. 408, 1917). Between 1917 and 1920 activity was curtailed by litigation and various groups controlled the mine for short periods. After about 12 years of inactivity, Early Gold Mining Company of Laredo, Texas reopened the mine under supervision of J. E. Rogers of Mariposa. Rogers operated the mine until 1939. In 1941, George Long of Mariposa worked the mine for a short period and in 1946, H. N. Hammond of Mariposa was operator. The present owners acquired the property in the late 1940's and have intermittently worked on the mine since that time except for a short period in 1951 when some ore was mined and shipped by H. H. Odgers and I. R. Toye of Midpines.

Two sets of east- and northeast-trending quartz veins cut hornblende quartz diorite. Most of the veins average between 2 and 3 feet wide, but widen locally to as much as 6 feet. The principal developed vein on Revel claim strikes roughly east and is vertical. The most thoroughly explored vein on the Early claim strikes N. 25° E. and dips 60° SE. Vein matter is chiefly quartz and pyrite with some chalcopyrite, argentiferous galena and light yellowish-brown native gold, but there are showings of scheelite in the vein matter and in the quartz diorite adjacent to the veins. Most of the high-grade scheelite ore is in stringers 2-4 inches wide.

The present extent of the mine workings could not be ascertained in September 1954 as all were inaccessible because of caveins and flooding. In 1914 there was a 16-foot shaft, surface trenches and 2 adits 50 and 60 feet long on the Louisa claim (Lowell, F. L., 1916, p. 586); an adit 800 feet long, overhand-stoped to a height of 150 feet, and a 40-foot crosscut-tunnel on the Early claim; and a shaft 150 feet deep with several hundred feet of drifts on the Revel claim. Considerable work was done on the claims since 1914 so the above list of workings undoubtedly is incomplete.

In September, 1954, exploration was progressing by use of bulldozer cuts which had uncovered several veins. Soil mantle is 18-20 feet deep in most places, but commonly carries considerable gold. Part of this was derived from the weathered upper portions of the vein and partly from old mill tailings. Some of the mantle material may prove rich enough to process.

Production records for the Early mine are incomplete but a total of $22,600 was recorded between 1909 and 1952. From 1909-1914, 478 tons of ore were mined which yielded 251.02 oz. of gold and 50 oz. of silver. From 1917 to 1920 a total of 697 tons of ore mined yielded 259.34 oz. of gold and 364 oz. of silver. For the period 1932-33, seven hundred tons of ore was processed which yielded 264.47 oz. of gold and 147 oz. of silver. Between 1934 and 1946 six hundred and fifty tons of ore yielded 156.34 oz. of gold, 207 oz. of silver and 1231 pounds of copper (U. S. Bureau of Mines records).

Feliciana Mine.   Location: Secs. 12, 13, T. 4 S., R. 18 E., secs. 7, 18, T. 4 S., R. 19 E., M.D. Accessible by way of State Highway 140 and 2½ miles of the graded dirt road up Trabucco Creek. Trabucco Creek road joins Highway 140 three miles north of Midpines. Ownership: Gold Ledge Mining Company, c/o Walter Gleeson, Pacific National Bank Building, San Francisco, California owns one patented claim, the Feliciana (20 acres) and 13 unpatented claims.

The Feliciana mine was discovered in the 1850s and was operated by Mexicans during the 1850s and early 1860s. In 1866 the mine was equipped with a 5-stamp mill and had a shaft 190 feet deep. Louis Trabucco of Bear Valley owned and operated the mill and had a part interest in the mine (Min. and Sci. Press, vol. 13, no. 8, p. 118, 1866). In the summer of 1866 fire destroyed the mill and later that year the mine was purchased by a San Francisco Company (Min. and Sci. Press, vol. 23, no. 18, p. 276). In December 1871 the mine was leased by J. Dolan of Bear Valley and reconstruction of the mill was started. By 1875 a 10-stamp mill had been built, a 300-foot tunnel driven and a 300-foot shaft sunk. In 1896 the mine was sold to E. S. Davis of Fresno. By 1914, ownership had passed to J. B. Campbell of Fresno (Lowell, 1916, p. 582). Thomas Doyle leased the mine from Campbell in 1917, after a long period of inactivity, but this activity was short lived, the mine having become idle by 1920 and ownership having passed to Mrs. Campbell. In 1925, the Feliciana Gold Mining Company, Inc. was organized with M. Farber as president and Pauline Farber as vice president, both of San Francisco (Laizure, 1928, p. 86). This company operated several other mines in the Briceburg district during the late 1920s, but had largely ceased operations by 1930. About 1931 the mine was acquired by the present owners and in 1933 the Gold Ledge Company began driving a long tunnel designed to intersect the vein at a depth of 900 feet. By September 1934 this tunnel had reached a length of 1186 feet but work on it was carried little further and was discontinued late in 1934 (Laizure, 1935, p. 33). In 1941 the mine was leased by Russell J. Wilson with William Bessler as mine superintendent. For a time ore was milled at the Buffalo mine, but in 1942 a mill was moved to the Feliciana from the Black Oak mine in El Dorado County (Eng. and Min. Jour., vol. 142, no. 10, p. 76, 1942). Closed down by War Production Board Order L-208, the mine has been idle since 1943 except for maintenance and assessment work.

The total recorded production of the Feliciana mine is approximately $159,500, chiefly in gold and silver but also from a little lead and copper. Between 1924 and 1934 a total of 3231 tons of ore yielded 2412.7 oz. of gold or an average of 0.74 oz. per ton. From 1941 to 1943 a total of 4428 tons of ore yielded 2873.1 oz. of gold and 465 oz. of silver or an average of about 0.65 oz. of gold per ton.

Well-defined and traceable for several thousand feet at the surface, the Feliciana vein averages between 2½ and 3½ feet wide, strikes N. 35-40° W. and dips 70-75° SW. It is exposd on rugged topography through a vertical distance of nearly 1000 feet. Vein matter consists of native gold, galena and chalcopyrite in a gangue of quartz and pyrite. A steeply dipping secondary vein intersects the main vein at an angle of 20° and the principal ore shoot has developed along this intersection (Laizure, 1928, p. 86). Two minor veins also join the main vein from the east or footwall side. These are largely unexplored.

Wall rocks are a series of northwest-trending, steeply tilted meta-sediments belonging to the Paleozoic Calaveras group. Slate, phyllite and graphitic and hornfelsic slates and schists are the principal rock types. The metamorphic series is cut by several small quartz diorite dikes.

Principal workings on the Feliciana claim are a 340-foot crosscut adit, with several hundred feet of drifts both north and south of the adit, and a shaft at least 190 feet deep. A winze from the crosscut adit connects with a lower level and drifts. The authors have no information on the extent of these workings. In August 1954, the adit could be entered for at least 100 feet and readily could be cleaned out beyond that point. The shaft was caved and the access road in poor condition. A 700-foot tunnel mentioned by Castello (1921, p. 115) could not be located and is presumed to be caved. This was reported to have several hundred feet of drifts. Most of the late work done on the mine has been from the lower tunnel on the Ace-in-the-Hole claim, the one driven in 1933-34. This is driven at an elevation of 3100 feet or 550 feet below the adit on the Feliciana claim. It is 1400 feet long and is connected with the older workings by an 80-foot raise. A second raise from the adit contacts the vein but is not connected with the upper workings.

French Mine.   Location: Secs. 9, 16, T. 4S., R. 17E., M.D., half a mile south of the Merced River and 1 mile east of the Pine Tree mine. Accessible by way of State Highway 49 and 1¼ miles of dirt road leading east from the water tanks of the Pine Tree and Josephine mill. Ownership: Pacific Mining Company, Crocker Building, San Francisco, California, owns 3420 acres which includes the French, French Pocket and Evans mines.

The French mine is a very old property active in the earliest days of mining in the county. Throughout most of its history the mine was held by the Mariposa Commercial and Mining Company and its predecessors and worked intermittently by various lessees. In 1897 the claims were jumped by a group headed by Bert McFadden and Robert Barnett because, through inaccurate surveys, they believed the land to be outside of Las Mariposas Grant (Min. and Sci. Press, vol. 75, no. 4, p. 80, 1897). The French mine and several others in the vicinity were acquired by the Pacific Mining Company in the early 1930s. Charles Sommers, et al., operated the French mine under lease in the middle 1930s. The last work done at the mine was in 1948.

The total recorded production of the French mine since 1903 is about $116,750. The early production may have been several times this amount. Ore taken out in 1903 averaged $44.16 per ton; in 1924 the average was $16.53 and in 1935-48 the yield was $13.99 per ton.

Two roughly parallel quartz veins cross the property. The principal vein averages about 20 inches wide, strikes N. 70° E. and dips 30° S. (Laizure, 1935, p. 33). Wall rocks are meta-pyroxene andesite greenstone of unknown age and slaty metasediments belonging to the Paleozoic Calaveras group. The principal working is a crosscut adit 155 feet long with drifts on the main level totaling 660 feet. Ore has been stoped almost continuously above the main level and there are seven stope-openings connecting with the surface. Two inclined shafts, each about 150 feet deep, connect with the main level. These are spaced about 50 feet apart. Both are in need of cleaning out. They connect with a lower level about 100 feet long driven about 38 feet below the main level. The total footage of workings aggregates 1048 feet.

Garibaldi (Blue Lead) Mine. Location: Secs. 4, 9, T. 3 S., R. 18 E., M.D., 1 mile west of U.S. Forest Service Kinsey guard station. Kinsey guard station is 12 miles from Briceburg on State Highway 140 by good graded dirt road. Ownership: Robert O. Greeves, Box 151, Columbia, California, owns one patented claim of 18.8 acres.

The Garibaldi mine is an old property dating back prior to 1897. Development of the mine has been hampered by water as the mine yields up to 600,000 gallons of water a day. Little has been recorded of the early history of the property. The principal period of activity was 1897-1902 when Garibaldi Mining Company operated the mine with S. R. Porter as superintendent. During this period a 325-foot inclined shaft was sunk parallel to the dip of the vein, a 40-foot crosscut to the vein was run and an unknown footage of drifts run on the vein from the end of the crosscut. Crosscut and drifts were run from the 300-level. Two 650-gallon capacity skips were kept running day and night to keep the mine unwatered. The Garibaldi Company operated a 5-stamp mill and a Huntington mill, processing 30 to 40 tons of ore daily (Min. and Sci. Press, vol. 84, no. 1, p. 11; vol. 84, no. 3, p. 39; vol. 72, no. 16, p. 503). Early in 1902, forty to sixty ounces of amalgam was being removed from the plates at the end of every shift, presumably from 10 to 15 tons of ore milled. This operation ended in March 1902 because of the expense and the technical difficulties experienced in handling the mine water (Min. and Sci. Press, vol. 73, no. 20, p. 704, 1902). In 1938 the mine was leased for a short time by Walter D. McLean of Coulterville. The tailings dump was worked under lease for a short period in the late 1930s by John P. McCormick of Sonora (Julihn and Horton, 1940, p. 136), but the mine has been idle since that time.

According to Julihn and Horton (1940, p. 135), the Garibaldi vein strikes a little north of east and dips 45° E. Early notes in the *Mining and Scientific Press* mention two parallel veins 1 to 15 feet apart separated by altered dike rock. Neither vein was exposed at the time of the authors' visit in August 1954. The veins follow close to the contact between dolomite and a series of schistose metasediments probably belonging to the Paleozoic Calaveras group. The principal rock types in the metasedimentary series are slate, graphitic schist, and black manganiferous chert. The dolomite, which occurs chiefly on the hanging wall side of the vein is blue-gray-to-black and somewhat platy. Most of the high-grade ore occurs as boulder-like to nodular masses in a black gouge found between dolomite and the altered dike rock. The black gouge zone, in many places simply a blue mud, is 8 to 10 feet thick. Much of the trouble with water came from cavernous dolomite on the hanging wall side of the vein, but a pump handling 500 g.p.m. will keep the mine dry.

Early work on the mine was done from two shafts of unknown depth and numerous open cuts. Work done on the mine from 1900 to 1902 has already been described. One of the shafts, probably one sunk prior to 1900 was open to a depth of about 100 feet in August 1954. Other workings were caved.

Geary (Geare) Mine. Location: SE¼, sec. 30, T. 4 S., R. 18 E., M.D. Adjoins the Nutmeg mine on the south and is about a quarter of a mile west of the Golden Gate and Coronado claims of the Golden Key group. Accessible via the graded Whitlock Creek-Sherlock Creek road from State Highway 49, a distance of 5½ miles by road from Mariposa. Ownership: Permit Mining Company, 1063 Howard St., San Francisco, California, owns one patented claim of 20 acres.

Little has been recorded concerning the early history of the Geary mine. It apparently was first worked by a series of open cuts from the surface as there are several hundred feet of almost continuous shallow workings. The property was purchased in 1898 by Terrill and Schroeder who sank a 100-foot inclined shaft, ran several hundred feet of drifts, and erected a small stamp mill (Min. and Sci. Press., vol. 76, no. 4, p. 86; vol. 76, no. 20, p. 517, 1898). The mill had a capacity of 14 tons in 24 hours. The amount and tenor of the ore mined at this time has not been recorded. In 1904 the mine was managed by James Peck of San Francisco and by 1914 ownership had passed to a Mrs. Potter. From 1918 to 1938 the owner was listed as J. S. Potter of San Francisco. It was explored during 1938 by two lessees, H. L. Womacks of Springdale, California, and Charles E. Farson of Mariposa. Under these operators a small tonnage of ore was shipped which yielded 0.38 oz. of gold and about 0.1 oz. of silver per ton. Little has been done on the mine since 1938.

The Geary vein, the same one that crosses the Nutmeg mine, is 3 to 4 feet wide, strikes N. 20° W. and dips 80° NE. From the disposition of the workings there apparently are several cutter veins crossing and at least one minor vein paralleling the main vein. Wall rocks are sheared, somewhat platy pyroxene andesite greenstones of unknown age.

According to Julihn and Horton (1940, p. 152) the main ore shoot was developed by two shafts 160 and 70 feet deep, respectively, these being about 300 feet apart. The California State Mining Bureau Mines Register for 1904 lists one shaft 100 feet deep and 300 feet of drifts. In 1954, one partly caved shaft surmounted by a wooden head frame could be identified. This apparently is the shallower of the two mentioned by Julihn and Horton. No equipment was on the property and there was no evidence of recent activity at the mine.

Gold Bug Mine. Location: Secs. 28, 33, T. 3 S., R. 17 E., M.D., in Solomon Gulch 1 mile west of the Black Bart mine. Accessible by dirt road from Date Flat on the Coulterville-Kinsey road via Dogtown and Buckhorn Peak. Ownership: N. D. Madden and K. I. Goulder, c/o Bay Construction Company, Ltd., Shelly Building, Vancouver, British Columbia, own several unpatented claims.

The Gold Bug mine produced several thousand dollars worth of gold from near-surface pockets in the 1880s and 1890s during the height of mining activity in the Cat Town district. It was operated for a short time in 1908 by Fred Shaw of Kinsey, in the late 1930s by N. D. Madden and in 1940 under lease by S. S. Escobar. One ore shoot mined between 1937 and 1940 contained 225 tons of ore which ran slightly more than 3.5 oz. of gold and 0.75 oz. of silver per ton.

The vein roughly follows the contact between slate of the Paleozoic Calaveras group, which forms the hanging wall side of the vein, and massive greenstones of unknown age. The vein, which varies from a few inches to several feet wide, is accompanied in places by albitite and diorite dikes, similar to other mines in the Cat Town and Flyaway Gulch district. According to Julihn and Horton (1940, p. 142) the high-grade pockets tend to occur when the vein is entirely in slate and where there is abundant calcite.

A 200-foot adit driven on the vein is the principal working. It penetrates the vein to a vertical depth of 70 feet. The ore shoot mined was 40 to 50 feet long, measured in the direction of strike of the vein, and 110 feet deep on the dip of the vein. In this shoot the ore averaged more than $115 per ton (Julihn and Horton, 1940, p. 142).

Gold King (Martin-Walling, Sunshine) Mine. Location: Sec. 11, T. 3 S., R. 17 E., M. D., in the Gentry Gulch district on the ridge crest half a mile southwest of the Hasloe mine. Ownership: Russell G. Rowe, Rt. 7, Box 1499, Modesto, California holds several claims adjoining the Lovely Rodgers mine on the southeast.

The Gold King mine was discovered in the 1860s and was known throughout its early history as the Martin-Walling or Martin and Walling mine. Through the first few years of operation it was reported to be fabulously rich. It was purchased in 1878 after some years of idleness by Douglass and Stevenson who developed two ore shoots and took out some ore. In 1879 it was reported sold by Joshua Hendy to eastern people along with the Bondurant for $80,000 (Min. and Sci. Press, vol. 39, no. 1, p. 5). By 1880 ownership had passed to P. P. Mast who rehabilitated the 300-foot shaft and did considerable mining results of which have not been recorded. Some exploration and rehabilitation work was done on the mine under the name Gold King during the 1930's through an adit and raise below the old workings, but no recent work has been done on it.

Two veins cross the property. The principal vein on which the main shaft is sunk strikes roughly north and dips 36° east at the surface. Poorly exposed at the surface, the vein appears to range from 2 to 4 feet wide. Vein matter is quartz with free gold, pyrite and tetrahedrite. The second vein strikes more nearly to the northwest, dips 60-70° northeast and is separated from the main vein by about 200 feet of slate and quartz-mica-graphite phyllite and schist. Vein matter is chiefly fractured quartz and fractured wall rocks.

The main vein has been developed by an inclined shaft, sunk on the vein, over 300 feet deep having 2 levels. The shaft is equipped with hoisting equipment and track but the mine is watered and partly caved 20 or 30 feet below the collar. No recent work has been done on it. The west vein has been superficially developed by shallow shafts.

Golden Key (Austin) Group. Location: Secs. 29, 32, T. 4 S., R. 18 E., M.D., on Whitlock Creek 5 airline miles northwest of Mariposa. Accessible from State Highway 49 by 5 miles of graded dirt road by way of Whitlock and Sherlock Creeks.

Ownership: Golden Key Mining Company, c/o J. E. MacDonald, Box 105D, Gilroy, California owns 6 patented claims, the Coronado No. 2, Dusenberry, Golden Gate, Hayseed, Haywire Fraction and Regan, and 3 mill sites.

Mining activity on the Golden Key claims dates back at least as far as the 1880s. In the late 1880s and early 1890s the Hayseed mine and the nearby Triumph and Farmers Hope (Miners Hope) mines were being operated by J. B. Helm and Son and considerable specimen ore was reported mined in 1890. In 1896 the Hayseed and Triumph claims were sold by the Helms to Angus McIntosh for $10,000 (Min. and Sci. Press, vol. 62, no. 9, p. 204, 1896). About the same time H. C. Austin of Fresno was operating the Regan mine from an 80-foot shaft (Storms, 1896, p. 222). By 1902 ownership of the Austin group of claims passed to T. R. Lombard, Coronado, California, and A. M. Kitchen of Chicago, and D. A. Connolly of San Francisco was named superintendent (Min. and Sci. Press, vol. 85, no. 5, p. 64, 1902). In 1903 the Austin Group Mining and Milling Company was organized with H. C. Austin as the principal owner and W. H. Cavin as general manager. This company intermittently operated various of the claims for more than 12 years. William Dolph operated the Hayseed and Miners Hope claims for a time in 1903 (Min. and Sci. Press, vol. 75, pp. 796, 944, 1903). Acquired by the Golden Key Mining Company during the early 1920's the mine was intermittently operated by them until 1949. The properties have been idle since 1949.

Two principal veins, the Regan and Hayseed, strike roughly north and dip east, the Hayseed at 40° and the Regan at 65°. These crop out about 50 feet apart and their surface courses are roughly parallel. The Regan vein is believed to be the extension of the Miners Hope vein and the Hayseed an extension of the Spread Eagle vein (Julihn and Horton, 1940, p. 151). These veins are intersected approximately at right angles by four lesser veins, the Golden Gate Nos. 1, 2 and 3 and the Coronado No. 2. The cutter veins dip south at angles between 35° and 40°. Much of the specimen ore taken from the mine has come from the cutter veins but several good sized ore shoots have been developed in the main veins. Wall rocks throughout the claims are meta-andesite greenstone of unknown age.

According to Julihn and Horton (1940, p. 151) the Golden Key group of claims had yielded $130,000 up to 1938 through a mined depth of only 135 feet. Total recorded production to 1949 is estimated at more than $154,000, but the production figures are difficult to check because of the many different names under which production was reported. The amouñt of gold produced before 1884 was not recorded. Ore produced between 1933 and 1949 apparently averaged about $12 per ton. The highest grade ore from sustained mining was recorded in 1912-13 when 343 tons of ore yielded 600.33 oz. of gold and 39 oz. of silver (U. S. Bureau of Mines records).

The principal workings on the Golden Key property are a long crosscut adit called the Greenstone tunnel and the inclined Hayseed shaft serving four levels. The Greenstone tunnel, more than 700 feet long, was driven to intersect the cutter veins and ultimately to connect with the workings from the Hayseed shaft. Its portal is close to the Sherlock Creek road and it is driven approximately S. 80° W. across the Golden Gate and Coronado claims. There are considerable drift and slope workings from this tunnel. The Hayseed shaft, sunk on the Hayseed claim and vein, has levels at vertical depths of 110, 220, 250 and 300 feet. More than 650 feet of drifts have been run from these levels, mostly north of the shaft. On the 300-level south of the shaft was an ore shoot 73 feet long and 5 feet wide in which ore averaged $10-$12 per ton (Julihn and Horton, 1940, p. 151).

Near the south end of the Regan claim is a partly caved shaft reported to be 310 feet deep. About 360 feet north of this shaft is the Arndke shaft 80 feet deep, also caved, and there are two other caved shafts on the Regan vein believed to be 190 and 110 feet deep, respectively (Julihn and Horton, 1940, p. 151). The Greenstone tunnel was the only working accessible to the authors in August 1954.

Granite King and Live Oak (Buckeye, Billings) Mine. Location: SW¼ sec. 3 and NW¼ sec. 10, T. 6 S., R. 18 E., M.D., 3 miles southwest of Mormon Bar and ¾ of a mile south of the old Yosemite Highway through Mormon Bar. Ownership: Edith McElligott, 1404 Poplar Ave., Fresno, California, owns a patented property of 16 acres, parts of the old Granite King and Live Oak claims.

The history of the Buckeye and Live Oak mine dates back beyond 1870. In 1871 rich ore from the upper, oxidized part of the vein was being mined by a man named Hambleton (Min. and Sci. Press, vol. 23, no. 16, p. 244, 1871). By 1902 ownership of the mine had passed to the Krogh Manufacturing Company of San Francisco who sold the property late that year to McCrae and Herman. (Min. and Sci. Press, vol. 84, no. 14, p. 194, 1902). The Buckeye mine appears on the list of mines of the Mariposa Commercial and Mining Company and apparently was acquired by that company early in the 1900s. In 1909 the mine was leased by John Hamm, C. P. Pratt and J. W. Pratt (Min. and Sci. Press, vol. 92, no. 22, p. 732, 1904). These operators produced 22 tons of ore which yielded $174.52 or an average of $7.93 per ton at the old price of gold (Logan, 1935, p. 188). By 1928 ownership had passed to O. S. Evans of Hornitos (Laizure, 1928, p. 129) but little or no mining was done in the 1920s. In 1938 the mine was reopened a partnership of M. T. McElligott, Guy Noble and Howard Campbell who operated the mine until 1941. During this period 9017 tons of ore was milled which yielded 2295 oz. of gold and 916 oz. of silver or an average of 0.243 oz. of gold per ton. The mine has been idle since 1941.

The main vein in the Granite King mine is 2 to 5 feet wide, strikes N. 40° E. and dips 35° to 50° southeast. Wall rocks are coarse-grained

FIGURE 21. Headframe, ore bin, and shop-building at the Granite King and Live Oak mine in the Buckeye district 3 miles southwest of Mormon Bar. Lying close to the south boundary of Las Mariposas grant, the mine was managed for a time by Mariposa Commercial and Mining Company. Although not among the leading producers in the county the mine has a substantial production of ore, chiefly of sulfide type.

hornblende-biotite granodiorite and a finer-grained granitic dike rock containing a greater proportion of dark minerals than the granodiorite. According to Julihn and Horton (1940, p. 134), the dike intrudes the vein fissure and in places has been brecciated and mingled with vein matter. Ore minerals are native gold, auriferous pyrite, galena and sphalerite in a milky quartz and brecciated wall-rock gangue. Milling ore apparently averaged from 0.2 to 0.3 oz. of gold per ton with occasional pockets of higher grade ore.

The principal workings are an inclined shaft at least 80 feet deep and several hundred feet of drifts. Several vertical shafts in the vicinity of the mill are flooded and of unknown depth. The main shaft is equipped with a serviceable headframe, ore bins, hoist, skips and track. It was flooded within 30 feet of the collar in September 1954 and there was some caving below the collar.

The partly dismantled mill consists of a 16-inch jaw crusher, 4 stamps, amalgamating plates, a small ball mill and concentrating tables. Repair shops and blacksmith shops are in good condition.

Greens Gulch and Greens Gulch Extension.   Location: Secs. 12, 13, T. 5 S., R. 17 E., M.D., on south side of the Mt. Bullion-Hornitos Road three-fourths of a mile west of Mt. Bullion. Accessible by paved road via Highway 49 and Greens Gulch. Ownership: Mariposa Commercial and Mining Company, c/o Eileen Milburn, Mariposa, California, owns the remaining parts of the Las Mariposas Grant which was largely liquidated in the early 1950s.

The Greens Gulch mine was well known through the early history of Las Mariposas Grant and was probably discovered during the 1850s. The principal periods of activity were in the 1860s and from 1901 to 1914. In the fall of 1941 the Green's Gulch mine, along with the Mt. Ophir, Princeton, Louis Mt. View and Ortega, were optioned by J. K. Wadley of Texarkana, Arkansas, and some exploration was done under the direction of Charles Greenamayer. No work has been done on the mine since 1941.

A typical Mother Lode mine, the Greens Gulch workings are on a broad vein striking N. 50-55° W. and dipping steeply east. The main vein is intersected by a northeast-trending cutter vein and much of the ore apparently was at the intersection of these veins. The vein matter is principally milky quartz containing native gold and auriferous pyrite. It locally contains quartz-ankerite-mariposite rock. The Mother Lode in this area consists of a series of northwest-trending, en echelon veins, rather than a single lode. The Greens Gulch vein is probably an extension of the main vein at the Princeton mine. Wall rocks are slate of the Upper Jurassic Mariposa formation.

Workings were caved and inaccessible in August 1954 but there is a line of stopes, several shafts and tunnels. There must be extensive drifts from the size of the dumps and the quantity of ore mined in the early 1900s.

Mariposa Commercial and Mining Company records (1900-1915) show that the Greens Gulch mine produced 10,625 tons of ore which yielded $119,140.70 or an average value of $10.25 per ton at the former price of gold. During the same period the Greens Gulch Extension produced 10.25 tons of ore from which $313.99 was realized, or an average of $13.96 per ton. These figures indicate an average tenor of ore in excess of half an ounce of gold per ton.

Hasloe (Funk, Coward, Hasloe and Centuary, Gentry Gulch) Mine. Location: Secs. 1, 2, T. 3 S., R. 17 E., M.D., in Gentry Gulch 5½ airline miles southeast of Greeley Hill and 8 airline miles east of Coulterville. Accessible by good dirt road from the surfaced Coulterville-Greeley Hill road by way of Date Flat and Gentry Gulch. Ownership: Walter D. McLean, Joseph Dupret, Jr., Ralph J. Jacobs and R. S. Hudgson, c/o Walter D. McLean, Coulterville, California, own two patented claims of 41.12 acres and two unpatented claims.

The Hasloe mine was first located by Thomas Palmer in 1857 and purchased that year by Samuel Funk. Samuel Funk was killed in a mill accident about 1860. From 1860 to 1869 the property was successfully worked by H. G. Coward who completed the 5-stamp mill started by Funk and increased it to 10 stamps. Then known as the Coward mine, the ore consistently averaged $40 per ton and considerable specimen ore was taken out (Browne, 1868, p. 32; Min. and Sci. Press, vol. 26, no. 24, p. 373, 1823). In the fall of 1873 the mine was sold by Coward to the Hasloe Mining Company and Robert Hannigan was made superintendent (Min. and Sci. Press, vol. 26, no. 16, p. 245, 1873). This company operated the mine until 1875 or 1876, finding several

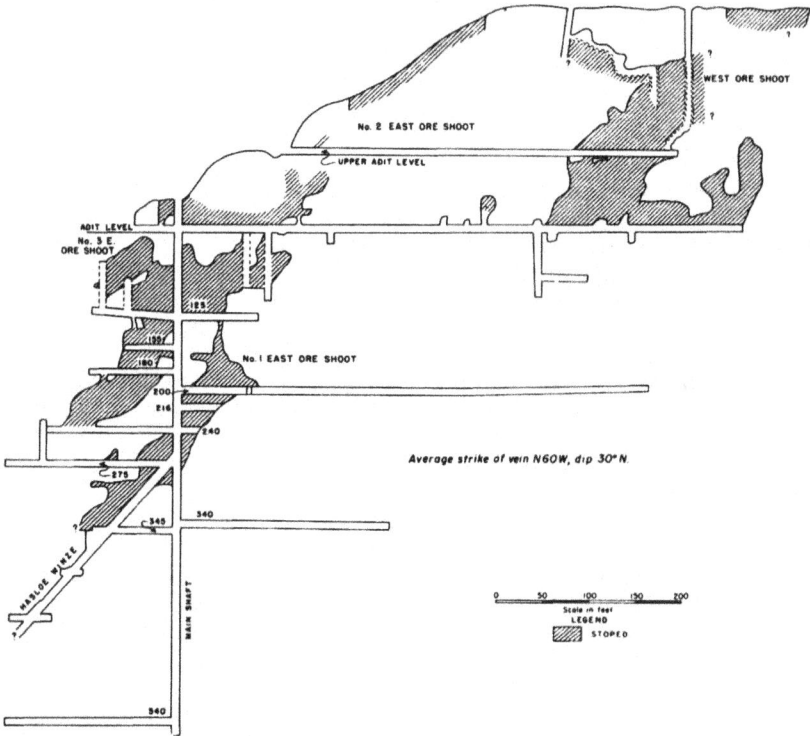

FIGURE 22.   Section in the plane of the Hasloe vein showing disposition of the workings and positions of the principal ore shoots. The mine had a production of about $3,000,000. *Drawing courtesy of Walter McLean, Coulterville.*

high-grade ore shoots. Some work was done on the property late in 1877 and in 1878. The *Mariposa Gazette*, March 22, 1879 states that the Hasloe had produced $800,000 to a depth of 300 feet. J. S. Morgan and Sons took over the property about 1880 and operated the mine until 1890. The mine is supposed to have produced over $3,000,000 under this management. After 5 years of inactivity the Hasloe was reopened and during 1895 and 1896 produced 435.38 oz. of gold from an unknown tonnage of ore. After a short period of inactivity an unsuccessful attempt was made to reopen the mine in 1898. In 1909, or 1910 a company headed by H. P. Dalton of Alameda acquired the property and it was operated by Hal G. Kennedy, Dalton's brother-in-law, until August 1911. A small tonnage of ore running 0.7 of an oz. of gold was produced during this period. Gentry Gulch Consolidated Mines Company had control of the Hasloe, Bandarita, Star and Texas Hill mines in the late 1920s (Laizure, 1928, p. 88). The present owners acquired the property in 1935 and operated it for two periods 1935-39 and 1942-48. It has been idle since 1948 and the workings below the adit level are flooded, but the present owners were preparing to reopen the mine in the summer of 1956.

Production of the Hasloe mine up to November 1, 1873 was reported in the *Mining and Scientific Press* (vol. 27, no. 20, p. 309) to be $200,000, but a previous issue of the *Mariposa Gazette* places the production at $600,000 up to 1866. No records are available for the period 1874 to 1910 but there was a substantial production of high-grade ore in 1874-75-76. According to U. S. Bureau of Mines records 3284 tons of ore was taken from the mine between 1910 and 1948 from which 1703.68 oz. of gold and 232 oz. of silver were extracted, or slightly more than $30,000.

The Hasloe vein averages about 30 inches wide but locally is as narrow as 6 inches and as wide as 5 feet. At the surface the vein strikes N. 50° W., and does not deviate more than 5° over a distance of 850 feet. It dips 28-30° northeast. Vein matter is milky quartz mingled with wall-rock gouge and in places ribboned by parallel inclusions of slate. Ore minerals are native gold, tetrahedrite and a little galena and arsenopyrite. Tetrahedrite is commonly associated with concentrations of gold and is believed to be an indicator of good ore. According to Browne (1868, p. 32) two ore shoots were developed during the early years of mining. These were about 100 feet long, pitched 40° east and were worked to a depth of 170 feet. Ore shoots have averaged 0.5 and 0.6 oz. of gold per ton but occasional pockets of specimen ores were found. Most of the gold occurs near the border of the vein, particularly on the footwall side (Julihn and Horton, 1940, p. 139). Wall rocks are principally slate and hornfelsic quartz-biotite rock derived from shale and siltstone. A diorite dike has been intruded for 200 feet along the vein fissure in the vicinity of the west ore shoot (Julihn and Horton, 1940, p. 139).

Principal workings on the Hasloe claim are a 500-foot inclined shaft and a 700-foot drift adit. Most of the ore above the adit level has been stoped. From the shaft there are 1750 feet of drifts on 6 levels. The disposition of the workings may be seen in the accompanying diagram.

Hite (Hite and Wynant, Hites Cove, Arkell) Mine. Location: Secs. 22, 26, 27, T. 3 S., R. 19 E., M.D., on the South Fork of the Merced River 1 airline mile south of Incline. Accessible by trail from Incline and from Jerseydale. A road was being constructed to the mine from Incline in 1954. Ownership: Cyrus Bell, 160 South Fairfax Ave., Los Angeles 36, California, owns the patented Priest and Coleman, Hite, Giltner, McConley, Old Dominion and Summit claims and several mill sites.

The Hite mine has been one of the most colorful and most productive gold mines in Mariposa County, and the authors have attempted to trace its history through old files of the *Engineering and Mining Journal* and *Mining and Scientific Press*. It was discovered in 1862 by John R. Hite and made Hite a relatively rich man in the 17-year period during which it was operated by Hite and Company. Development of the mine was entirely by hand-methods, and arrastra milling. Hite was able to get a small stamp-mill driven by water power into operation by 1866 and the capacity of the mill was doubled by 1868. The mine was surveyed for patent in 1866. During the 17-year period of operation by Hite and Company, Hite was the principal director of operations although Peter Wynant was a partner with Hite for a short period in the early 1860's. Under Hite's management, 2 long crosscut adits were driven to the vein. The upper adit intersected the vein between 200 and 250 feet from the surface. The lower adit intersected the vein at

FIGURE 23.   Adit of the Horseshoe mine in the Bull Creek district 10 airline miles northeast of Coulterville. One of but few mines recently operated in Mariposa County the small, high-grade shoots of ore have yielded nearly $14,000 during recent development work. *Photo by Mary H. Rice.*

FIGURE 24.   Mill building at the Horseshoe mine. *Photo by Mary H. Rice.*

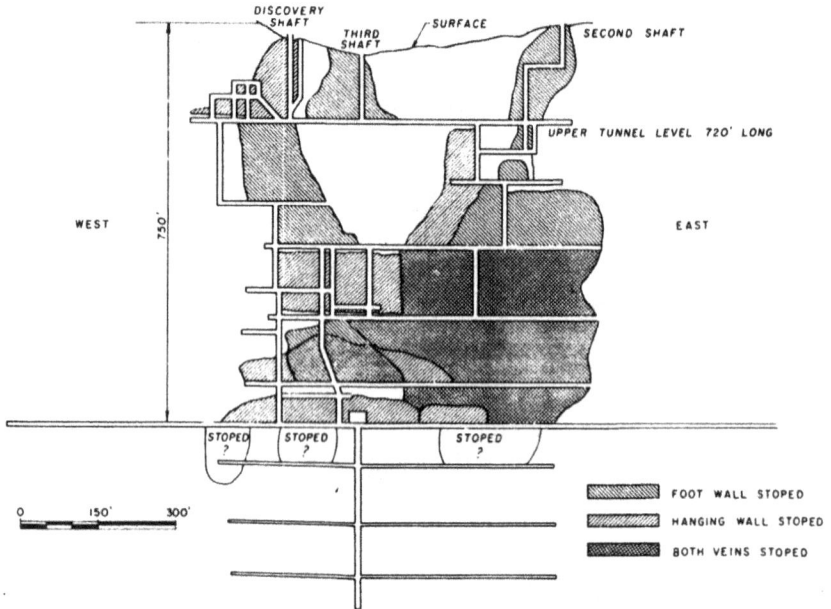

FIGURE 25. Longitudinal projection of the workings of the Hite gold mine drawn in the plane of the vein. Located in the Hite district close to the South Fork of the Merced River, the Hite mine has produced at least $3,000,000—mostly prior to 1900. Two long crosscut adits trending toward the observer connect the lateral workings to the surface. The lower crosscut was the main means of entry during the later years of operation. *Diagram reproduced through courtesy of Francis H. Frederick.*

a vertical depth of between 700 and 800 feet or an inclined depth of over 900 feet. During the last few years of operations by Hite and Company a winze or underground shaft was sunk from the end of the lower adit to a depth of 330 feet and three drift levels were run from this working.

In August 1879 the mine was sold by Hite for between $600,000 and $650,000 to a corporation headed by W. S. Clark and J. R. Bothwell, Hite remaining as a trustee in the new corporation. Considerable mining and development was done by the company, particularly on workings from the winze. Operation of the mine by this group ended in 1882 with a dispute and prolonged litigation. The litigation was not settled until 1887. During that year the mine was reopened under management of Judge Walker but the operation was short lived. In 1883 the mine was managed for a short time by the McCaw brothers. In 1895, W. S. Chapman and associates organized the Hite Gold Mining Company and the property was bonded to them for $100,000. This venture also ended in litigation after minor production. A. H. Ward and H. H. Todd purchased the Hite mine in 1899 for $46,000 and organized the Hite Cove Gold Mining Company. Little was accomplished under this management. In 1903, Thomas Churchill of San Francisco purchased the property for an undisclosed sum and leased it to the Arkell brothers. Early in 1905 the Yosemite Mining Company made an unsuccessful attempt to open the Hite mine and later that year the properties were purchased by W. M. Kirkhoff and associates of Minneapolis. The mine was active, apparently under lease, in 1909 and 1912 and some production was made during the latter year. By 1916, ownership had passed to J. S. Spillman of San Francisco who held the property until the late 1930s. About 1938 it was acquired by Minerals Engineering Company of Los Angeles. This company did some work but no production was recorded. In 1953-54 a road was cleared across very difficult terrain from Incline almost to the mine and it may be reopened in the near future.

The total production of the Hite mine has been estimated at various figures between $2,750,000 and $3,000,000. (Min. and Sci. Press, vol. 81, no. 10, p. 287; Julihn and Horton, 1940, p. 145) U. S. Bureau of Mines figures show that about $55,000 in gold and silver was produced from the mine between 1896 and 1912 but no precise figures are available for the period 1862 to 1896. Published reports in the Mining and Scientific Press show that $250,000 was realized during 1878 and similar amounts for several preceding years. A large percentage of the ore mined prior to 1879 ran more than $40 per ton at the old price of gold. The lowest grade of ore mined was taken out between 1904 and 1909 when 4625 tons of ore yielded 788.85 oz. of gold and 849 oz. of silver or an average of only $3.68 per ton. In 1912 a small tonnage of ore was mined which yielded nearly $3\frac{1}{4}$ oz. of gold per ton.

At depth the Hite vein takes an almost straight course N. 70° W. through the Priest and Coleman, Hite and Giltner claims, and dips 75° to 80° north. Through the Hite Central claim, which adjoins the Giltner on the southeast, the strike of the vein swings to about N. 50° W. At the surface and to a depth of 600 feet on the Hite claim, the Hite vein splits into two branches separated by a lenticular horse of slate and schist 600 feet long and 50 feet thick at the thickest place. Vein matter in the two parts of the Hite vein varies from a few feet to more than 12 feet. Along the borders of the slate-schist horse the vein is 25 feet thick (Storms, 1894, p. 170).

A second vein striking N. 60°-70° W. and also dipping steeply north traverses the McConley and Old Dominion claims. Where observed this vein was 2-5 feet thick. A third vein striking about N. 10° E. and essentially vertical crosses the Summit claim.

The vein fissures cut metasedimentary rocks of the Paleozoic Calaveras group. Dark slate, graphitic schist, quartzite, metasandstone and dense, massive quartz-biotite hornfels are the principal wall rock types. Bedding, where discernible, is parallel to the schistosity and cleavage, has a general strike of N. 50° W. and dip of 75°-80° NE. The metasediments are cut by a variety of small dikes and sills of granitic rocks ranging in character from white aplite to dark biotite-hornblende granodiorite.

Inasmuch as most of the veins of the Hites Cove district crop out on the south slope of a steep, west-trending ridge, they have been developed almost exclusively by crosscut adits connecting with drift-levels. There are several relatively short adits on the McConley and Old Dominion claims but the principal workings are the two long adits on the Hite claim driven by Hite. The upper adit is 720 feet long, and has east drifts reaching to the Hite Central claims and west drifts for an unknown distance. The lower adit, a little more than 500 feet below and almost due south of the upper level, is 1400 feet long. It has several thousand feet of drifts on the adit level. At the end of this adit is a winze or underground shaft 330 feet deep with 3 levels, presumably 100 feet apart. The *Mining and Scientific Press* (vol. 58, no. 26, p. 611, 1894) mentions east and west drifts from the bottom of the inclined winze and a crosscut and drift driven from the footwall branch of the vein into the 14-foot wide hanging wall branch of the vein from the end of the east drift. Very little other work had been done on the hanging wall branch of the Hite vein up to 1895 but the ore apparently was of considerably lower grade than in the hanging wall side. In various places, on the lowest level the footwall vein apparently pinches out and no effort was made to follow it deeper. Most of the ore in the footwall branch of the vein above the lower adit probably has been stoped out but records are fragmentary.

Horseshoe (Bower Cave) Mine. Location: NE ¼, sec. 23, T. 2 S., R. 17 E., M.D., on the southwest side of Jordon Creek ¾ of a mile northwest of Bower Cave or 8 airline miles northeast of Coulterville. Accessible via the Greeley Hill-Smith Creek-Bower Cave road from Coulterville, a distance of about 13 miles. The road is paved as far as Dudley Hill. Ownership: Joseph Dupret Jr. and Irene Dupret, Star Route, Box 21B, Cazadero, California, own one patented claim of 20 acres. Walter D. McLean of Coulterville is mine manager.

The Horseshoe mine, known in early days as the Bower Cave mine, was discovered about 1867 probably by Andrew Rocca. In 1879 it was sold by a Dr. Austin to P. P. Mast who was operating the Bandarita and Martin Walling mines. Most of the early work on the mine was done by Mast between 1879 and 1881. Ore was milled at the Martin-Walling mill (Min. and Sci. Press, vol. 39, no. 26, p. 413, 1879; vol. 43, pp. 236, 252, 1881). By 1904 ownership of the mine had passed to Rodgers and Loomis of Springfield, Ohio (Aubery, 04, p. 11). Some work was done on the mine in 1910 by Francis B. Loomis. In 1913 the mine was bonded to T. G. Mudgett of San Francisco but no production was recorded. In July 1951 the present owners reopened the mine and it was put into active operation in November 1953. Since 1951 the mine has produced $13,800 almost entirely from development work.

There are two nearly parallel veins on the Horseshoe claim striking N. 68°-72° E. and dipping 65° SE. to vertical. The north vein, approximately 40-70 feet from the south vein, varies from a few inches to 16 inches wide and the south vein varies from 2 to 4 feet wide. Wall rocks are graphitic slate and hornfelsic quartz-biotite-graphite rock derived from clay shale and siltstone. Cleavage in the metasedimentary series strikes NW and dips SW. Vein matter consists chiefly of bluish to milky quartz, commonly ribboned with thin sheets of slate, and wall rock gouge. Ore minerals are native gold, pyrite, chalcopyrite, arseno-pyrite, galena and sphalerite. Millheads assay from $30 to $50 per ton in gold.

The principal workings consist of a crosscut adit about 200 feet long, which intersects the north vein at a depth of about 125 feet, 1500 feet of drifts and secondary crosscuts from the adit level, and an under-ground vertical shaft or winze now being sunk on the south vein at its intersection with the crosscut. In August 1955 this shaft was down 50 feet and had 130 feet of drifts on the one sub-level.

The ore is hand-trammed from the mine to a 20-ton ore bin, passed through a jaw crusher and Huntington mill. It is ground to pass a 40-mesh screen before being passed over amalgamation plates and concentrating tables and sent through flotation cells. Sponge gold is sent to the mint and the sulfide concentrate to the Selby, California, smelter. Three men operate the mine and mill. Six tons of ore are processed per day.

Independence (Twin Springs) Mine.   Location: NE ¼ sec. 23, T. 3 S., R. 18 E., M.D., on the Jenkins Hill road half a mile (airline) east of the Briceburg-Kinsley road but about 3½ miles from the Briceburg road via the unimproved Jenkins Hill road. Ownership: Charles Jordan, 633 Green Ave., San Bruno, California owns one unpatented claim of about 20 acres, but ownership was disputed in August 1955 by other individuals.

The authors were unable to gather any information on the early history of this property. At the time it was reactivated under lease by Delaney and Rohrback in 1934 the workings were extensive. One shoot mined between 1932 and 1934 ran between $80 and $140 per ton (George Matlock, personal communication, 1955). The claim is operated intermittently by the listed owner and by other individuals who also claim ownership.

The Independence vein strikes N. 55° W., dips 75-80° NE. and averages about 18 inches wide. Vein matter is ribbon quartz with pyrite, galena, and native gold. Wall rocks are slate and schist of the Paleozoic Calaveras group intruded by granitic dikes. The dikes commonly are schistose and well altered.

The principal working is a crosscut adit 300 feet long driven northeast to the vein and striking it 50 to 60 feet below the surface outcrop. From the end of the crosscut a drift has been run about 200 feet northwest and 100 feet southeast developing 300 feet of backs averaging at least 50 feet high. Much of the vein between the crosscut and the surface has been stoped, but no work has been done below the crosscut level.

Iron Duke Mine.   Location: NW¼ sec. 4, T. 4 S., S. 16 E., SW¼ sec. 33, 3 S., 16 E., M.D., in the northwestern part of Hunter Valley about 5 miles by paved and graded dirt road from the intersection of the paved Bear Valley-Hornitos road. Ownership: Mrs. Louise M. Broad, 3469 Twentieth St., San Francisco, California, owns five unpatented claims, the Iron Duke, Commercial, Pedro, Protection and Protection No. 1.

The Iron Duke mine was discovered prior to 1865, probably about the same time as the Oakes and Reese mine, which adjoins it on the northwest. During the early history of operations a 100-foot shaft was sunk at the junction of the principal Blue Lead vein and an intersecting cutter vein. During the sinking of the shaft a small pocket of ore was found which yielded 190 oz. of gold (Julihn and Horton, 1940, p. 119). The property was worked from time to time by lessees but the principal periods of exploration were 1934-35 under E. F. McTarnahan and 1938-

41 under J. O. Gillice. It has been idle, except for assessment work, since 1940. In July 1954 the underground workings were inaccessible because of flooding. There was no headframe but a hoist house, compressor house, blacksmith shop, mill building, watchman's cabin and some milling equipment were on the property.

The Iron Duke claim is oriented on the Blue Lead which passes through the adjoining Iron Duke and Carson mines as well as several others in the district. This vein strikes N. 30-35° W., dips east at an average of 61°, averages 4 feet wide, and is traceable for 1500 feet (the entire length of the claim). It is joined at angles between 45° and 90° by a series of cutter veins spaced about 250 feet apart. These enter the Blue Lead from the west but do not cross it (Gillice, J. O., private report). Most of the ore has been taken where the cutter veins join the Blue Lead or from the cutter veins themselves, but some small ore shoots have been found in the Blue Lead on the adjoining Oakes and Reese property. The two Protection claims and the Commercial claim cover extensions of the cutter veins followed west from the Iron Duke claim. Several of the cutter veins have been traced west of the Blue Lead for more than 600 feet.

The Blue Lead follows a fissure which in general lies between basalt and andesite greenstone on the west and tuffaceous slate and chert on the east. The cutter veins are found entirely in greenstone. The wall rocks are part of the Upper Jurassic Amador group (Taliaferro, 1943, pp. 282-284) which is predominantly a volcanic assemblage.

The principal working is single compartment and manway shaft 200 feet deep inclined 61°. At the 185-foot level a drift was run south 500 feet to Cutter Vein No. 2 and a secondary drift was driven west 100 feet on Cutter No. 2. This secondary drift encountered an ore shoot and a 30-foot raise was driven, but no stoping was done (Gillice, J. O., private report, p. 6). One powder box full of ore taken from a pocket in this shoot yielded $524. The main drift was continued south to the junction point with Cutter Vein No. 3 and a 75-foot drift was run west on that secondary vein (Julihn and Horton, 1940, p. 119). In addition to the main workings there is a partly caved 35-foot shaft and numerous shallow surface workings in the vicinity of Specimen Hill on the Iron Duke claim.

According to Gillice (private report) $16,000 in gold was taken from shallow workings on Specimen Hill, $3800 was taken from a small stope at the junction of the Blue Lead and Cutter Vein No. 1 at a depth of 104 feet, and $2000 was recovered during driving of the raise on Cutter Vein No. 2. U. S. Bureau of Mines records indicate a gold production of $2000 for the period 1932-1941. Except for occasional pockets of bonanza rock the average milling ore runs about $10 per ton at a gold price of $35 per oz. Sulfide concentrates, chiefly marcasite and arsenopyrite, rarely contain more than $50 per ton (Julihn and Horton, 1940, p. 119).

Jumper (Jumper and Bear) Mine. Location: N½ sec. 30, T. 4 S., R. 17 E., M.D., on top of Bear Valley Mountain 2½ airline miles west of Bear Valley. Accessible via 1 mile of unimproved dirt road from the surfaced Bear Valley-Hornitos road. Ownership: Not determined. Last operator was C. H. Burt of Bear Valley. Property originally consisted of two claims, the Jumper and Cold Springs or Bear embracing about 35 acres.

The Jumper mine was discovered in the 1850s, worked intermittently to 1917 and almost continuously in a small way from 1917 to 1937. Ore, though not present in large shoots, ran from ½ to 1 oz. of gold per ton. Most of the work from 1917 to 1940 was done by C. H. Burt of Bear Valley. There has been no recorded production from the mine since 1940.

According to Laizure (1928, p. 91) there are two nearly flat, laminated quartz veins one above the other, varying from a few inches to more than 4 feet wide which carry very finely divided native gold. These apparently strike N. 75-80° E. but are well obscured by soil mantle. The lowest mill run made in the late 1920s gave a recovery of $19 per ton at a gold price of $20.67 per oz. Production records for the period 1925-28 show a yield of nearly 1 oz. of gold per ton. Between 1919 and 1922 over 621 oz. of gold and 75 oz. of silver was obtained from only 43 tons of ore.

In September 1954, there were 3 inclined shafts less than 100 feet apart watered and partly filled with debris. One of these is at least 180 feet deep (Laizure, 1928, p. 91). There are also two vertical shafts open to a depth of at least 100 feet, three other caved shafts and numerous partly caved stopes and open cuts. There was a small Gibson mill and a steam engine on the property when it was last visited in 1954.

Juniper (Juniper and Patricia, Patricia and Charles) Mine. Location: SW¼ sec. 19 and NW¼ sec. 30, T. 4 S., R. 17 E., on Bear Valley Mountain 2½ miles west of Bear Valley. Accessible via half a mile of unimproved dirt road from the surfaced Bear Valley-Hornitos road. Ownership: Pereno O. Zirker, Pearl Crowell, Ella B. Edwards and Hallett B. Hammatt, 832 Franklin St., Monterey, California own a patented claim of 20 acres and several unpatented claims. The property originally included the Juniper, Independence, Starlight (Patricia) and Mt. View lode claims and the Independencio placer claim. One of these claims apparently was renamed the Charles.

As nearly as can be determined from meager early records, the Juniper group of claims was probably first located in the 1850s about the same time as the adjacent Jumper mine. The mine was known to be active in 1882 with Stephen Arthur as superintendent and Samuel C. Bates as principal owner (Min. and Sci. Press, vol. 45, no. 12, p. 180, 1882). Later C. C. Schlageter of Mariposa came into possession of part of the claims and others were acquired by the Zirker family. The last period of operation was 1933-37 when both properties were leased by George L. Reed of Bear Valley and worked under the direction of Warren Dutton and D. R. Jones. There has been no apparent recent activity at the mine.

There are two sets of veins in the Juniper group of claims. The veins on the Juniper and old Mountain View claims strike roughly north but are somewhat sinuous and dip east and angles between 45° and 60°. Those on the Starlight (Patricia) and Independence claims strike generally northwest but are somewhat arcuate and dip northeast at angles similar to those of the other vein system. Some vein offshoots from the north-trending system strike east of north and tend to be steeper than the main veins. Vein widths vary from 1 foot to 5 feet. Wall rocks are massive and schistose greenstone derived from pyroxene andesite. They are of Upper Jurassic age.

The Juniper vein is opened by an adit level 500 feet long which reaches a vertical depth on the vein of 130 feet at the south end. There are several partly caved stopes or airshafts from this level to the surface giving a total footage of workings of about 800 feet (Laizure, 1935, p. 34). On the Starlight or Patricia claim there is a 320-foot vertical shaft with several hundred feet of drifts. These workings were partly caved and inaccessible in August 1955.

Data on the production of the mine is incomplete but the total tonnage of ore is relatively small. At least 1000 tons were milled between 1919 and 1940 with returns running from 14.4 oz. of gold and 4 oz. of silver per ton to 0.35 oz. of gold per ton. The average tenor of ore was about 1.04 oz. of gold and a fraction of an ounce of silver per ton.

Lafayette and Lafayette Extension (Ghirardelli) Mine. Location: Sec. 15, T. S., R. 16 E., M. D. near the south end of Hunter Valley west of the paved Hunter Valley road and ¾ of a mile north of the Pyramid mine. Ownership: Louis and Gabriel Queriolo, 861 San Ramon Way, Sacramento 21, California, owns two claims, part of a parcel of patented argicultural land.

The Lafayette mine is an old property active chiefly in 1866-67. In 1866 it was being operated by a San Francisco group headed by Samuel Miller with H. C. Treon as superintendent (Min. and Sci. Press, vol. 12, no. 18, p. 274, 1886). Late that year an inclined shaft had been sunk to a depth of 90 feet and a vertical shaft 70 feet deep. A 4-stamp mill was in operation. The ore yielded $25-$30 per ton. By 1867 ownership had passed to D. Ghirardelli of Oakland. The vein varied from 1 foot to 3 feet thick and was yielding $30-$40 per ton (Min. and Sci. Press, vol. 15, no. 15, p. 69, 1867; vol. 17, no. 3, p. 38, 1868). By 1920 ownership of the property had passed to Dr. C. A. Queriolo of Oakland and has remained in the Queriolo family to the present.

The Lafayette vein is narrow but ore shoots have been consistently of good grade. It strikes north and dips east at about 40°. The vein fissure has developed along or close to the contact of pyroxene andesite greenstone of the Logtown Ridge member of the Upper Jurassic Amador group and laminated tuffaceous sediments of the Hunter Valley chert member of the same group (Taliaferro, 1943, pp. 282-284).

Workings are inaccessible and there has been no recent work done on the mine.

Landrum (Barley Field) Simeon Landrum Mine. Location: Secs. 27, 34, T. 4 S., R. 18 E., M.D., 1 mile south of Colorado School just west of the Colorado-Sherlock Creek road and 5½ airline miles northeast of Mariposa. Ownership: George Matlock, Mariposa, California, owns the Old Standby claim containing 20 acres, surveyed but not yet patented.

The Landrum or Barley Field mine was discovered sometime prior to 1889, probably by Joseph Landrum. In 1899 it apparently was operated in conjunction with the Champion mine by a partnership of James Ridgway, Colonel Dunbar and Mr. Hay (see Champion II mine). The property originally consisted of 3 claims including the Simeon Landrum and Old Standby, but the others apparently were abandoned prior to 1928 when the property passed to Harry I. Maddox of Merced.

Most of the production of the mine was never recorded, but from 1897 to 1905 about 250 oz. of gold was produced from an unknown tonnage of ore (U. S. Bureau of Mines records). Some shallow exploratory work is done from time to time by the present owner but underground workings are difficult to support and the rather extensive old workings cannot be entered. The mine is characterized by pockets of very rich ore which are in places found in vein quartz and in others in disintegrated dike material.

The main Landrum vein of brecciated milky quartz strikes N. 35-40° W., dips 45° SW. and averages 2-4 feet wide. It is accompanied over much of its length by a clayey, much altered dike, mineralized in part, so that the ore-bearing zone is, in places, over 10 feet wide. Wall rocks are hornblende-biotite granodiorite and finer-grained feldspar porphyries. A second vein crops out 50-60 feet southwest of the main vein and roughly parallel thereto. In the few places it is exposed it is 4-5 feet wide and dips 50 to 65 degrees southwest. About 400 feet northwest of the end-line of the Old Standby claim and slightly en echelon to the Landrum vein is a 2-foot wide vein of bluish, glassy quartz carrying abundant chalcopyrite.

In the 1920's there was 90-foot shaft serving over 800 feet of drifts. In September 1955 these workings had been almost obliterated but the course of the vein is marked by a long series of shallow open cuts some of which may once have been stopes. All of the present work is done from shallow open workings. Owing to the softness of the vein and surrounding rock open-cut mining by steam shovel might be feasible in the dry season.

Live Oak and Governor (Stud Horse Flat group, Live Oak group, White Oak group) Mines. Location: Sec. 35, T. 3 S., R. 16 E., M. D. on the south side of Merced River Canyon 2 miles northwest of Bagby. Accessible by 3½ miles of jeep and foot trail from the Schoolhouse mine on the paved Hunter Valley Road. Ownership: North American Gold Mines, Inc., c/o Ada Stewart, 25 Delmar Ave., San Jose, California owns a group of unpatented claims and several mill sites, notably the Live Oak, Governor, Red Oak, Monitor, Grand View, Henrietta, Lincoln and Roosevelt claims.

Originally known as the Stud Horse Flat group of mines, the Live Oak and Governor probably were discovered in the 1890s or early 1900s. A 3-stamp mill, one of the early installations at the mines, burned in 1907 (Castello, 1921, p. 139). The first owner of record is John P. Carroll of Bagby who intermittently operated the property from about 1910 to the late 1930s. In 1914 workings consisted of a 140-foot shaft with two levels, 300 feet of drifts and a stope 250 feet long and 60 feet high; also several shallow shafts and short tunnels (Lowell, 1916, p. 598). The property was bonded for a short period in 1918 to a Los Angeles company, but no production was recorded (Castello, 1921, p. 139) and a 5-stamp mill brought on to the property was never erected. During the 1920s John Carroll ran a 225-foot crosscut adit southwest to the vein cutting it at a depth of 68 feet. From this crosscut a 250-foot southeast drift was run, a winze put down 20 feet and a raise put up to the surface (Laizure, 1928, p. 95). During the early 1930s a 250-foot double compartment shaft was sunk 25 feet southeast of the crosscut adit, the drift from the 68 or adit level was lengthened to 322 feet and drifts on the 225-foot level were run 250 feet northwest and 250-feet southeast of the shaft. There has been little activity at the mine other than assessment work since 1938.

The Live Oak vein strikes northwest and dips about 75° southwest. It can be followed at the surface for more than a mile, averages 3-4 feet wide and reaches a maximum width of 10 feet. A smaller cutter vein, the Roosevelt, intersects the main vein crossing the Roosevelt and Lincoln claims. Vein matter is milky quartz with native gold, galena, and pyrite. Wall rocks are pyroxene andesite greenstones of the Peñon Blanco member of the Upper Jurassic Amador group (Taliaferro, 1943, pp. 282-284). The greenstones contact the Mariposa slate about half a mile east of the mine.

Most of the workings have been described previously. The two shafts on the Live Oak claim are 212 feet apart, the more southerly caved above the 68-foot level, but the main (northerly) shaft in good condition. On the Governor claim are two adits, a lower, 112 feet long, and a second one 50 feet higher, 20 feet long with a 10-foot winze near the end (Julihn and Horton, 1940, p. 128).

The Carrolls estimated in 1932 that there was 14,000 tons of ore between the 68-foot and 225-foot levels from the main shaft that would average $10.50 per ton. There is no record of any of this having been removed, but recorded production from the mine between 1931 and 1934 indicated that some ore ran considerably less than $10.50 per ton. (U. S. Bureau of Mines records) Some sampling data are included in accounts by Julihn and Horton and Laizure, references previously cited. The authors were unable to get an accurate estimate of the total production from the Live Oak and Governor group of mines.

Long Mary Mine. Location: Sec. 17, T. 5 S., R. 17 E., M.D., ¼ of a mile north of the Mount Bullion-Hornitos road and 6 miles by paved road west of Mount Bullion. Ownership: Not determined. Most of the land in this area has reverted to agricultural status. Property was for many years a part of the holdings of the Mariposa Commercial and Mining Company.

The Long Mary mine has a long history dating back prior to 1900. Most of the recorded activity has been by lessees working with the management of Las Mariposas Grant. The principal periods of activity were 1900-1905; 1908-11 and 1914-1915. During this span of time the mine produced 9348.9 tons of ore which yielded $108,928.44 or an average of $11.65 per ton at the old price of gold. (Logan, 1935, p. 188) The entire surface installations including a 5-stamp mill were destroyed by a forest fire in 1927 (Laizure, 1928, p. 95). In 1939 an option to purchase the Long Mary and several other prominent Las Mariposas Grant mines was granted to the San Rosario Mining Company headed by A. R. Gordon of Washington, D. C., but the deal was never consummated. (Eng. and Min. Jour., vol. 140, no. 7, p. 62, 1939). There has been no production from the mine since 1915 and the workings are inaccessible.

The Long Mary vein strikes N. 85° W., dips 65° south and averages about 2 feet wide. Wall rocks are slate of the Upper Jurassic Mariposa formation. Vein matter is milky quartz with native gold and pyrite. The principal ore shoot was 350 feet long and at least 400 feet deep, judging from the shape of the stopes reported by Castello (1921, p. 121).

In 1920, the workings consisted of a 413-foot inclined shaft with 4 levels and an adit 263 feet long. There were 360 feet of drifts on the 200-level, 383 feet on the 300-level, 315 feet on the 350-level and 163 feet on the 400-level plus 900 feet of raises and a stoped area 400 feet high and 350 feet long (Castello, 1921, p. 121).

Louis and Louis Extension Mines. Location: Sec. 11, 5S., 17E., M.D., half a mile southwest of the Mt. Opher Mint historical marker on Highway 49. Adjoins the Mountain View mine on the southeast and is half mile northwest of the Greens Gulch mine. Ownership: Mariposa Commercial and Mining Company, c/o Eileen Milburn, Mariposa, California owns a parcel of land containing 240 acres that includes the Mt. View, Louis, and Louis Extension mines.

The Louis and Louis Extension are adjoining claims, on a conspicuous quartz vein of the Mother Lode system and probably were first opened up in the 1850s. Throughout its history it has been operated principally under lease from the management of the Mariposa Grant, most notably from the Mariposa Commercial and Mining Company. The principal periods of activity have been 1900-1903, 1905-06, and 1912. In this period the Louis claim produced 1240.14 tons of ore from which gold worth $27,691.33 was extracted or an average of $22.36 per ton at the former price of gold. The Louis Extension, last active in 1912, produced 10 tons of ore yielding $141.42 or $14.14 per ton (Logan, 1935, p. 188).

Through the Louis and Louis Extension claims the main vein strikes approximately N. 62° W. and dips 55° NE. As it reaches the Mt. View claim farther to the northwest the strike changes to about N. 47° W. According to Lowell (1916, p. 586) the vein exposed in the workings averages 4 feet wide. At the surface much of it is obscured by mantle. Wall rock is black slate belonging to the Upper Jurassic Mariposa formation. Vein matter is chiefly milky quartz containing native gold. According to Castello (1921, p. 121) the sulfide content is low. Castello mentions a second, presumably parallel, quartz vein also 4 feet wide but lists no further details. This could be any of several veins in the vicinity.

In 1920, the workings included a vertical shaft 375 feet deep with three levels at 100-foot intervals, several crosscuts and winzes and a 150-foot air shaft. There were 275 feet of drifts on the 100-level, 75 feet on the 200-level and 500 feet on the 300-level. Crosscuts totaled 775 feet and winzes 600 feet. None of the workings were accessible in September 1954. The site of the main shaft is marked by a large white dump.

Louisa (Louise, Louise Point) Mine. Location: Sec. 3, T. 3 S., R. 16 E., M.D., astride State Highway 49, ¼ of a mile south of Coulterville. Ownership: Walter J. Lautenschlager, 626 South Catalina St., Los Angeles 5, California, owns a patented property of 25.29 acres.

The Louisa mine is typical of the Mother Lode mines in northern Mariposa County. The prominent, massive, milky quartz reef and accompanying quartz-mariposite-ankerite vein matter crossing Highway 49 just below Coulterville make well-known landmarks in the southern part of the gold belt. Early records of the mine have been lost but it undoubtedly was one of the first claims to be located in the 1850s, the vein system being the most conspicuous in the Coulterville district short of the Peñon Blanco itself. The principal period of activity at the mine was 1894-1900 when the Merced Gold Mining Company, J. R. Gilbert, superintendent, spent more than $56,000 in improvements and development work (Eng. and Min. Jour., vol. 61, no. 3, p. 68, 1896). Milling was done at the Potosi mill 1½ miles to the west. Prior to acquisition of the mine by Merced Gold Mining Company the property was part of the 20,000 acre Cook Estate. A General Boyd of San Francisco owned the property for a short period about 1893-94 (Storms, 1896, p. 172). There has been no sustained work on the mine since 1900. No equipment is on the property and the workings are inaccessible.

The east branch of the Mother Lode passing the length of the Louisa claim is a broad, multiple vein more than 300 feet wide striking N. 40-45° W. and dipping 70-75° NE. The bulk of the vein consists of quartz-mariposite-ankerite rock derived in considerable part from hydrothermal alteration of serpentine. Several tabular to lenticular sheets of massive, milky quartz traverse the quartz-mariposite-ankerite rock in sub-parallel or braided fashion and the adjoining masses of vein matter are interlaced by stringers of milky quartz. In the upper, oxidized parts of the vein system the quartz-mariposite-ankerite rock has been converted to a spongy, rust-colored, earthy rock consisting chiefly of yellow oxides of iron intermixed with quartz stringers. Ore shoots of native gold and auriferous pyrite commonly are large but of relatively low grade. They generally are found in quartz-mariposite-ankerite rock on the hanging wall or footwall side of the massive quartz subveins. The bulk of ore produced ran less than $12 per ton at the former price of gold. According to Storms (1896, p. 172) when the vein matter was solid mariposite-ankerite rock without quartz or a brecciated appearance the gold content in the ore was too small to pay.

The vein system in the east branch of the Mother Lode lies along the contact of black slates of the Upper Jurassic Mariposa formation and massive pyroxene andesite greenstone probably belonging to the Peñon Blanco member of the predominantly volcanic Amador group, also of Upper Jurassic age. A good sketch of the vein system, as seen in plan, has been drawn by Storms (1896, p. 172). The rock marked diabase on the sketch is greenstone derived from pyroxene andesite.

In 1900, workings consisted of the vertical Nelson shaft 375 feet deep with 3 levels and an air shaft 150 feet deep. There were 275 feet of drifts on the 100-level, 75 feet on the 200-level and 500 feet on the 300-level, plus crosscuts totaling 775 feet and winzes totaling 600 feet (Laizure, 1928, p. 96).

Louisiana Mine. Location: Secs. 25, 26, T. 2 S., R. 17 E., 12 miles by road east of Coulterville via the Greeley Hill, McDiermid guard station (U. S. Forest Service) and Dutch Creek roads. Adjoins the Bondurant mine on the northwest. Ownership: Sarah W. Treat, c/o Mrs. Robert Sherman, 975 Roble Ridge Road, Palo Alto, California, owns a patented property of 30 acres. The Louisiana claim covers 3500 feet in the vein.

Discovery of the Louisiana mine was probably made in the early 1860s. In January 1866 the mine was purchased and placed in operation by the Heslep family (Min. and Sci. Press, vol. 12, no. 3, p. 40, 1866). Judge Heslep, W. C. Heslep and his son P. C. Heslep operated the property until the spring of 1867. During this period ore ran from $10 to $50 per ton plus the yield of a sulfide concentrate not readily reduced which assayed $100 to $500 per ton. Some interesting machinery was installed by the Hesleps and their predecessors including two mechanically driven sledge hammers made by Rex, Logan and Company, a walking-beam type of pumping engine (English-made) and a Hogan de-sulfurizing furnace burning hydrogen gas. The sulfide concentrates could be handled in this furnace for $1 per ton (Min. and Sci. Press, vol. 12, no. 3, p. 40, 1866). At the time the Hesleps took over the mine there was a crosscut tunnel 200 feet long cutting the vein at a depth of 100 feet and a connecting shaft from the surface. In the spring of 1867 management passed into the hands of Colonel Conley, terminating a short layoff. The Louisiana claim was patented in February 9, 1878 (Julihn and Horton, 1940, p. 136) and was active at least intermittently through the 1870s and 1880s. About 1891 three additional shafts were sunk and a 4-stamp mill installed. A. C. Cable superintended operations for J. J. Grove of San Francisco at the mine in 1896 (Min. and Sci. Press, vol. 73, no. 12, p. 242, 1896). In 1898 the owners gave up operation of the mine but it was leased for a short period by former employees who took out 600 or 700 tons of ore which yielded $10,330.98 (Julihn and Horton, 1940, p. 136). In 1911 the property was bonded to G. F. Dyer of Merced who apparently did some exploration and development work but no mining. By 1920 ownership of the mine had passed to C. P. Treat and the mine was put under option by A. T. Adams of the Bondurant mine (Castello, 1921, p. 126), but final sale was not consummated. The mine has remained in the Treat family since that time. In 1953 and 1954 some work was done on the north end of the claim by E. W. Ferguson of Coulterville.

Massive white quartz of the Louisiana vein is boldly exposed at the crest of a low ridge north of the road through the Bondurant property. It strikes N. 70° W., dips 55° NE. and ranges from 2 feet to more than 15 feet wide. The southwest or hanging wall side of the vein consists mainly of massive milky quartz in which gold values are low. The best ore apparently is concentrated in the 3- to 6-foot wide part of the vein nearest the footwall (Julihn and Horton, 1940, p. 137). The best grade of ore apparently came from the southeast half of the claim, judging from the concentration of workings there. Vein matter in the ore shoots consists of native gold, galena, sphalerite and pyrite, pyrite being most abundant. Julihn and Horton (op. cit.) stated that the ore would probably average $1\frac{1}{2}$ to $2\frac{1}{2}$ percent sulfides. Wall rocks are mainly black slate and graphitic mica schist, metasediments belonging to the Paleozoic Calaveras group. The vein is accompanied over part of its length by a badly altered ochrous granitic rock. According to Julihn and Horton the vein contains ore for more than 1200 feet of strike length, principally on the hanging wall side. Samples ran from $6 to $15 per ton, presumably at the present price of gold. The mine production figures have not been recorded.

There are five shafts of various depths on the property exclusive of the one started recently on the hanging wall side of the vein near the north end of the claim. The Number 2 shaft is reported to be 250 feet deep and the No. 3 shaft 150 feet deep (Julihn and Horton, 1940, p. 138). The various shafts could not be distinguished one from the other at the time of the authors' visit in 1954. There are 2 tunnels on the northwest part of the claim which could easily be cleaned out and an undetermined number of other adits on the southeastern part of the claim.

Lovely Rogers (Cherokee, Shimer) Mine. Location: Center sec. 11, T. 3 S., R. 17 E., M.D., in a saddle half a mile south of the Gentry Gulch road and 1 mile west of the Hasloe mine. Accessible by road via the Martin Walling or Gold King mine. Ownership: Mrs. Nettie Shimer Hauck, et al., Yosemite National Park, California, owns one partially patented claim of 12.57 acres.

The Lovely Rogers mine was discovered by a Cherokee Indian named Rogers in 1857 (Browne, 1868, p. 32). Rogers operated the mine by hand methods for several years, recovering gold from high-grade surface ore by crushing the ore in a hand mortar. In 1859 an 8-stamp steam mill and two arrastras were built, ore milled yielding $35 per ton. The mill apparently never paid its way and was sold at sheriff's sale in the early 1860s. During the 1880s and 1890s the mine was worked intermittently by the Shimer Brothers and ownership has remained in the Shimer family up to the present. There has been no recent work done on the mine.

According to Laizure (1928, p. 96) there are three more or less parallel veins on the Lovely Rogers claim. None of the veins exceed 3 feet in width. The principal vein is narrow and sinuous and the strike varies from N. 56° W. to N. 73° W. It dips about 70° northeast. Wall rocks are slate and massive quartz-biotite rock that is commonly horn-felsic. These are metasediments belonging to the Paleozoic Calaveras group. The metasediments are cut by several narrow, dark granitic dikes.

Principal workings are a shaft 100 feet deep and three adit levels, 400, 300 and 250 feet long, respectively. There is also a 70-foot crosscut and an 80-foot winze from one of the adit levels, several shallow shafts or pits on the northwest third of the claim and numerous open cuts along the vein over the southeast two-thirds of the claim.

Malone (Bear Creek) Mine. Location: Sec. 5, T. 5 S., R. 19 E., half a mile east of Bear Creek School and State Highway 140 and 1¼ miles south of Midpines. Ownership: Not determined.

The Malone mine was located in 1880 and carries the discoverer's name. In 1881 the shaft was down 60 feet with George McCaffery superintending the operation for Malone (Min. and Sci. Press, vol. 42, no. 18, p. 276). In 1883 the property was being operated under lease by J. Mitchell and Company. These operators had a 5-stamp mill working and were driving a tunnel (Min. and Sci. Press, vol. 46, no. 16, p. 268). The mine was intermittently active under adverse conditions through the 1880s constantly hampered by inadequate financing. McCormick and Hay were the operators in 1889 and Koch and Mitchell in 1890 (Min. and Sci. Press, vol. 61, no. 5, p. 70 and no. 13, p. 202). In 1914

W. H. Johnson operated the mine and 5-stamp mill (Min. and Sci. Press, vol. 108, no. 23, p. 945). By 1915 ownership had passed to the Golden Wreath Mining Company (Lowell, 1916, p. 587) but the mine had to be sold that year to satisfy debts incurred by the company. Between 1915 and 1919 the property was intermittently active, principally by Thomas Gordon and associates. Will Johnson worked the mine for a short time in 1919 (Castello, 1921, p. 122) under lease from Gordon, and three men also leased it for a while in 1935 (Laizure, 1935, p. 37). The property has been idle for many years. It is on patented agricultural land.

A series of five roughly parallel, narrow quartz veins crosses the Malone property. The main vein strikes N. 35° W. and dips 35-40° SW. It varies from a few inches to 3½ feet wide. The other veins are narrower and less persistent. The veins cut a small, roughly circular intrusion of granitic rock three-quarters of a mile in diameter which has intruded schistose and massive greenstone. The eroded upper parts of the veins apparently yielded considerable placer gold prior to 1880. Vein matter is ribbon quartz containing thin sheets of altered wall rock, abundant auriferous pyrite and some free gold (Preston, 1890, p. 300-302). The principal ore shoot is over 350 feet long and pitched steeply northwest. It has been stoped to a depth of about 90 feet but the total depth of the shoot has not been reached. According to Preston (1890, p. 302) the ore averaged about $14 per ton at a price of $7.50 to $18.00 per ounce.

Workings consist of a 200-foot inclined shaft, which reaches a vertical depth of 140 feet, 2 airshafts 30 and 75 feet deep, respectively, and a drift adit and connecting crosscut tunnel 600 feet long.

Malvina Group (Potosi, Mahoney, Douglass, D. Cook, Merced). Location: Secs. 4, 9, 10, T. 3 S., R. 16 E., and secs. 32, 32, T. 2 S., R. 16 E., M.D., along Maxwell and Black Creeks just west of Coulterville. Ownership: Walter J. Lautenschlager, 626 So. Catalina St., Los Angeles 5, California, owns 14 patented claims and several mill sites aggregating over 300 acres.

The Malvina and adjacent claims were first located in 1852 (Goodyear, 1888, p. 347). According to news items in the *Mining and Scientific Press* much of the early work was done by the Maxwell Creek Mining Company more popularly known as the French Company. F. L. A. Pioche was the principal owner in the company. In 1867 most of the claims on the Malvina vein were sold to Douglass, Chamberlain and Company of San Francisco, Gilbert Douglass being the principal partner. By the end of that year four shafts had been sunk but the portion of these shafts was not recorded. A 440-foot adit was also driven which followed the vein for 330 feet (Logan, 1935, p. 183). By 1876 the Douglass Company had erected a 20-stamp, steam-driven mill on or near the Potosi claim. About 1878 the claims on both the Malvina and Mary Harrison veins were sold to Seth Cook and Company. This company rapidly acquired properties in the Coulterville district which ultimately aggregated 22,000 acres. In the spring of 1880 the Cook Company commenced driving a tunnel from Black Creek south for more than 3,000 feet to get under the supper workings of the Malvina claims. This was completed in 1881 and was used for many years as a haulageway for ore from the Malvina and adjoining claims to the Potosi mill.

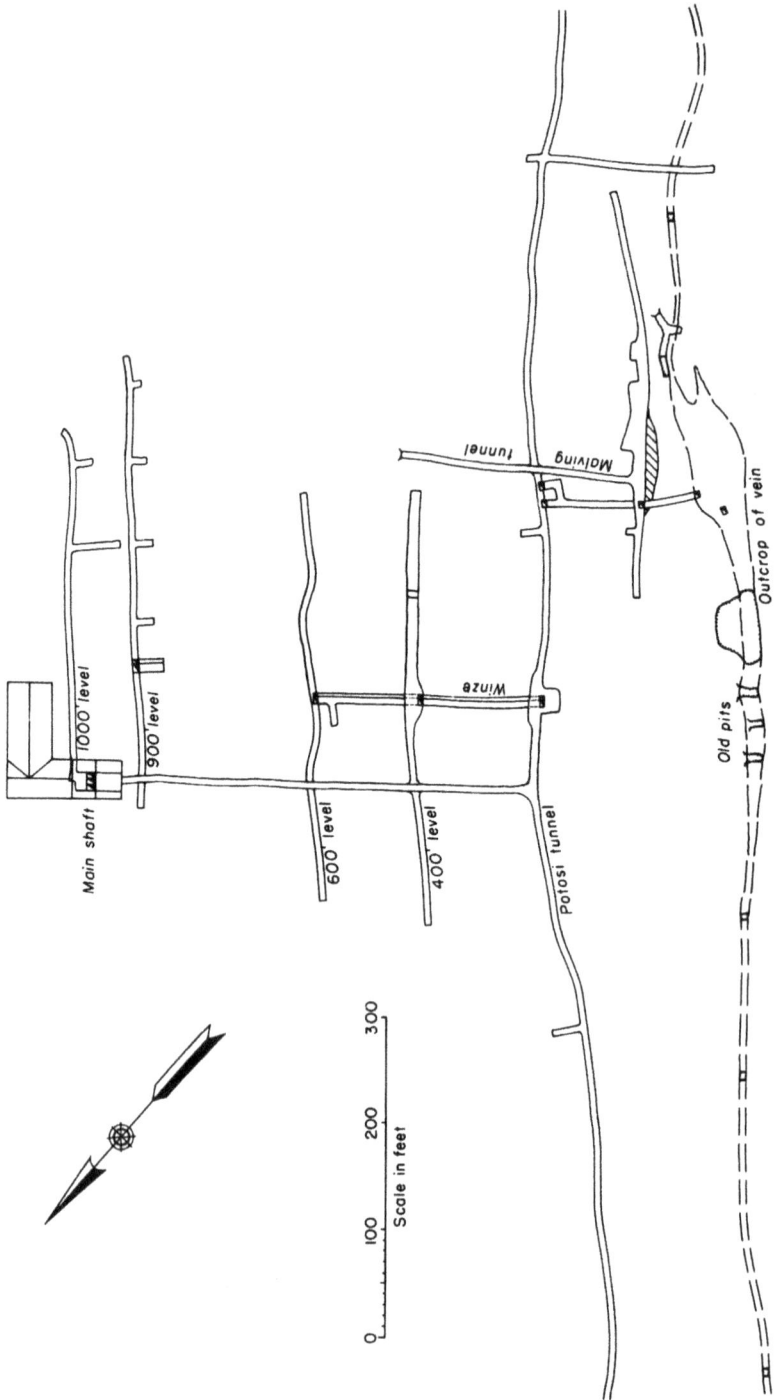

FIGURE 26. Plan drawing of the Malvina mine on the Mother Lode near Coulterville. Reproduced from fig. 43 of the U. S. Bureau of Mines Bulletin 424 (Julihn and Horton, 1940).

Operations by the Seth Cook and Company apparently ceased about 1882 and the properties remained idle until about 1894 when the Cook Estate was acquired by the Merced Gold Mining Company. This company was headed by John W. Mackay, John P. Jones and Alvinza Hayward. Captain Couch and Captain Ward were their mine superintendents. The Merced Company was unable to keep the grade of ore sufficiently high to pay expenses. The Malvina mine was closed down on August 1, 1897 and the Mary Harrison later that year. The Malvina group remained idle until 1910. For a short period that year the D. E. Lutes Company operated the Mahoney claims. In 1911 H. P. Daulton of Alameda did some work on some of the claims on both the Malvina and Mary Harrison veins but was unsuccessful in making any sustained production (Mining and Sci. Press, vol. 102, no. 18, p. 640, 1911). By 1920 ownership had passed to A. S. Bigelow and G. M. Hyams of Boston and some work was done at the mine superintended by C. I. Mentzer of Coulterville (Castello, 1921, p. 123). According to Castello the ore averaged about $4 per ton. In 1933 the Boston California Mining Company, headed by Charles H. and Eric Segerstrom of Sonora, began unwatering the Malvina mine to sample 250,000 tons of .2 oz. per ton ore believed to be present from old mine records (Julihn and Horton, 1940, p. 104). By November 1934 the 1010-foot shaft had been reconditioned and exploration revealed a substantial quantity of $10 ore. From 1938 to 1942 the company milled 121,093 tons of ore which yielded 13,197 ounces of gold, 1460 oz. of silver, 4988 lbs. of lead and 24,548 lbs. of copper or an average of $3.82 per ton in gold. The properties have been idle since 1942.

The Malvina group of claims is on the Malvina vein which is generally recognized as the west branch of the Mother Lode in the Coulterville area. The vein is almost continuously traceable from the Rittershoffen claim on the south to the Mahoney claim on the north, a distance of about 2½ miles. The strike of the vein is N. 45-50° W. and the dip is 56-75° NE. In the Potosi Tunnel the Malvina vein is only 2 feet thick for much of the exposed distance (Goodyear, 1888, p. 347). Elsewhere it reaches thicknesses up to 20 feet, and the average thickness is about 10 feet. According to Goodyear (1888, p. 347) the Potosi tunnel on the Malvina No. 1 claim developed an ore shoot 400 feet long from which 3,000 tons of ore were taken which averaged $7 a ton at the old price of gold. Ore minerals are principally native gold and auriferous pyrite in a ribbon structure of milky quartz and slate. Quartz-mariposite-ankerite rock is locally present. Wall rocks are chiefly sandy slate although Logan (1935, p. 183) mentions a strip of greenstone occurring on the hanging wall side of the vein and the vein has formed only a few hundred feet from the contact between slate of the Mariposa formation and pyroxene andesite greenstone of the Peñon Blanco member of the Amador group, both formations of Upper Jurassic age. The Mariposa slate occupies a synclinal trough faulted off on the northeast by the main branch of the Mother Lode thrust. West of the Malvina vein and north of the Potosi claim there are conglomeratic beds near the base of the Mariposa formation containing quartz and slate pebbles. These beds were first described by H. W. Fairbanks in 1890. The Mariposa formation is in gradational depositional contact with the greenstones of the Amador group in this vicinity.

The principal workings and the ones most recently used (1938-42) are a 3-compartment, 1010-foot vertical shaft with levels at 200, 400, 600, and 900 and 1,000 feet on the Malvina claim and a 3,000 foot haulageway tunnel extending from the Malvina to the Potosi claims. There are more than 2287 feet of drifts and 1225 feet of crosscuts on the five levels. There is also an older inclined shaft 875 feet deep sunk on the Malvina claim, several airshafts and tunnels on the Potosi claim and numerous other workings on claims north of the Potosi, particularly on the Mahoney. None of these was readily accessible in September 1954. Most of the known ore has been stoped above the 600-level on the Malvina and Potosi claims. There was formerly a shaft at least 350 feet deep on the Potosi claims near the old mill site which had 9 levels in 1944 (Lowell, 1916, p. 594). The upper part of the shaft was filled to mill level in September 1954.

The total recorded production of the Malvina group is approximately $867,800 in gold, silver, lead and copper. Including the early production of the mine, of which there is no precise record, the claims must have produced more than $1,000,000. By far the largest proportion of this total was amassed in the periods 1897-1903 and 1938-1942. The average tenor of ore during these periods was $4 per ton although nearly half the recorded production was made during a time when gold only brought $20.67 per ounce. Much of the ore mined in the 1800s would have yielded $7-8 per ton at the present price of gold.

Marble Springs (Compromise and Eubanks) Mine. Location: Secs. 30, 31, T. 2 S., R. 18 E., in the Bull Creek district 1½ miles west of Bull Creek School and 2 miles south of Bower Cave. Accessible by good partly surfaced, partly graded dirt road from Coulterville or by graded dirt road from Briceburg on Highway 140. Ownership: George B. Glenn, 3134 East 10th St., Oakland 1, California, owns two patented claims, the Compromise and Eubanks totaling 41.3 acres.

The Marble Springs mine was discovered in either 1850 or 1851 (Browne, 1868, p. 32; Eng. and Min. Jour. Vol. 101, no. 4, p. 201, 1916). It was equipped with a 5-stamp mill using wooden stamps, one of the earliest mills built in California. According to Browne (1868, p. 32) the first owner failed to make the mine pay, and sold it to a partnership who operated about 8 years at a loss, even though the ore ran $25 per ton. After a period of idleness of several years duration the mine was purchased and reopened about 1866 by H. G. Coward. A new mill was built, the old mill having been removed. Some specimen ore was taken out by Coward that ran as high as $900 per ton and the mine was noted for the production of gold-bearing quartz for jewelry (Browne, 1868, p. 32). By 1869 the mine had again become idle, fire having destroyed the mine buildings. By 1879, ownership had passed to A. G. Black. In 1881, the mine was reopened, presumably by an English company, and a new mill and hoisting works were built. The shaft, which was 100 feet deep in 1879 was deepened to 400 feet, several hundred feet of drifts were run, and a substantial quantity of high-grade ore was taken out using Chinese labor. The amount produced during the period 1881-84 was not recorded. By 1885, the mine again had fallen idle because of a high-grading dispute between miners and Superintendent Lawrence (Min. and Sci. Press, vol. 51, No. 7, page 120, 1885). From about 1894 to 1916 the mine was owned by various

FIGURE 27.   Installations at the Marble Springs mine in the Bull Creek district 7 airline miles northwest of Briceburg. Fully equipped and operated as recently as 1952, the Marble Springs mine has a long history of production totaling at least $200,000.

members of the Hilliard family, first by M. Hilliard and then by P. J. and W. W. Hilliard. The Hilliards, or their lessees, took out about $11,500 in gold and silver from the mine between 1898 and 1901 (U. S. Bureau of Mines records). The mill and hoist buildings were again destroyed by fire in 1909. In 1910 the Marble Springs Mining and Milling Company, F. M. Bernou, manager, secured a bond on the Marble Springs and North Fork mines but was unsuccessful in putting them into operation. In 1912, J. A. Flink secured an option to purchase but no deal was consummated until 1916 when a new stock company, The Marble Springs Gold Mining Company, was organized with R. C. Haywood and J. H. Flink as principal stockholders (Eng. and Min. Jour., vol. 101, no. 4, p. 201, 1916). This company continued to function under various managements and lessees until about 1943, the principal production periods being 1918-20, 1930, and 1935-43. In 1950 the Glenn-Steintorf Company took over the mine, completely renovated it, built a new mill head frame and hoist house, sunk the shaft 80 feet deeper, ran 250 feet of drifts and several raises. Operations were temporarily suspended in 1952 and Mr. Glenn since has become the sole owner.

The Compromise and Eubanks veins are segments of the same vein which strikes N. 10-15° E. and dips 35-47° E. It varies from 2 to 5 feet thick. Vein matter is milky to bluish quartz, in places carrying thin seams of wall rock, containing native gold, galena, chalcopyrite, pyrite and probably a little tetrahedrite and arsenopyrite. Old accounts in the Mining and Scientific Press and by J. Ross Browne (1868, p. 32) indicate that the main ore shoot was 400 feet long near the surface, widening and then narrowing somewhat at depth, and that there were several other shoots one of which was over 100 feet long. The main ore shoot pitched moderately south. The vein is almost continuously stoped above the 300-level for 600 total feet south of the shaft. (Perrin, Tom, personal communication, 1954). Wall rocks are chiefly massive, blue-black hornfelsic quartz-biotite rock with some inter-layered greenstone and occasional slaty and quartzitic strata. These metasediments are part of the Paleozoic Calaveras group. The vein system is intersected and accompanied by a fine-grained greenish granitic dike rock approximating quartz diorite. Dikes of this rock generally strike the vein from the hanging wall side (Castello, 1921, p. 123) but are not continuous the full length of the vein.

During the last period of operation, ore taken largely from the 400-level averaged about $10.70 per ton in gold with considerable lead and a little silver and copper. Much of the ore mined prior to that date ran better than $25 per ton at the former price of gold. Between 1936 and 1943, a total of 1767 tons of ore milled ran $24.62 per ton at the present price of gold. Incomplete records indicate that the total production of the mine must be at least $200,000.

The principal working is a 470-foot inclined shaft with four levels and several thousand feet of drifts. There is also a drift adit a short distance south of the main shaft which is 130 feet long. The vein has been stoped discontinuously for an aggregate distance of nearly 3000 feet, between the surface and the 300-level. In September, 1954 the shaft was full of water to within 20 feet of the collar. The modern mill, which includes two jaw crushers, a ball mill, rake classifiers, four flotation cells, a cone concentrator and a small Gibson mill, could quickly be put into shape for operation.

Mariposa Mine. Location: Secs. 23, 24, T. 5 S., R. 18 E., M.D., in the southeastern edge of the town of Mariposa. Ownership: Mrs. Frank E. Galhagher, 211 Twenty-sixth St., Merced, California. Under lease to George Adams of Mariposa.

The following historical summary is drawn principally from Bradley (1954, p. 32), Logan (1935, p. 184), Browne (1868, p. 28) and files of the *Mining and Scientific Press* and *Engineering and Mining Journal*.

Discovery of the Mariposa mine is generally credited to Kit Carson and two associates in the spring of 1849. As early as July 1849, Palmer-Cook and Company were crushing ore from this mine in a stamp mill. Credited by some as being the first stamp mill in California, it is certainly among the first 2 or 3 built in this state. According to Browne (1868, p. 28) the mine had produced $200,000 before litigation was settled which gave Fremont undisputed title in 1859. From 1859 to 1864 it was successfully operated by a lessee named Barnett. In 1863, Fremont sold Las Mariposas grant to a New York syndicate which, in 1864, took over direct operation of the Mariposa mine. The 40-stamp mill of the Greens Gulch mine was moved to the site of the Mariposa mine in 1864 and during that year the gross yield of the mine was $84,948 from ore that averaged $25 per ton (Browne, 1864, p. 29). The operating syndicate was dissolved about 1868 but the mine continued in operation until 1870 at which time the shaft was down 275 feet and the vein was described as 4 to 5½ feet wide at that depth. Although some rich ore was found on the lower level in 1870, the average grade of ore evi-

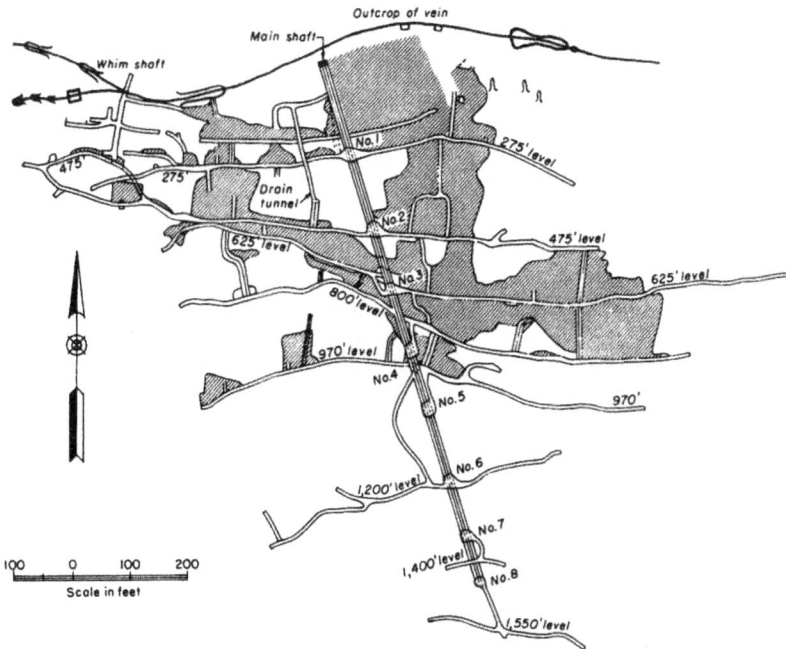

FIGURE 28. Plan drawing of the Mariposa mine workings showing stoped areas. Reproduced from fig. 55 of U. S. Bureau of Mines Bulletin 424 (Julihn and Horton, 1940).

dently fell off from $25 in 1864 to $10 or $11 per ton. The mine had produced $1,500,000 up to the time of its closing in 1870 (Logan, 1935, p. 184). In 1887 the Mariposa Commercial and Mining Company was organized by San Francisco financiers Hayward, Flood, et al., but nothing was done with the Mariposa mine until 1897 when the company was reorganized by another San Francisco group headed by Fred Bradley. Work of rehabilitating and re-equipping the mine began in September 1899 and was completed in 1900, with William Dodge as foreman and F. T. MacGuire as resident manager. From December 1900 to the end of 1915, when mining was halted by adverse wartime conditions, the mine produced a total of 112,379 tons of ore yielding $693,-205—an average of $6.17 per ton at the former price of gold. A new inclined shaft was sunk to a depth of 1550 feet (measured along the incline) and several thousand feet of drifts was run. In 1927 the State Highway Commission produced 35,000 tons of crushed rock of unannounced value from dumps at the Mariposa mine. Between 1927 and 1932 lessees worked the extension of the Mariposa vein, taking out about 1500 tons of low-grade ore and several small high-grade pockets. F. E. Gallagher reactivated and re-equipped the Mariposa mine in 1952 and intermittently produced small tonnages of ore mined during cleanout and development work. By July 1955, the main shaft had been unwatered and cleaned out to a depth of 800 feet and a new 100-150-ton-per-day-capacity mill put into operation. The death of Mr. Gallagher early in 1956 suspended activities at the mine. The Mariposa mine has been among the five most productive in Mariposa County, known production to date aggregating $2,395,405.

The Mariposa vein east of the main shaft strikes N. 70° W. and dips 60-70° SW. Between the main and old shafts the vein branches with one branch striking off at N. 50° W. Vein widths vary from 2 to 8 feet wide east of the junction line of the two branches. The branches do not average much greater than 2 feet wide. Vein matter is milky quartz, commonly ribboned toward the borders by thin inclusions of wall rock, with native gold, pyrite and arsenopyrite (Browne, 1868, p. 28). According to Browne there are high-grade pockets in the branch veins that nearly always contain arsenopyrite. He reported two pockets from which $30,000 and $15,000 was obtained. East of the vein intersection the vein matter uniformly contained gold and ore shoots are ill-defined. Several theories of vein structure are discussed by Julihn and Horton (1940, p. 155-157).

Workings of the Mariposa mine include the old, now unusable, shaft 475 feet deep, located 350 feet west of the main shaft, and the three-compartment main shaft 1550 feet deep (measured along the incline) sunk in the 1900's. It has 8 levels with drifts as follows (Laizure, 1928, p. 99).

Level	East	West
275	390	400
475	400	637
625	700	600
800	322	410
970	396	412
1200	250	400
1400	60	50
1550	200	130

Most of the ore has been stoped out above the 800-level and there are a number of raises and crosscuts as well as a drainage tunnel. A diagram of the workings may be seen in the accompanying figure.

The mine is equipped with a 100-150-ton-per-day mill including jaw-crusher, bank of five stamps, ball mill, jig, Knudsen bowl, amalgamating barrel, and rake classifier. Mill heads run from $8 to $15 per ton, mostly in free gold. Pyrite concentrate runs $60 per ton.

Mary Harrison Group. Location: Secs. 3, 10, 11, T. 3 S., R. 16 E., M. D., 1 mile south of Coulterville on Highway 49. Ownership: Walter J. Lautenschlager, 626 South Catalina St., Los Angeles 5, California, owns 6 patented claims, the Balance, Choteau, Dahlia, Ely, Sheridan and Venture, and several mill sites totaling nearly 80 acres.

The history of the Mary Harrison group of mines closely parallels that of the Malvina group, the two groups being operated for many years by the same company. Discovered about 1852, the mine was operated by the French Company, also known as the Maxwell Creek Mining Company, through the early years of its development. F. L. A. Pioche was the principal owner (Goodyear, 1888, p. 346) in this company. In 1867 the claims on the Mary Harrison vein were sold to Douglass, Chamberlain and Company of San Francisco. By August 1878 the old inclined shaft on the Mary Harrison claim was down 450 feet (Min. and Sci. Press, vol. 33, No. 14, p. 221; Browne, 1868, p. 34). Later that year the mine was sold by the Douglass interests to

Seth Cook and Company. About November 1894, the Merced Gold Mining Company took over the Cook Estate mines and for the next 2 years did extensive development work on the Mary Harrison group of claims. The Merced Company discontinued work on other claims on the Mary Harrison and Malvina veins in 1897 and from that year to 1904 the Mary Harrison mine was the sole producer for the company. According to the historical marker now at the site of the old boiler house the Mary Harrison mine produced $1,500,000 from ore averaging $7 to $12 per ton at the old price of $20.67 per ounce. According to Logan (1935, p. 185) between $330,000 and $400,000 was produced between 1895 and 1903, mill returns in 1897 being $5 per ton. Milling was done at the Potosi mill throughout most of the history of the mine, ore being hauled to the mill over narrow-gauge railroad by steam locomotive. D. G. Kidder, later Mariposa County Assessor, was the last superintendent. A little development work was done on the Mary Harrison mine in 1911 by H. P. Daulton of Alameda but there was no recorded production. Sometime prior to 1920 the Merced Gold Mining Company was dissolved and although the properties have since passed through numerous ownerships no further mining has been done. All buildings were destroyed by fire in August 1926 (Laizure, 1928, p. 100).

The Mary Harrison vein is part of a well-defined system belonging to the main or east vein of the Mother Lode. It is clearly traceable from the county line on the north to the Virginia mine on the south, after which its character is less well-defined. Cross sections of the vein are well exposed in road cuts along Highway 49. The general strike of the vein is N. 50° W. with dips from 60-75° NE. It is characterized by a prominent, massive sheet of milky quartz 4 to 20 feet thick between irregular, ribboned masses of quartz-mariposite-ankerite-pyrite rock. Because of decomposition of pyrite with formation of sulfuric acid groundwater the upper parts of the quartz-ankerite-mariposite-pyrite rock are commonly converted to porous, earthy masses of yellowish brown oxides of iron and silica. Some of the high-grade ore was taken from this sort of vein matter but other similar material proved of very low grade. Adjacent to the vein, wall rocks, which are chiefly slate, serpentine and greenstone, have been impregnated with pyrite in large irregular patches, but such material was evidently too low in gold content to constitute ore. The entire width of vein matter varies from 50 to more than 100 feet wide and there are extensive horses of schistose serpentine caught in the vein matter. Along most of the length of the Mary Harrison claim the hanging wall side generally consists of black slate and schist assigned to the Paleozoic Calvareas formation (group) by Turner and Ransome (1895, maps). The footwall side is serpentine along much of the length of the vein on the Mary Harrison claim but sheets of black slate and greenstone are in contact with it at some points farther north. The greenstone appears to be derived from pyroxene andesite of Upper Jurassic age and the footwall slate is probably part of the Upper Jurassic Mariposa formation. The vein occupies a profound thrust-fault fissure along which there has been displacement aggregating many thousands of feet.

The most extensive ore shoot was encountered on the footwall side of the central quartz sheet. It had a strike length of 300 or more feet and was stoped chiefly between the 400 and 700 foot levels. The vein width embraced by the shoot varied from 3 to 10 feet. Ore was not obtained from the central sheet of massive quartz but some ore was found on the hanging-wall side of the quartz sheet. Large areas of quartz-ankerite-mariposite rock are barren of values and the bulk of the vein matter was too low grade to mill. North of the Mary Harrison group of claims vein matter reaches a width of 300 feet. Although occasional pockets of high-grade ore were found that touched off stories of high-grading activity (various issues of *Mining and Scientific Press*) most of the ore was of milling grade and yielded $5 to $12 per ton at the old price of gold. There is no indication of the probable quantity of ore left in the lower levels of the mine.

Principal workings are a 1200-foot vertical shaft, the most northerly, with levels at 100-foot intervals, a second inclined shaft, also with levels at 100-foot intervals and an airshaft of unknown depth. These shafts are within 250-300 feet of each other in a nearly straight line, and are open to depths of several hundred feet. According to Logan (1935, p. 185) there are more than 5000 feet of drifts. None of the shafts are timbered nor are they accessible without hoisting equipment. There are numerous other workings of unknown extent on the Dahlia, Venture, Ely and Choteau and Sheridan claims but these are largely caved and inaccessible. There has been no recent activity on any of the claims.

Mexican I Mine. Location: NE¼ sec. 23, S.E. ¼, sec. 28, T. 4 S., R. 17 E., 1½ mi. south of Bear Valley. Ownership: Not determined; probably Ellen T. Simpson, et al.

The Mexican mine was discovered sometime during the 1850s and was worked by Mexicans up to the time Fremont gained undisputed title to Las Mariposas grant in 1859. Over $50,000 is supposed to have been taken from a shallow pit prior to the time ownership passed into Fremont's hands (Julihn and Horton, 1940, p. 130). In 1908 a pocket containing $62 was taken out and a 150-foot shaft was sunk (Mariposa Mining and Commercial Company records). In 1934 the mine passed into control of Hickman and Dolman but little or no work was done on the Mexican mine by the company. Between 1934 and 1940 water for the Dolman mill was pumped from the shaft of the Mexican mine.

Striking N. 35° W. and dipping 70° NE., the Mexican vein follows the approximate contact between slate of the Upper Jurassic Mariposa formation and pyroxene andesite greenstone of probable Upper Jurassic age. Greenstone forms the west or footwall side of the vein and slate the hanging wall side. The vein system is discontinuous and is made up of a series of subparallel members and intersecting cutter veins. The upper part of the vein apparently contained pockets of high-grade ore formed by oxidation and removal of valueless sulfide minerals. There is little indication that rich ore was found at depth, but the vein was never thoroughly explored because of the large amount of water to be dealt with in the workings.

Mockingbird (Talc, Lacy) Mine. Location: Sec. 27, 4 S., 18 E., M. D., 1 mile south of Colorado School and 5½ airline miles northeast of Mariposa. Accessible by 4 miles of good dirt road from Summit Inn on Highway 140. Ownership: David W. and Maxie S. Dukes, 612 Blackburn, Watsonville, California own 1 patented claim of 20 acres.

The Mockingbird mine, known prior to 1900 under the names Talc and Lacy, was discovered sometime prior to 1890. It is characteristically a pocket mine noted for specimens of wiry and arborescent gold (Preston, 1890, p. 304). Through the 1890s it was operated principally by George Lacy of Mariposa (Storms, 1896, p. 219). In the early 1900s the property passed into the hands of J. A. and C. J. Schroeder of Mariposa and remained in the Schroeder family until the 1950s. No work has been done on the property in recent years. Much of the work on the mine was done by the Weston Brothers prior to 1920 (Castello, 1921, p. 126).

Gold occurs at the borders of a nearly vertical, chloritic, altered greenstone dike which strikes N. 40-45° W. More precisely it occurs where gently dipping quartz stringers cutting the dike contact the slate wall-rocks (Preston, 1890, p. 304). The altered dike, averaging approximately 3 feet wide, consists in most places of a talcose, chloritic, clayey mass including quartz stringers a fraction of an inch to several inches wide. During the early years of mining the gold was sluiced or hydraulicked.

According to Castello (1921, p. 126) there is also a northwest-trending quartz vein containing free-milling ore. He describes a shaft 100 feet deep with drifts and a tunnel several hundred feet long. The property was not visited during this investigation.

Mount Buckingham (Vanderbilt, Sunset III, Crown Point) Group. Location: Secs. 1, 2, 11, 12, T. 5 S., R. 19 E., M.D., near the southeast end of Mount Buckingham above Snow Creek, 1 mile west of Darrah. Ownership: Helen M. Ketler et al., 3325 Kempton Ave., Oakland, California owns one patented mining claim, the Vanderbilt and about 300 additional acres of patented mining and timber land.

The Mount Buckingham group of mines was first located by William Buckingham in 1850 (Castello, 1921, p. 127). Prior to 1869 an 8-stamp, water-powered mill was built near the mine (Min. and Sci. Press, vol. 18, no. 6, p. 86, 1869). In April 1871 George Bernhard completed an 80-foot tunnel that exposed a 2-foot vein showing free gold. In 1883 a crosscut adit was started that had reached a length of 400 feet by 1885 (Min. and Sci. Press, vol. 46, no. 13, p. 221, vol. 50, no. 26, p. 416). Hall and Starr were the owners in the early 1880s. Ore was described as running $10 to $15 per ton. By 1889, ownership had passed to a man named MacDonald who is reported to have recovered $2,000 in one cleanup, representing 20 days of milling (Min. and Sci. Press, vol. 58, no. 13, p. 234). In 1893 Judge Condon of Mariposa was listed as superintendent but the mine was largely idle that year. In 1894 the mine was owned by a partnership in which George Beebe was the principal partner. In 1895, W. S. Chapman, of the Hite mine, purchased an interest in the Mount Buckingham and the Sunset Mining Company was organized with George Beebe acting as superintendent. A new 10-stamp mill was built on Snow Creek half a mile east of the mine (Storms, 1896, p. 224). By 1904 ownership had passed to Mrs. Eliza J. Starr of San Francisco (Wilkinson, 1904, p. 13). In 1916 a landslide damaged the adit and buried 200 tons of ore lying on the dump. From about 1912 to 1916 the mine was intermittently operated by the Mt. Buckingham Gold Mining Company of Mariposa, J. L. Diven of Mariposa, superintendent during much of this period (Lowell, 1916, p. 589). By 1920 ownership had passed to J. L. Diven and Mrs. C. A. Morgan but by 1928 Mr. Diven was listed as sole owner (Castello, 1921, p. 128; Laizure, 1928, p. 102). The properties apparently were idle during the 1920s but were operated for a short time in 1930-31, producing a small tonnage of ore running slightly less than 0.2 oz. of gold per ton. There has been no recent work done on the mine and there is little record indicating the total production.

Five more or less parallel quartz veins have a general northwest strike and dip southwest at angles between 70 and 85 degrees. Vein widths vary from 2 to 25 feet. One of the veins is accompanied by a granitic dike in which gold has been deposited. Others associated with granitic dikes are relatively barren of gold. The veins cut a roof-pendant of chiastolite-mica schist, most of the surrounding rock being granitic. Vein matter consists of glassy and milky quartz, ore shoots carrying abundant pyrite as well as some native gold. Much of the ore milled in the 1880s returned from $12 to $15 per ton with pyrite concentrates running as high as $900 per ton. Most of the ore encountered in later development work was considerably lower in grade, much of it running only $3 to $4 per ton. Ore mined in 1930-31, totaling 252 tons, averaged about 0.2 oz. of gold and 0.2 oz. of silver per ton.

The principal working is a 500-foot crosscut tunnel with a connecting 250-foot raise, 30-foot winze and 2 large stopes. There are several open pits and minor workings.

Mount Gaines (Barfield, Frenchman, Bearfield) Mine. Location: Secs. 35, 36, T. 4 S., R. 16 E., M.D., on a tributary to Burns Creek 4½ airline miles northeast of Hornitos. Accessible via 5½ miles of the surfaced Hornitos-Bear Valley road. Ownership: J. W. Radil, 444 California St., San Francisco, Calif., owns over 300 acres of patented mining and agricultural land.

The first mining done in the vicinity of the Mount Gaines properties was placering of Burns and Eldorado Creeks carried on from 1853 to 1873 (Julihn and Horton, 1940, p. 122). Lode mining began about 1868 on pockets and narrow stringers, vein matter being crushed in arrastras. In the early 1870s a 400-foot vertical shaft was sunk on the Frenchman claim about 2000 feet northeast of the present main shaft. These workings yielded about $100,000 according to Julihn and Horton (1940, p. 122). These authors state that "beginning in 1880, three successive mills were built on the Mount Gaines property, two having been destroyed by fire and the third by flood." Shallow surface workings supplied most of the ore crushed in these mills, but $150,000 was recovered from ore mined on the Barfield (also called Bearfield) claim. About 1881 the Mount Gaines mine was purchased, along with a large group of mines in the vicinity of the Number Nine mine, by the Yosemite Mining Company in which M. Huling, a Pennsylvania oil man was the chief owner (Mining and Scientific Press, several entries, 1881). Four thousand dollars was reported as the purchase price. This company operated the mine intermittently or leased it to various groups until about 1904 when it was sold to the Consolidated Mining and Processing Company of Los Angeles, W. T. Carter, secretary. Prior to this sale in 1904 considerable ore had been milled from the Barfield claim.

FIGURE 29.   Mine installations at the Mount Gaines mine in the Hornitos district, observer facing northwest. The Mount Gaines mine has an estimated production of $3,590,000, mostly in sulfide ore running $18-$20 per ton in gold.

FIGURE 30.   The Mount Gaines vein as exposed in the "I" stope 50 feet above the 500-level. The width of the vein is about 4 feet. The very shallow dip of the vein and the schistose character of the greenstone wall rocks are clearly shown. *Photo by courtesy of Francis H. Frederick.*

FIGURE 31.   Part of the Mount Gaines mill as it was in the spring of 1941, showing several banks of stamps and three flotation units. *Photo by courtesy of Francis H. Frederick.*

The main shaft, started by the Yosemite Company in 1881 was down 300 feet by 1897. At that depth the vein was reported to be 5 feet wide and the ore to average $25 per ton. The Consolidated Mining and Processing Company failed to get the mine into production and after a period of inactivity of nearly 3 years the Mount Gaines Mining Company was organized in 1906 with A. R. Gaines of Los Angeles as the principal owner and manager. Under this management, which operated the property until 1911, the shaft was deepened to 1322 feet and many thousands of feet of drifts were run. Mine production from 1900 to 1911 totaled nearly $1,000,000 (Julihn and Horton, 1940, p. 122). Operations by this company ended in indebtedness and litigation in 1911. In April 1914, a group of Denver and Los Angeles men headed by A. M. Gillespie attempted to reopen the mine but was unsuccessful. About 1917 another company, the Mount Gaines Gold Mining Company, was organized with G. W. Crotts and Serona E. Crotts as principal owners and W. J. McCray as superintendent. Considerable expenditure was made for equipment but indebtedness and litigation hampered the various managements and no material mining was accomplished during the 1920s.

In May 1934 the mine was leased to International Mining and Milling Company of Los Angeles and soon was put into operation under the direction of Nelson L. Wagner. This company went bankrupt in 1939 with the Mount Gaines mine its only productive and profitable asset. From 1939 to 1949 the mine was operated under the control of the Federal Bankruptcy Courts, Mr. J. P. Hart of Reno, Nevada acting as operating trustee. A. V. Udell was manager part of this time and later

John L. Dynan. The operating profits from the mine for that 10-year period were substantial and were used to pay the very high litigation costs connected with the various bankruptcy claims and procedures. Mine development was kept at a minimum during that period and it was largely through lack of development that the mine ceased to be profitable. A small production was recorded in 1951-52 in a mill cleanup but the mine was idle until 1956 when the present owner began exploration and development works.

Although production records are not complete the gross production of the Mount Gaines mine is estimated to be at least $3,590,000. From 1932 to 1947 the ore averaged more than half an ounce of gold per ton in addition to substantial quantities of silver, copper and lead, giving an average of approximately $19.79 per ton. The Mount Gaines mine ranges among the 5 most productive mines in Mariposa County.

The Mount Gaines mine is on a northeast-trending vein system that at the surface is not very well defined. Much of the early work on the veins was done on pockets and stringers of sulfide-bearing quartz and the main branch of the vein system was discovered at depth. In the main vein, which is sinuous, the best ore tends to be concentrated on the east sides of arcuate irregularities, particularly along the parts of the vein having the gentlest dip. (Frederick, Francis H., personal communication, 1956). Shoot lengths vary from 40 feet to 450 feet long. The main vein strikes about N. 35° E. and dips southeast at an average dip of 20°, but dips vary from 10° to 30°. The vein system has a known length of about 9000 feet and has been developed for a strike length of more than 2000 feet and a depth of over 1300 feet, measured on the incline. It varies from a few inches to more than 15 feet wide averaging about 5 feet wide. Vein matter is chiefly milky quartz, but some ore occurs in quartz veinlets in fractured, slaty to chloritic green-

FIGURE 32.   A gathering of mining men in front of Hornitos saloon, photo taken about 1895. Standing, left to right, Messrs. Henry Nelson, Smith Thomas, John Branson, Dennis (?), B. A. Sheppard, G. Gagliardo, J. D. Craigham, and Mose Rodgers. Seated, left to right, Messrs. R. Bancroft, Al Sylvester, Collier (?), J. Spagnoli, Bailey, Tom Thorne, Robert Arthur, and Tom Williams (?). *Photo by courtesy of Francis H. Frederick.*

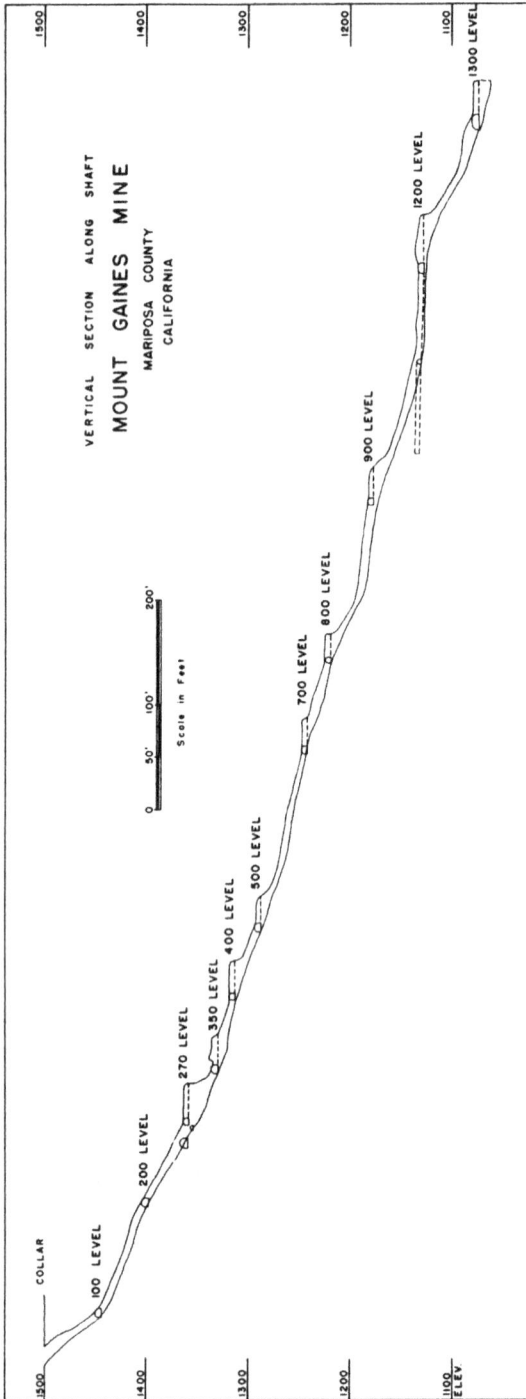

FIGURE 33. Vertical section along shaft, Mount Gaines mine. *Courtesy of J. W. Radil.*

stone along the footwall. Ore shoots contain about 3 percent sulfides. Ore minerals include native gold, galena, sphalerite. chalcopyrite, pyrite, arsenopyrite, barite, proustite and argentite, according to Julihn and Horton (1940, p. 123). Deposition of pyrite preceded that of galena; sphalerite and gold were deposited later than the pyrite. No high-grade ore was observed on the dump at the time of the authors' visit in September 1954. Wall rocks are chiefly massive pyroxene-andesite greenstone belonging to the Peñon Blanco member of the Upper Jurassic Amador group, but there are some slaty and horfelsic metasediments in the section. Shearing and hydrothermal alteration along the vein have caused widespread development of schistose, chloritic, slate-like material, particularly on the footwall side of the vein. Hornblende granodiorite dikes apparently have penetrated the vein fissure in many places as there is considerable material of this kind on the dumps. Julihn and Horton (1940, p. 122) describe a prominent dike on the hanging wall side of the main vein.

The disposition of the principal workings may be seen in the accompanying diagrams. The main inclined shaft is 1322 feet long, measured on the incline, and has 13 drift levels at 200, 270, 350, 400, 500, 600, 700, 800, 900, 1000, 1050, 1100 and 1200 feet. There are more than 11,000 feet of drifts and 1200 feet of raises. The principal development has been northeast of the main shaft. There is also a tunnel driven N. 20° E. which was not accessible in September 1954. The main shaft and workings were watered within 20 feet of the collar of the shaft. The mill and hoisting equipment were partly dismantled and rehabilitation of mine and buildings will require considerable capital outlay.

Mount Ophir. Location: Sec. 12, T. 5 S., R. 18 E., M.D., 1 mile northwest of Mount Bullion close to the south side of Highway 49. Ownership: Mariposa Commercial and Mining Company, c/o Eileen Milburn, Mariposa, California, owns a large tract of land, a portion of the old Las Mariposas Grant, which includes the Mount Ophir, Louis, Mountain View and Greens Gulch mines.

The Mount Ophir mine is one the most accessible and conspicuous of the Mother Lode mines in Mariposa County. Massive white quartz of the Mother Lode crowning the summit of Mount Ophir makes a prominent landmark along Highway 49. Close by are the ruins of the Mount Ophir mint which manufactured octagonal gold slugs from locally mined gold in 1850-51. Discovered in 1849 or 1850, it was worked extensively up to the time Fremont gained title to Las Mariposas Grant in 1859 and intermittently thereafter until 1914. Three months of operation in 1860 grossed $43,000 from 1845 tons of ore for an average of $23.20 per ton (Julihn and Horton, 1940, p. 112). Browne (1868, p. 29) states that the yield in 1864 was $12,540. From 1864 to 1901 little or no work was done on the mine. The reorganized Mariposa Commercial and Mining Company, which took over the Las Mariposas Grant in 1897, reactivated the property in 1901. Between that year and 1914 a total of 2366 tons of ore was mined which yielded $29,252.75 or an average of $12.37 per ton at the old price of gold. Although the mine was examined by George R. Barnett of San Francisco in 1935, under option to purchase, and although available records indicate that the mine was not worked out, there has been no production since 1914. The estimated total production of the Mount Ophir mine is between $250,000 and $300,000 (Julihn and Horton, 1940, p. 112) but the recorded production is $85,703.

At the Mount Ophir workings, the Mother Lode strikes N. 55-60° W. and dips 55-65° NE. Vein matter crowning the hill is at least 30 feet wide. According to Lowell (1916, p. 590) the vein averaged 5 feet wide in the workings. Only stringers are visible in the lower tunnel which has been driven at the contact of schists of the Paleozoic Calaveras group, which forms the hanging wall, and serpentine which forms the footwall. Both of these units have come in along a fault-zone cutting the Mariposa slate. Veinlets of quartz penetrate the serpentine walls and locally form stockworks. There is very little mariposite-ankerite rock in this segment of the Mother Lode system. Ore shoots apparently are confined to the footwall and hanging wall sides of the vein, the central milky quartz being relatively barren of ore minerals. The larger ore shoots apparently were concentrated on the footwall side. One pocket of ore was found in a vein on the footwall side of the main vein in 1913 that yielded $711.65 (Laizure, 1928, p. 98). Ore mined in the early days apparently averaged a little better than $23 per ton whereas ore mined in the 1900's averaged a little over $12 per ton, both these figures at the old price of gold. According to a report made in 1922 by Frank Eichelberger (Julihn and Horton, 1940, p. 112) the workings at that time contained 4000 to 5000 tons of developed ore averaging 4 feet wide that would mill $9 to $11 per ton. He further stated that nearly all the development work had been carried on on the footwall side of the vein with little or no exploration of the hanging wall side.

Principal workings are two adits, one about 115 feet above the other, both driven on the vein from the north side of the hill. The lower tunnel, 700 feet long, reaches a vertical depth of 250 feet below the top of Mount Ophir. This tunnel was accessible in August 1955 although in need of cleaning out. The upper adit is 320 feet long but is partly caved. A third adit, mentioned by Lowell (1916, p. 591) could not be located and presumably is caved. At the summit of Mount Ophir are two open stopes or air shafts each at least 100 feet deep which could easily be cleaned out as accessways. According to Lowell, there were 1500 feet of drifts, 360 feet of raises, 80 feet of crosscuts and one large stope in 1916.

Mountain King (Omparisa, Calender and Calendonia) Mine. Location: Sec. 31, T. 3 S., R. 18 E. and sec. 6, T. 4 S., R. 18 E., M.D., on the north side of the Merced River Canyon 1 mile east of Quartz Mountain and 6 airline miles east of Bagby. Accessible by 5½ miles of good dirt road from Briceburg on Highway 140. Ownership: 13 of the claims are owned by J. W. Radil, 444 California St., San Francisco 4, California and 5 claims are owned by Ralph E. Dailey, 1165 Twenty-second St., Merced, California.

Very little concerning the early history of the Mountain King mine has been recorded. It was being developed in 1899 by Egenhoff and Merritt who were running a tunnel designed to tap the vein 1500 feet from the surface. In 1904 the properties were bonded to the Omparisa

Mining Company headed by H. C. Austin for $40,000. (Min. and Sci. Press, vol. 88, no. 12, p. 122, 1904). H. A. Kunz was one of the first superintendents under this company which started a crosscut tunnel to intersect the 5 parallel veins cropping out higher on the hill. A 5-stamp mill was completed in March 1905 and this was increased to 10 stamps in October the same year. Several high-grade ore-shoots were found in 1909 and the mill was increased to 20 and then to 40 stamps. William S. Thompson was superintendent 1910-1911. After a short period of inactivity the Mountain King Mining Company was organized in 1914 and took over the holdings of the older company. Alexander Hamilton was superintendent in 1915 and B. C. Austin in 1917. Operations by this company ceased in 1922 because operating costs rose above mill returns. The mine has been idle ever since except for a little leasing activity between 1922 and 1925.

From 1909 to 1925 the mine produced 162,800 tons of ore yielding 39,833.51 oz. of gold and 18,042 oz. of silver for a total of $825,425.65 and an average value per ton of $4.51 (at the former price of gold). Estimated total production of the mine is nearly $1,000,000.

Five persistent veins as well as a number of smaller ones cross the Mountain King properties at the surface. Four of the 5 main veins have been identified in underground workings and a fifth was discovered during development work that does not reach the surface. The principal veins are the Mountain King, Flat O1, Back or No. 3, McFadden and Big Flat Footwall. Production thus far has come only from the Mountain King and Flat O1 veins. The McFadden, Mountain King and No. 3 veins are roughly parallel, strike N. 20-25° W. and dip 65-80° N.E. The Big Flat Footwall vein strikes southwest and dips northwest. The Flat O1 vein appears to diverge from the steeper Mountain King vein about 800 feet from the surface at about the 200-level. The best grade of ore taken from the Mountain King mine came from between the 200- and 300- levels of the Flat O1 vein during the period 1910-12.

Three ore shoots, numbered 1 to 3, east to west have been found and worked in the Mountain King vein. The No. 1 oreshoot is known to extend from the surface, at the collar of the Egenhoff shaft to a known inclined depth of 1000 feet and was still well defined at that depth when abandoned to work the No. 2 shaft. The No. 2 shoot has been explored between the 850- and 2000-levels where it is known to merge with the No. 1 shoot near the 2000-level. The No. 2 shoot is 200 feet long, averages 46 inches thick on the 2000-level and averages $9 per ton at the present price of gold. The third and most westerly ore shoot dies out above the 850-foot level but is still well defined near the bottom of the workings on the 2000-level.

Mountain View I Mine. Location: Sec. 11, T. 5 S., R. 17 E., 2 airline miles northwest of Mt. Bullion. Adjoins the Louis mine. Accessible by unimproved dirt road from Greens Gulch via the Louis mine. Ownership: the mine is part of a tract of land containing several hundred acres, part of Las Mariposas Grant, still held by Mariposa Commercial and Mining Company, c/o Eileen Milburn, Mariposa, California.

The Mountain View mine was discovered prior to 1859 and its history has been similar to many other Las Mariposas Grant mines. It apparently was characteristically a pocket mine or at least had small ore shoots, and most of the recorded production has been from ore of fairly high grade. One of the principal periods of production was 1900-1915 when 376.16 tons of ore mined yielded $17,792 or an average of $47.52 per ton at the old price of gold. No other production records are available. No mining has been done since 1915.

Both the Mountain View and Louis properties are crossed by the same vein, which is generally considered to be part of the Mother Lode system. This vein is roughly parallel and en echelon to the vein passing through the Mount Ophir mine, the two veins being about 3200 feet apart. Crossing the Mountain View claim the main vein strikes N. 48° W. veering to about N. 65° W. crossing the Louis property. The dip is 60-65° northeast and average width is about 2½ feet. Wall rocks are black slate of the Upper Jurassic Mariposa formation. Strike of the slaty cleavage is roughly parallel to the strike of the vein. Vein matter is milky quartz with pyrite and native gold. Inclusions of thin sheets of slate give a ribbon-like structure to some parts of the vein.

Workings consist of 2 adits driven on the vein totaling about 500 feet, 400 feet of crosscuts, 160 feet of raises and a stope 50 feet long and 100 feet high (Lowell, 1916, p. 591). Milling was done at the Princeton mill throughout the last period of operation of the Mountain View mine.

Nellie Kaho. Location: Sec. 4, T. 5 S., R. 17 E., sec. 33, T. 4 S., R. 17 E., M.D., 3 miles south of Bear Valley and just west of Cow and Calf Road. Ownership: Harold Hansen, Mariposa, California, owns a property consisting of about 20 acres. Under lease (1956) to Harmon and Gault.

Prior to the late 1930s the Nellie Kaho mine was one of many old prospects which had never been well explored. Taken over by a group of leasers including the present owner, in June 1937, about $1400 was recovered from surface workings (Julihn and Horton, 1940, p. 126). By 1938 a 100-foot shaft had been sunk which connected with a 50-foot south drift and a 50-foot east crosscut on the 100-level. The advent of World War II ended this period of operation. Clyde Diffenbaugh of Bear Valley did some work on the mine in 1951-52 and the present operators have worked the property intermittently since 1954.

The quartz vein, one of the Mother Lode system, is about 18 inches wide and crops out for a distance of 100 feet at the surface. It strikes northwest and dips about 80° SW. Wall rocks are slate and greenstone. Ore runs from ¼ to ½ percent sulfides, 30 percent of the precious metal value coming from the sulfide concentrate. This averages 6 to 7 oz. of gold per ton. The grade of ore mined in the late 1930s ranged between $10 and $20 per ton with occasional pockets of higher-grade ore.

Workings consist of 2 shafts about 50 feet apart. The main shaft, the more southerly, is 200 feet deep and is inclined about 85° W. There is a 50-foot level, a 125-foot level and a 200-level. Recent work has been on the 200 level. The operators plan to ultimately deepen the north shaft, now about 50 feet deep, and connect the 2 workings. In 1938 the workings made 8000 gallons of water per day.

Number Five (Monte Carlo) Mine. Location: Secs. 2, 11, T. 5 S., R. 16 E., M.D., 5 miles northeast of Hornitos and 1 mile northeast of the Number Nine mine. Accessible from the paved Hornitos-Bear Valley road by 1½ miles of good dirt road via either the Number Nine or Mount Gaines mines. Ownership: not determined.

The Number Five mine is an old property consisting of the Number 5 and Jack Quartz claims and the fractional Standby and No. 5 Extension claims. A small production was recorded in 1899, but the operators' names were not recorded. The principal period of activity was 1909-1914, operators being the Number Five Mining Company, J. J. Le Tourneau of Duluth, Minnesota, president and E. S. O'Brien of Merced, secretary. Various superintendents under the management were Hugh Branson (1909), Martin Sutherland (1910) and Carl J. Smith (1911). A 260-foot inclined shaft was sunk with a 300-foot drift on the 100-level and a 56-foot crosscut on the 200-level (Min. and Sci. Press, vol. 99, no. 2, p. 41, 1909). In 1914 the property was operated for a short time under bond by the Nevada Mineral Extraction Company in which S. W. Parker of Berkeley was the principal owner and C. H. Gage was superintendent. A 50-ton Lane mill was installed (Eng. and Min. Jour., vol. 98, no. 20, p. 894, 1914). The mine was operated under lease for a time in 1921-22 and a small production was recorded. No material amount of mining has been done since 1922 and the land has reverted to agricultural status.

There are at least three veins on the Number 5 property, the Number 5, Number 2 and south extension of the Prescott. The Number 5 vein, the most prominent and the most southerly of the group, trends N. 75-80° E. and dips about 65° north. It ranges from 5 to 30 feet wide and consists principally of milky quartz with abundant pyrite. Surface showings of gossan are common. The Number 2 vein, apparently the one crossing the Jack Quartz claim, strikes N. 80-85° E. and dips steeply north. It is much narrower and less well defined at the surface than the Number 5. The southern extension of the Prescott vein apparently has been the most extensively worked. It strikes N. 20-30° E. on the Number 5 property and swings more nearly to the north on the Prescott claim. Ranging between 1 and 4 feet wide, it dips toward the east at angles varying from 20 to 30 degrees and blankets much of the hillside area. Judging from the numerous surficial workings there must have been numerous parallel blanket stringers in addition to the main Prescott vein. All of the soil mantle in the vicinity of the veins has been placered. The veins cut a variety of wall rocks, chiefly slate, spotted slate, schistose greenstone, hornblende schist and hornfelsic slate probably belonging to the Upper Jurassic Amador group (Taliaferro, 1943, p. 282). These are cut in places by irregular intrusions of porphyritic granitic rocks, chiefly granodiorite porphyry.

The main working on the Prescott vein is a long northwest-trending crosscut adit with extensive connecting drifts and stopes. Laizure (1928, p. 104) mentions a main shaft 260 feet deep and a second shaft 200 feet deep but does not state the location of these workings. There are numerous workings on the Prescott vein which could be either shafts or stope openings. All of the workings are in need of cleaning out and those on the Prescott vein are flooded below the main adit level.

Number Nine (Bill Jones and McCall, Yosemite, Ginaca) Mine. Location: Sec. 10, T. 5 S., R. 16 E., M.D., 2 airline miles northeast of Hornitos. Accessible via 1¼ miles of improved dirt road from the paved

Hornitos-Bear Valley road. Ownership: W. R. Plunkett and Ulysses M. Peyrellade, 1278 Twenty-sixth Ave., Oakland, California, own three patented claims totaling about 41 acres.

The Number Nine mine was probably discovered in the 1850s and mined in a small way from surface pits and shallow underground workings. The first owners of record were Major Hardwick and son who reopened the mine 1873 and erected a mill. Prior to erection of the mill in 1873-74 ore was crushed at the Mt. Gaines mill where one run of 2,000 tons yielded $7.375 per ton (Min. and Sci. Press, vol. 29, no. 4, p. 53, 1874). In 1880 the mine was sold along with several others to Marcus and J. W. Hulings of Oil City, Pennsylvania, for $77,000 (Min. and Sci. Press, vol. 41, no. 24, p. 372, 1880). At that time the ore in the mine was sampled and was supposed to average $9 per ton. A 30-stamp mill was erected which crushed 40-45 tons of ore in 24 hours. Only $1-$2 of the total gold proved to be free-milling and the first year of operation proved a failure. In 1881 a new milling procedure, which recovered sulfide concentrates, was introduced and several years of profitable operation ensued in which recovery varied from $6 to $9 per ton and profits from $2-$4 per ton. J. Frank Thorn was superintendent during this period. A $15,000 chlorination plant was erected in 1883 but this evidently was not successful. The mine closed down soon afterward and nothing further was done until 1896 when it was held for a short time by W. S. Chapman and Associates of San Francisco. Ownership had passed to Moses L. Rodgers after the Hulings left. The Chapman management failed to get the mine into production and it lay idle until 1903 when some cleanout, exploration and development work was done by Harmon and Stevens under the direction of W. F. Stevens. About 1909 the Number Nine Gold Mining Company was organized, initially with New York capital but later reorganized by San Franciscans. Richard O'Brien superintended mine operations from 1909 to about 1911. During this time the old inclined shaft was re-timbered to the 250 level and then abandoned. In 1910 a new vertical shaft, the Keys was started which ultimately reached a depth of 140 feet (Laizure, 1928, p. 104). Between 1911 and 1919 there was a little intermittent activity at the property but no sustained mining (James Peck managed the mine for the Rodgers estate). About 1919 the Number Nine Gold Mining Company was reorganized with L. A. Ginaca of San Francisco as president. This company moved in new equipment, unwatered the mine and was preparing to re-open when litigation cancelled further work until 1924. From 1924 to 1930 activity was confined largely to rehabilitation and development work and only a few ore shipments were made. B. S. McArthur was superintendent in 1928. Between 1930 and 1936 there was a moderate production of low-grade ore. There has been no activity since 1936 and all equipment has been removed.

According to Laizure (1928, p. 104) twelve veins have been discovered on the Number Nine property. The main system strikes north to N. 20° W. and dips 25-30° E. It is crossed by a secondary, nearly vertical system of east-trending veins. The most prominent vein crops out on the east side of a low ridge northwest of the Key shaft, blanketing the hillside for perhaps an acre. Most of the early work apparently was done on this vein as a great many shallow workings have been driven into it. Vein matter consists chiefly of massive to vuggy milky quartz 3 to 15 feet thick, with very erratically distributed masses of auriferous pyrite. Wall rocks consist of quartz-biotite schist, gneiss and hornfels with some small intrusions of altered granodiorite. Although not well exposed, some of the ore appears to have developed in altered dike material on the footwall side of the vein. The width of the main vein in the vicinity of the Number 5 shaft varies from 38 to 44 inches. The vein exposed in the drift adit 150 feet from the end was 18 inches wide.

Six shafts and a 350-foot drift adit were identified on the Number Nine property. Four of these shafts are approximately on a line bearing N. 20° W. The most northwesterly of this group is the No. 9 at least 125 feet deep but long unused. The most southeasterly is the No. 5 shaft 387 feet deep measured on the incline. The shaft serves 3 levels of which the 300-level is most extensive. The No. 9 and No. 5 shafts are about 880 feet apart. Two 30-foot deep shafts are about evenly spaced in this interval. All these shafts are inclined from 20° to 28° northeast. The Key vertical shaft is 150 feet deep with the lowest working level at 125 feet. This shaft is located about 500 feet northeast of the Number 5 shaft. The Key Extension shaft, also vertical is located close to the Mount Gaines mine road about 1000 feet north and slightly east of the Key shaft. All of the workings have been inactive for a long time and are in need of rehabilitation.

Nutmeg Mine. Location: Sec. 30, T. 4 S., R. 18 E., M.D., on the west side of Whitlock Creek 1 mile north of the Permit mine and 4 airline miles northeast of Mount Bullion.

Ownership: Sam Le Barry, c/o Permit Mining Company, Midpines, California owns one unpatented claim.

The Nutmeg mine is on a north extension of the vein system that passes through the Milburn, Permit and Geary mines. Little has been recorded of the early history of the mine but it was discovered prior to 1900. A moderate production is recorded from 1938 to 1951, a total of 5239 tons of ore having yielded 1204 oz. of gold, and a little lead and copper, and 331 oz. of silver. The average yield per ton has been between $8 and $8.50 although some pockets of ore run as high as 3 oz. per ton. The mine is still worked intermittently by the Permit Mining Company under the direction of H. H. Odgers.

The main vein is about 3½ feet wide, strikes N. 20-25° W. and dips steeply east. Wall rocks are massive and sheared pyroxene andesite greenstone. Four ore shoots have been developed which average 25 feet long (Julihn and Horton, 1940, p. 152). The principal working is a 106-foot vertical shaft with more than 160 feet of drifts on the 50-foot level and 300 feet on the 100-level.

Oakes and Reese (Grand Prize and Badger) Mine. Location: Secs. 32, 33, T. 3 S., R. 16 E.; sec. 4, T. 4 S., R. 16 E., M.D., in the northwestern part of Hunter Valley 2 miles east of Exchequer reservoir on Temperence Creek. Accessible via the paved Hunter Valley road, 11 miles north and slightly east of Hornitos.

Ownership: W. A. Hayes, Martin S. Heller and J. P. Warren, 1900 Leimert Boulevard, Oakland 2, California own 2 patented claims, the Grand Prize and Badger, totaling 26.15 acres.

The Oakes and Reese mine, discovered in 1863, was one of the famous early producers in Mariposa County. It has produced between $500,000 and $600,000 (Julihn and Horton, 1940, p. 117), most of this total coming from the period 1863-1870. The operator in 1865 was J. W. Adams and Company and in 1870 Robinson and Company. According to Castello (1921, p. 131) the mine was closed down in 1870 because of a lawsuit over water rights and because one of the ore-shoots was mined out. The property was operated under lease in 1870-1871 by Dan Jones (Min. and Sci. Press, several entries). Much of the ore mined in the 1860s ran much higher. Early milling was done in arrastras. A 10-stamp mill was built in 1866 which had been increased to 28 stamps by 1868 (Raymond, 1870, p. 24). The vein had been opened to a depth of 200 feet by 1868 (Min. and Sci. Press, vol. 16, no. 9, p. 134).

Except for intermittent small mining by lessees and occasional development work the mine remained closed until June, 1937. Some work was done in the late 1920s by the Oakes and Reese Mining Company, a New York firm (Laizure, 1928, p. 131). During 1938 a small tonnage of low grade ore was milled by W. A. Hayes of San Francisco, Thomas Henry of Hornitos acting as superintendent. No high-grade pockets were found and the mine shut down in 1939, since when the property has been idle. The shaft was caved, and all equipment removed in July, 1954.

The Oakes and Reese mine is located on a well-developed vein of bluish quartz, called the Blue Lead, that strikes N. 35-40° W. and dips 70-80° northeast. The Blue Lead is intersected, approximately at right angles, by a series of nearly vertical cutter veins or laterals which enter the main vein from the east or hanging wall side, and which in general carry more high-grade ore than the main vein. The two principal laterals, the Potts and Floyd, are approximately 400 feet apart. The main vein averages 4-5 feet wide and the laterals between 1 foot and 2 feet wide. A third lateral vein crosses the Blue Lead about 300 feet south of the Floyd lateral, and there is a fourth still farther south. The Blue Lead has developed on or close to the contact between the Hunter Valley chert beds and massive greenstones, both of Upper Jurassic age. The chert series contains thin-bedded tuffaceous shales and siltstones as well as chert. The principal ore body was a pipe-like shoot of high-grade developed at the junction of the Blue Lead with the Potts lateral. The high-grade ore graded into lower-grade material long the footwall side of the Blue Lead. According to Julihn and Horton (1940, p. 117) the ore from the shoot averaged about 2.5 oz. of gold per ton and the best of it 5 oz. or more.

The main shaft was sunk to an inclined depth of 450 feet at the junction of the Blue Lead with the Potts lateral. From the shaft levels were run at depths of 130, 200, 300 and 450 feet, the total amount of drifting being only about 800 feet (Julihn and Horton, 1940, p. 117). The Blue Lead vein has been stoped for a strike length of 60 feet, north of the shaft, and to a depth of 200 feet. The Potts lateral has been stoped for a length of 282 feet from the shaft on the 130-level and 147 feet on the 300-level (Julihn and Horton, 1940, p. 117).

There are two other shafts 160 and 70 feet deep, respectively, both sunk at the junction points of laterals and the Blue Lead. Drifting from these shafts is much less extensive than from the main shaft. The last work done on the mine in 1938 was the extension of the 200 level from the main shaft south toward the Floyd lateral.

Oro Rico (Peñon Blanco) Mine.   Location: Secs. 19, 20, 29, T. 2 S., R. 16 E., M.D., just south of the Tuolumne County line on Highway 49. Ownership: Oro Rico Mines Company, c/o A. D. Vencile, Room 14, 1584 West Washington Blvd., Los Angeles, California, owns the patented Old Judge, South Judge, Little Judge, Peñon Blanco and North Peñon Blanco lode claims and the unpatented Star and Stevenson claims totaling more than 100 acres.

The prominent white quartz ridges cropping out on the Peñon Blanco (White Cliff) are among the best known landmarks along Highway 49. Oro Rico properties cover more than 2 miles of the Mother Lode between the county line and Coulterville. Because of its prominence, the vein was one of the earliest to be worked in Mariposa County. The Peñon Blanco claim, over 5800 feet long, is the longest on record. As early as 1868 there were two shafts and several tunnels (Browne, 1868, p. 35) and several ore shoots 2 to 4 feet thick had been discovered. At that time the ore averaged $10-12 per ton. The first application for patent was made by the Oro Rico (Peñon Blanco) mine under the act of 1866, according to Browne. By 1896 there were well over 1400 feet of workings on the 'property, then owned by A. H. Ward of Alameda (Storms, 1896, pp. 221-222). Ward built a 20-stamp mill and worked various parts of the property until about 1910 when it was bonded to J. C. Wilson and associates of San Francisco (Eng. and Min. Jour., vol. 89, no. 16, p. 839), who did some work on the mine between 1910 and 1912. The property was tied up for much of the time between 1912 and 1928 because of litigation, and little or no mining was done. During this period ownership passed to A. E. Tower of San Francisco. In 1934, J. C. Kempvanee and associates unwatered the mine and did some exploration and development work but little or no work has been done since that time. The total production, although incompletely recorded, is small.

The Mother Lode, as it crosses the Oro Rico property, consists of a single massive vein striking N. 30° W., dipping 52° northeast, and reaching a thickness of over 250 feet. It is intersected in several places by small lateral veins. Great thicknesses of vein matter consist of quartz-ankerite-mariposite rock although the included quartz lenses that cap the ridges are by far the most conspicuous parts of the vein. Over much of the vein length the ankeritic portions have been altered by acid groundwater to spongy masses of iron oxide and silica boxwork. Ore shoots contain pyrite, chalcopyrite and native gold (Storms, 1896, p. 221). Toward the southern end of the mine property the Mother Lode lies close to the contact between slate of the Upper Jurassic Mariposa formation and serpentine, also of Upper Jurassic age. Farther north it is wholly within a large serpentine mass.

In comparison with the size of the vein the ore shoots in this segment of the Mother Lode proved small and of relatively low grade. The two shoots mentioned by Browne (1868, p. 35) on the Peñon Blanco claim were 2 and 4 feet thick and averaged $10-12 per ton near the surface. Inasmuch as these were considerably enriched through removal of worthless material during alteration in the oxidized zone, it is probable that the grade of ore lessened with depth.

In 1868, workings consisted of a 175-foot crosscut adit driven from the west or footwall side, a 285-foot crosscut adit driven from the east or hanging wall side and two shallow shafts 2000 feet apart (Browne, 1868, p. 35). All of these apparently were on the Peñon Blanco claim. By 1896, there were 2 adits 25 and 100 feet long, respectively, and several open cuts on the Old Judge claim; a 200-foot adit on the North Peñon Blanco claim; and more than 1000 feet of tunnels, drifts and shafts on the Peñon Blanco claim (Storms, 1896, p. 221). According to Logan (1935, p. 186), the Oro Rico Mines Company drove a 600-foot crosscut which reached the vein at a depth of 320 feet and more than 1000 feet of drifts were run from this adit. From the floor of the northwest drift a winze was sunk to a depth of 450 feet and 4 levels of unknown length were driven. In August 1955, five workings were still recognizable on the Peñon Blanco claim, two adits and a shaft toward the south end of the claim on the east side of the ridge and one adit and shaft toward the north end on the west side of the ridge. None of these workings would be accessible without some cleaning out and none were entered.

Ortega Mine. Location: Sec. 19, T. 5 S., R. 17 E., sec. 24, T. 5 S., R. 18 E., M.D., on Highway 140 midway between Nigger Hill and Guadeloupe Mountain. Adjoins the Sorrel (Sarle) mine on the north. Ownership: Frank A. Cassaccia, Mariposa, California owns a tract of about 125 acres which includes the Ortega mine. Formerly part of Las Mariposas Grant.

Gold was discovered at the Ortega mine in the early 1850s by a Spanish Californian named Ortega. He worked the vein extensively until Fremont gained title to Las Mariposas Grant in 1859. Ten or more shafts were sunk on the vein over a distance of 1485 feet (Julihn and Horton, 1940, p. 131), all but two of these being north of Highway 140. The deepest of the shafts, cleaned out and reconditioned in 1936-37 was found to be 167 feet deep measured along the incline. Milling was done mainly in arrastras but some ore was milled in a 2-stamp mill by a millman named Turner. After Fremont dispossessed Ortega's group in 1859, no further work was done until 1900 when some ore found lying on the dumps was milled at the Princeton mill and yielded between $14 and $27 per ton (Julihn and Horton, 1940, p. 131). In 1934, a partnership of Price, Willmer, Givens and Givens leased the mine from the Mariposa Commercial and Mining Company. This group cleaned out the No. 3 shaft to a depth of 180 feet and ran a drift 125 feet south. Good ore was found below the old Mexican workings 10 tons of which yielded $91.60 per ton. However, the quantity of ore developed proved small and the partnership was forced to withdraw because of insufficient funds to properly develop the mine. In 1936 John Q. Finfrock leased the property and operated it until 1940. He cleaned out and rehabilitated the No. 4 shaft, mined considerable new ore and processed some old broken ore found on the property. In 1940 the property was again taken over by Mariposa Commercial and Mining Company and put in charge of D. Mullins. In 1941 the Ortega mine was one of a group optioned to J. K. Wadley of Texarkana, Arkansas and placed under the management of Charles Greenameyer. World War II prevented continuation of this enterprise and no further work had been done at the Ortega mine up to September 1955.

Two veins have been found on the Ortega property. The main vein averages about 3 feet wide, strikes about N. 20° W. and dips 50-60° W. It is a continuation of the vein which crosses the Sorrel mine south of Highway 140. Vein matter is coarse milky quartz containing sphalerite, pyrite, chalcopyrite, galena and arsenopyrite as well as native gold. At various places along its length the vein fissure has been intruded by pegmatite, aplite and dark granitic dikes. The enclosing rock is coarse-grained biotite granodiorite. A second vein was discovered in the late 1930s about 500 feet east of the No. 4 shaft which strikes N. 30° W. (Julihn and Horton, 1940, p. 132). This vein has not been well explored nor has the possible point of junction of the 2 veins been discovered.

A line of caved shafts and stopes extending northwest from Highway 140 marks the position of the old workings. None of these were accessible in July 1954 and the various shafts could not be identified. The shafts were evidently numbered from north to south, no. 2 and no. 3 shafts lying 170 feet apart. The No. 4 shaft, worked by Finfrock in the late 1930s, was 167 feet deep and had 3 levels at 100, 147 and 167 feet, respectively. According to Julihn and Horton all of the ore above the 100-level south of the shaft has been stoped. Ore was found in place on the 147-level from within 60 feet of the shaft to 347 feet south of the shaft and ore has been found continuously from 30 feet north of the No. 3 shaft to the No. 4 shaft. The length of one ore shoot must be at least 527 feet with a width varying from $2\frac{1}{2}$ to 4 feet. Disposition of other workings indicates that the vein must have been almost continuously mineralized for over 1000 feet. The workings in the vicinity of the No. 4 shaft draw 24,000 gallons of water a day (Julihn and Horton, p. 132, 1940).

The early production of the Ortega mine was not recorded. Julihn and Horton state that "a total of $19,796 was recovered by Finfrock from ore taken from the levels once operated by Ortega and from remains of their surface dumps. Of this total, $10,154 was derived from 691 tons of newly mined ore, an average recovery of $14.69 per ton, and 346 tons of old gob yielded an estimated $5 per ton. In milling, 9700 pounds of concentrates were recovered which assayed $171.54 per ton."

Our Chance (Clark Mines, Albert Austin group) Mine. Location: Sec. 29, T. 4 S., R. 18 E., on the Sherlock Creek road, 4 airline miles northeast of Mt. Bullion. Adjoins the Diltz mine on the west. Ownership: Not determined.

The following discussion is drawn chiefly from Julihn and Horton (1940, pp. 147-148), the last authors to visit the mine while it was operating. The Our Chance mine apparently was first worked by Albert Austin who is supposed to have recovered more than 2000 oz. of gold from pockets near the exposed surface of the vein. Austin used a hand mortar and an arrastra. A pocket taken out by Arthur Clark, the last operator, yielded about $17,000. Clark operated the mine from 1932 to about 1946, mining and milling a total of 2563 tons of ore yielding 1357.77 oz. of gold and 223 oz. of silver (U. S. Bureau of Mines records) or an average of about $18.60 at current metal prices. The mine is noted for its specimens of crystallized and arborescent gold.

The Our Chance vein strikes northwest and dips about 27° northeast, the width averaging 14 inches but varying between 4 inches and 4 feet. Above the vein is a dike of dark granitic rock 4 to 6 feet thick. Wall rocks enclosing the vein and dike are pyroxene andesite greenstones of unknown age. Vein matter is banded quartz with pyrite, chalcopyrite, arsenopyrite, galena and native gold, in places associated with calcite and manganese oxides.

The principal working in 1938 was an 800-foot adit driven northeast to the vein with 2500 feet of connecting drifts, winzes, etc. At the end of the adit is a 500-foot southeast drift which has a 200-foot winze sunk 200 feet from the end of the adit. At a depth of 100 feet a level has been run from the winze 100 feet to the southeast and 70 feet to the northwest. Above the drift from the main adit ore has been stoped to a height of 50 or 60 feet between the adit and winze for a length of 175 feet. Between the winze and the end of the drift there is a 300-foot stope about 100 feet high. Very little timbering is required. The mine was idle in October 1955 and there was no equipment on the property.

Permit (Boulder, Kockel, Bulldog, Bullpup) Mine. Location: Sec. 23, 31, T. 4 S., R. 18 E., M.D., on Whitlock Creek 3 airline miles northeast of Mt. Bullion. Accessible via the Whitlock-Sherlock Creek graded dirt road. Six miles north of Mariposa. Ownership: Permit Mining Company, c/o H. H. Odgers, Midpines, California, owns 5 unpatented claims, the Aladdin, Aladdin No. 1, Aladdin No. 2, Permit and Permit Extension.

According to Castello (1921, p. 109) the old Bulldog and Bullpup claims were located in 1894 but an entry in the Mining and Scientific Press (vol. 60, no. 14, p. 232, 1890) indicates that the Bulldog mine operated that year under N. J. Farrens, producing a few tons of low grade ore. In 1904 (Wilkinson, 1904, p. 10) the owner of the Bulldog was listed as Czerney and Company of Merced. In 1916 the Bulldog claim was operated by Theodore E. Kockel of Lyons Gulch and the Bullpup by Jack Czerney of Merced. At that time there was a shallow shaft on the Bulldog claim and an inclined shaft 165 feet deep with 100 feet of drifts on the Bullpup claim. By 1928 the mine had been renamed the Permit and consisted of 2 claims, the Boulder and Permit, owned by Theodore Kockel of Mariposa (Laizure, 1928, p. 110). In 1932 the mine was acquired by the Permit Mines, Inc. of Mariposa. At that time the directors of the company were A. J. Walters, J. J. Fahnlender, Mabel R. Brocke and Theresa J. Thompson. More recently the Permit Mining Company has been operated by E. B. Cook of San Francisco and H. H. Odgers of Midpines. A small production was recorded in the 1930s and again in the late 1940s. From 1898-1900, a total of 241.87 oz. of gold was produced from the Bulldog and Bullpup claims. In 1912 the Kockel mine (Bulldog) produced 6.87 oz. of gold from 28 tons of ore (U. S. Bureau of Mines records). The mill is fully equipped and the mine is operated intermittently.

The main vein on the Permit property is on the same system that extends from the Milburn mine north to the Nutmeg mine. At the Permit mine it strikes N. 10-20° W., is nearly vertical and is 2-4 feet wide. According to Laizure (1935, p. 40) there are two other more or

FIGURE 34.   Installations of the Permit mine, Sherlock Creek district.
Photo by Mary H. Rice.

less parallel veins. He describes a west vein 3-4 feet wide, a center vein 2 feet wide and an east vein 2 feet wide. In 1935 there was a 70-foot shaft on the west vein, a 70-foot shaft with a 500-foot drift on the center vein and a 74-foot shaft on the east vein. At that time ore averaged $15 to $20 a ton. The main vein observed by the authors apparently is the center vein mentioned by Laizure. Wall rocks are pyroxene andesite greenstones of unknown age. In 1954 the mine was fully equipped with hoist house, compressor house and mill but there was no one on the property and the equipment was not accessible.

Pine Tree and Josephine Mine. Location: Secs. 8, 9, 16, 17, T. 4 S., R. 17 E., M.D., on Highway 49, 2 miles north of Bear Valley. Ownership: Pacific Mining Company, Crocker Building, San Francisco, California, owns about 5⅛ square miles of the northern end of the former Las Mariposas Grant which includes the Pine Tree and Josephine, French, Evans and Queen Specimen mines.

History: Along with the Princeton and Mariposa mines the Pine Tree and Josephine mine is among the best known and most chronicled in the southern part of the Sierran gold belt. Conspicuously located astride Highway 49 overlooking the Merced River it has a prominence and background of historical romance equalled by few other gold mines. The Pine Tree and Josephine vein system was discovered in 1849, probably soon after discovery of lode-gold at the Mariposa mine. Originally separate mines, the Pine Tree and Josephine have been worked jointly for so long that they have emerged as a single property.

The original locators and operators of the mine, whose names have been forgotten, lost their rights and title to the Pine Tree and Jose-

FIGURE 35.   Tailings dump and mine buildings at the Pine Tree and Josephine gold mine on Highway 49 between Bear Valley and Bagby. The principal mine opening is a long adit driven away from the observer from a point behind the shop buildings. The mine has a production of about $4,000,000. *Photo by Mary H. Rice.*

phine mines through a court decision in 1859 giving John C. Fremont control of the vast Las Mariposas Grant of 44,387 acres. In spite of the court decision an armed struggle for possession of the mines took place in 1859 and Fremont's followers were actually besieged on the fortified Pine Tree property by miners then in possession of the Josephine mine. Fremont ultimately gained undisputed possession of all the grant mines, but his management was never very successful. Logan states: "the grant properties became subject to one promotion scheme after another and comparatively little gold was produced (considering the possibilities) until 1900. . . ." In 1862 a disastrous flood damaged many of the grant mines. In 1863 a group of Fremont's creditors in New York formed a stock company capitalized for $10,000,000 but this company went into receivership in 1864. In spite of the financial vicissitudes of Las Mariposas Grant at large, the Pine Tree and Josephine mines operated almost continuously through the 1860s to about 1875. The elaborate Benton mill with its much publicized Ryerson process of dry-milling and steam-activated amalgamation, was built about 1865 just south of the present town of Bagby on the Merced River. This was ultimately enlarged to 80 stamps (Raymond, 1871, p. 28). In 1873 the Mariposa Land and Mining Company succeeded the former Mariposa Company. Under this management a long tunnel was driven south on the vein which ultimately reached a length of 3300 feet and extended under the Queen Specimen mine workings to a depth of 1200 feet. It was the intention of the company to extend this tunnel clear to Mariposa, a distance of 12 miles, but the ore shoots

FIGURE 36. Large tailings dump consisting of vein matter that has been crushed during recovery of the precious metals. Tailings were piped down from the Pine Tree and Josephine mill shown in figure 30. The dump is on the Merced River east of Bagby. *Photo by Mary H. Rice.*

encountered proved disappointing and the strength of the company was gradually dissipated in litigation and internal disputes. The grant was finally purchased at sheriff's sale by a man named Donahue. In July 1887, the present Mariposa Commercial and Mining Company was organized by San Francisco financiers headed by Hayward, Flood, Hobart, Jones and Mackey, who bought the grant from Donahue. This group engaged in little or no mining and little was done at the Pine Tree and Josephine mine until after 1898 when the company was reorganized with Fred Bradley as president. Sustained mining began about 1900 and was successfully continued until about 1915, about a third of a million dollars being produced from the Pine Tree and Josephine property. Another period of inactivity ended in 1933 with acquisition of the northern end of the grant by Pacific Mining Company. Between 1933 and 1944 this company operated continuously and achieved by far the greatest production ever realized by the mine. Operations ceased because of wartime conditions and the mine has never resumed production because of high operating costs. The property is by no means exhausted of ore and simply awaits more favorable economic conditions.

Geology: The Pine Tree and Josephine mine workings follow ore shoots in various parts of a single, immense, multiple-vein system, the Mother Lode here averaging at least 125 feet wide. It strikes N. 30-35° W. and dips 55-60° northeast. In general the vein system occupies a thrust fault of large displacement bordered on the east by serpentine and on the west by the Mariposa slate. Both of these rock units are believed to be of Upper Jurassic age. A typical cross section of the

Mother Lode in the Pine Tree workings is in the vicinity of the No. 2 East and No. 3 West crosscuts on the level of the New Pine Tree Tunnel (see Logan, 1935, p. 188). Southwest to northeast across this section the character of the vein is as follows:

Footwall side—black Mariposa slate_____thickness unknown
Josephine "vein"—quartz lenses and stringers in brecciated slate with ore
   shoots _____ 18 feet
Massive carbonate rock—ankerite—mariposite-quartz rock probably de-
   rived by replacement of serpentine_____ 60 "
Old Pine Tree "vein"—ribbon quartz with ore shoots (largely stoped)___ 10 "
Milky (bull) quartz—massive, barren of ore or low-grade_____ 12 "
Hanging Wall Pine Tree vein—ribbon quartz with ore shoots_____ 8–14 "
Serpentine and talc schist_____ 10 "
Greenstone or metadiorite_____ 25 "
Hanging wall side—serpentine_____thickness unknown

One version of the structure of the Mother Lode at the Pine Tree and Josephine mine to a depth of about 1200 feet is shown facing page 108 of Julihn and Horton (1940). In the ore shoots ore minerals are chiefly pyrite, arsenopyrite, and native gold with minor chalcopyrite, sphalerite, galena, niccolite and millerite. Copper has been the only metal recovered other than gold or silver. The nickel content of the concentrates ranges from 0.35 to 1.35 percent and during 1936 averaged 0.77 percent (Julihn and Horton, 1940, p. 110). Erythrite and danaite, both containing cobalt, have been found in the mine but presumably are rare and well disseminated. Thus far there has been no attempt to recover the nickel or cobalt.

According to Storms (1900, p. 144), the Old Pine Tree and Josephine veins in the southern parts of the Josephine mine are very close together—in places less than an inch apart, whereas they diverge to the north and are 60 feet or more apart. The Mother Lode pinches materially between the Pine Tree mine and Bagby as well as to the south toward Bear Valley. Between Bear Valley and Mt. Bullion the Mother Lode is represented by several en echelon veins rather than by a single wide multiple vein. The northeast-trending veins represented by the French and Lucky Boy veins do not intersect the Mother Lode in the vicinity of the Pine Tree and Josephine mine.

Workings: The workings of the Pine Tree and Josephine mine are connected by a single raise driven, during the last stages of operation, from the Pine Tree Tunnel level to the English Trail Tunnel level of the Josephine mine. The April adit in the Pine Tree mine and the September adit in the Josephine mine are only about 200 feet apart and at approximately the same elevation. The River Tunnel extends under the Queen Specimen mine and if extended would connect with main (MacKenzie) underground shaft in the Pine Tree workings theo-retically at the 1150-level. The accompanying diagram shows the disposi-tion of the workings.

The most southerly working is the McMurray shaft, about 60 feet deep which has no lateral workings of note. The Josephine workings begin with the Josephine and the Hunt and Oyler adits about 550 feet north of the McMurray Shaft, the last about 275 feet above the Josephine adit. Connecting with the Josephine adit is the Septem-ber adit, nearly 1200 feet farther north, and there are three others, the No. 1, No. 2 and English Trail in between. A winze from the Josephine adit has 3 levels, the 100, 300 and 500 and there is one drift level (the Black drift) between the Josephine and Hunt and Oyler adits connected by winzes. The total depth developed on the vein in the Josephine workings is 820 feet and the total length of workings is about 7700 feet. The principal ore shoot is developed for at least 600 feet.

The Pine Tree workings consist of 7 adits spread over a distance of about 1500 feet. The principal working is the New Pine Tree adit about 1200 feet long and the inclined MacKenzie shaft sunk from the New Pine Tree adit approximately 600 feet from the portal. The MacKenzie shaft serves levels at 150, 270, 430, 550, 700 and 850 feet and there are 3 levels above the New Pine tree adit at depths of about 150, 200 and 250 from the surface. The workings aggregate over 8 miles.

Summary of Production: The estimated production of the Pine Tree and Josephine mine is over $4,000,000 (Bradley, 1954, p. 32) and the recorded production is about $3,386,000. Prior to acquisition by Fremont no production records are available, due chiefly to secrecy connected with disputes and the final litigation. Between 1860 and 1863 forty-five thousand tons of ore was mined which yielded $350,000, an average of $7.77 per ton (Logan, 1935, p. 187, quoting Browne). In 1864 production was $67,940. For the fiscal year ending in July 1870 the Mariposa and Pine Tree and Josephine mines yielded $170,000 of which nearly $100,000 (Browne, 1868, p. 30) must have come from the Pine Tree and Josephine. No records are available for the period 1865 to 1875 but the yield must have been substantial. Little or no mining was done between 1875 and 1900. From 1900 to 1915 the production was $371,748 from 20,968 tons of ore for an average of $12.40 per ton. Between 1933 and 1937 production was $989,174 from 170,943 tons of ore or an average of $5.78 per ton, and $271,392 was produced from 55,021 tons of ore in 1938 (Pacific Mining Co. records). From 1939 to 1944 a total of 248,481 tons of ore yielded 35,215 oz. of gold, and 8219 oz. of silver for a gross of about $1,242,000, an average of $5.00 per ton.

Princeton, New Princeton and Princeton Extension.   Location: Secs. 13, 18, T. 5 S., R. 18 E., M.D., at the southern outskirts of Mt. Bullion on the paved Mt. Bullion-Cathay road. Ownership: Mariposa Commercial and Mining Company, c/o Eileen Milburn, Mariposa, California owns a tract of land containing several hundred acres that includes the Princeton and several adjoining mines. Formerly part of Las Mariposas Grant.

The Princeton mine was discovered in 1852 (Browne, 1867, p. 29) and has had a history similar to other large mines of Las Mariposas Grant (see Mariposa and Pine Tree and Josephine). It has the greatest recorded production of any mine in Mariposa County and the size of the workings is exceeded only by the Pine Tree and Josephine mine. At various times it has been the largest producer in the state. During the early years of its history the ore yielded about $70 per ton, to a depth of about 100 feet, but the bulk of the ore mined ran between $4 and $7 per ton at the old price of gold. By 1867 it had reached a depth of 560 feet as measured on the incline, and had been explored along the strike for 1200 feet (Browne, 1867, p. 41-42; 1868, p. 30).

The New Princeton mine, adjacent to the Princeton to the southeast was discovered and developed between 1871 and 1875 (Raymond, 1871, p. 30). Two inclined shafts were sunk on this location 200 feet apart.

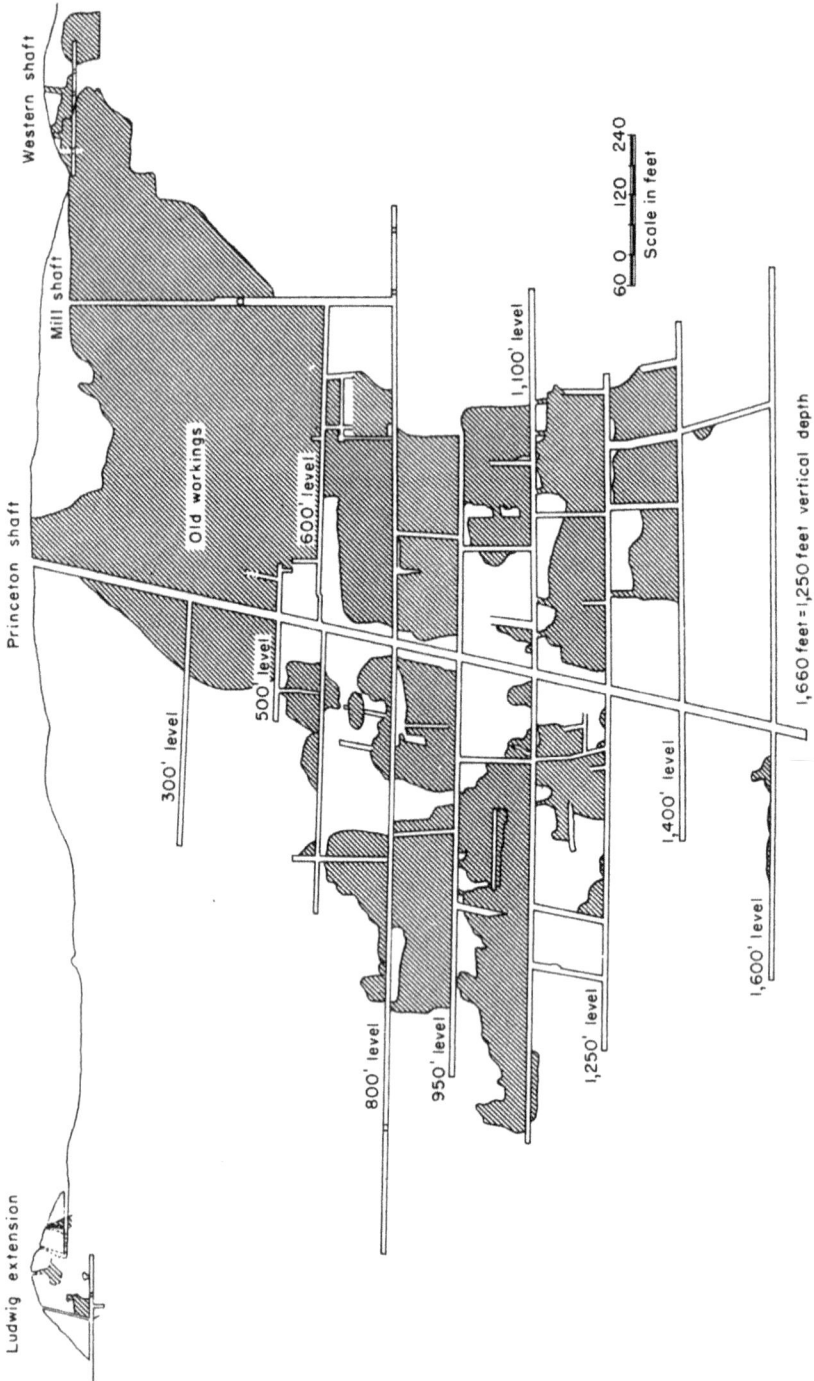

FIGURE 37. Longitudinal section in the plane of the vein of the Princeton gold mine showing stoped areas. *Reproduced from figure 46 of U. S. Bureau of Mines Bulletin 424 (Julihn and Horton, 1940).*

According to Knopf (1929, p. 84) a shaft 300 feet deep was sunk at the New Princeton between 1921 and 1924. It had levels at 150 and 300 feet.

Between 1875 and 1900 the mine lay idle but mining was resumed and vigorously pursued between 1900 and 1911 under the direction of John H. MacKenzie. The shaft was deepened to 1600 feet (inclined) and lateral workings greatly expanded. During this period most of the ore in the main shoot was stoped to a depth of 1200 feet (Lowell, 1916, p. 595). A small tonnage of ore was mined in 1915 after which the mine again lay idle until 1921. That year T. E. Kelso and W. H. Holmes contracted to purchase the Princeton mine and held the property until 1927 when a forest fire destroyed the surface buildings (Laizure, 1928, p. 111). Soon after this the property reverted to the Mariposa Commercial and Mining Company. Under Kelso and Holmes 3944 tons of ore was mined which averaged about $3.16 per ton at the old price of gold. W. J. Loring was their superintendent in 1924.

Lessees produced a small tonnage of high-grade ore in 1930-31 and 3828 tons of ore and tailings were milled between 1935 and 1941, but no sustained mining has been carried on at the Princeton mine since 1927.

The Princeton vein, believed to be part of the Mother Lode system, occupies a thrust-fault fissure cutting slate of the Upper Jurassic Mariposa formation. The slate contains thin beds of dark-colored graywacke and is cut by dikes of sheared, fine-grained granitic porphyry. Some of these interesect the vein. The trace of the vein is not well defined at the surface and lacks the massive, multiple characteristics of the Mother Lode at the Pine Tree and Josephine and Mount Ophir mines. The vein is en echelon to the vein passing through the Mount Ophir but is more or less in line with the veins passing through the Greens Gulch, Louis and Mountain View mines. It is probable that the Mother Lode in this vicinity consists of several separated veins rather than one or two major persistent features. The Princeton vein system strikes N. 54-57° W. and dips 45 to 60° northeast, averaging perhaps 50°. Vein matter, which is ribboned, milky quartz carrying numerous parallel sheets of included wall-rock, varies from 4 to 8 feet wide. Parts of the vein which constitute ore carry a considerable amount of pyrite. Native gold is seldom visible but minor amounts of galena, sphalerite and tetrahedrite are commonly seen (Knopf, 1909, p. 85). At the New Princeton workings the slate is severely bleached and hydrothermally altered near the vein.

The main shaft of the Princeton mine, now caved and inaccessible, is 1600 feet deep, measured on the incline, and reaches a vertical depth of about 1250 feet. It has 9 levels at 300, 500, 600, 800, 950, 1100, 1250, 1400 and 1600 feet. The longest level is the 800, approximating 2100 feet. There is a total footage of drifts approximating 11,500 feet together with over 3000 feet of crosscuts and raises. The old Phillips shaft, about 350 feet southeast of the main shaft was approximately 600 feet deep. On the New Princeton part of the property the 6 x 10 ft. main shaft (east shaft) is 300 feet deep and has levels at 100 and 300 feet. The west shaft is 60 feet deep. The New Princeton shaft was open to a depth of about 75 feet in August 1955.

There are various estimates as to the total production of the Princeton mine. A total of $4,397,743 is well authenticated. Storms (1900, p.

143) conservatively estimated the production to 1900 at $3,000,000 and production since that time has been $1,397,743, calculated from records of the Mariposa Commercial Mining Company and U. S. Bureau of Mines. Raymond (1871, pp. 30-31) places the early production at $4,000.000 to $5,000,000 and Knopf (1929, p. 84) places the total production to 1929 at $5,000,000. The mine has been characteristically one of large ore bodies of relatively low grade, although the characteristic ore mined prior to 1915 would yield $10.50 per ton at the present price of gold.

Pyramid (Castagnetto I) Mine.   Location: Secs. 14, 15, 23, T. 4 S., R. 16 E., M.D., on the west bank of Cotton Creek half a mile from the paved Hunter Valley road or about 9½ miles from Hornitos. Ownership: Lloyd A. Mason, Hornitos, California owns about 200 acres of patented agricultural land which includes the Pyramid mine.

The Pyramid mine was discovered prior to 1900, probably by Daniel Castagnetto. It was operated for a time in 1915-16 by C. II. Burt and Dodge of Bear Valley (Castello, 1921, p. 110). In 1925 the mine was purchased from the Castagnetto estate by George K. Allen of Piedmont (Laizure, 1928, p. 83). A mill was installed and some development work was done but there was no sustained production. In the middle 1930s it was operated by the Pyramid Gold Mining Company in which Eugene B. Gratton of San Jose and Lloyd Mason of Hornitos were the principal officers (Julihn and Horton, 1940, p. 121). From 1933 to 1942 a total of 5117 tons of ore was mined from which 4062.33 oz. of gold and 544 oz. of silver were taken for an average value of $27.86 per ton. The total production of the mine must be at least $200,000.

At the surface the main vein at the Pyramid mine strikes N. 60-65° W. and dips 45-71° southwest. According to Julihn and Horton (1940, p. 121) the vein steepens to 80° within 190 feet of the surface. Vein matter is mainly milky quartz with a little white calcite and includes leaves and fragments of wall rock. Ore minerals are native gold, pyrite, galena and sphalerite. About 95 percent of the gold is free-milling. Sulfide concentrates amount to only ½ percent of the ore (Julihn and Horton, 1940, p. 121) but contain 5 to 6 oz. of gold per ton. Wall rocks are chiefly pyroxene andesite greenstones belonging to the Upper Jurassic Amador group with thin strata of tuffaceous slate. Julihn and Horton (1940, p. 121) state that diorite forms the hanging wall of the mine, but no diorite was seen by the authors at the surface and none was on the dumps. A narrower parallel vein has been worked to some extent about 600 feet south of the main shaft.

Workings consist of four shafts two of which apparently have not been used since early days. The main working in the 1930s was a 6 x 10 foot shaft 190 feet deep with three levels at 65, 105 and 175 feet. From this shaft there is a 90-foot north drift and a 240-foot south drift; 238 feet of drift north and 85 feet south on the 105-foot level and 215 feet of north drift on the 176-foot level (Julihn and Horton, 1940, p. 121). A second shaft is 165 feet deep, but the depths of the others are not known. None were accessible to the authors in August 1954. The compressor hoist and hoist had been damaged because of sliding ground. The 5-stamp mill appeared to be in fair condition. No recent work has been done at the property.

FIGURE 38. Mine installations at the Quail gold mine on the north side of Indian Creek Canyon 5 miles southeast of Greeley Hill. The mine is entered by several adits, openings of which are not shown in the photo. Mine workings were being put into condition for operation during 1955. The Quail mine has a long record of productivity extending back prior to 1873 and totaling over $400,000.

Quail (Alvina, Hartford) Mine. Location: Secs. 15, 16, T. 3 S., R. 17 E., M. D., on the north side of Indian Gulch about 7 miles by fair dirt road southeast of Greeley Hill. Accessible via McDiermid guard station and Date Flat. Ownership: H. E. Moerlien, San Martin, California owns four patented lode claims, the Juniper, Sunset, Bonanza and Mammoth, 7 unpatented claims, the Violet, Starlight, Morn Sight, Fairview, and West Lode 1, 2 and 3, plus several mill sites aggregating over 380 acres.

The Quail mine was discovered prior to 1873 and has a total production estimated at slightly more than $400,000 (Julihn and Horton, 1940, p. 140). The first owners of record were Hanbidge and Gonigall who made a strike of good ore in 1873 (Min. and Sci. Press, vol. 26, No. 10. p. 149, 1873). These operators had driven at least one tunnel by 1873 and were in process of driving a raise for ventilation. By 1891 ownership had passed to Francisco Bruschi and considerable production was maintained. Ore was said to average $15 per ton in native gold and sulfide concentrates ran $800-$900 per ton (Min. and Sci. Press, vol. 62, no. 12, p. 180; vol. 62, no. 18, p. 276). Various members of the Bruschi family operated the property intermittently through the 1890s. In 1899 the property was bonded for a short time to Kendall Brothers and Whitney at which time the working tunnel was 700 feet long and an 18-inch wide shoot of ore was being mined. Later in 1899 the mine was sold by D. Bruschi to the Quail Mining and Milling Company, a group made up of Sonora, California, businessmen (Min. and Sci. Press, vol. 78, no. 18, p. 488; vol. 78, no. 21, p. 565). A 4-stamp mill was moved onto the property from the Louisiana mine and it was renamed the Hartford mine in 1900. The *Engineering and Mining Journal*, vol. 68, no. 12, p. 293, 1899, states that the mine included 3 tunnels, one 800 feet long, one 1300 feet long, and the third of unstated length, and that an ore shoot $2\frac{1}{2}$ feet wide was yielding ore running $20 per ton. Between 1899 and 1901 a small production of gold was recorded that year. Operations were carried on through 2 working tunnels and 4 raises. The vein varied between 18 and 60 inches wide and ore averaged $7.50 per ton at the old price of gold (Eng. and Min. Jour., vol. 71, no. 9, p. 283, 1901). There is no record of activity at the mine between 1901 to 1915. Various lessees worked the property under agreement with the Bruschi Brothers between 1915 and 1920. Some ore was milled which ran between $9 and $12 per ton at the prevailing price of gold (Castello, 1921, p. 134). A 10-stamp mill was destroyed by fire in 1917. Castello states that the workings, in 1920, consisted of a tunnel about 1000 feet long, two shafts 150 feet and 55 feet deep, respectively, and several stopes and raises. In 1915 one of these stopes was reported to be 200 feet long and 70 feet high (Lowell, 1916, p. 595). Between 1920 and 1935 the mine lay idle until reactivated late in 1935 by J. E. King and associates of Sonora. About 1937 the property was purchased from J. Bruschi by Quail Gold Mines, Inc., Jerome L. Drumheller of Spokane, Washington, in charge and Otto D. Rohlfs of Coulterville, general manager. Between 1937 and 1942 a total of 7161 tons of ore was milled which yielded 444 oz. of gold and 394 oz. of silver or an average of approximately $2.22 per ton. Again idle through the war years, the Quail mine was leased in 1949 from Jerome Drumheller by the Golden State Mining Company, a partnership of George Marshall, Clarence and Stanley Silvia and H. E. Moerlien. More recently ownership has passed to H. E. Moerlien of San Martin, California, who has been retimbering two adits and preparing to reopen the mine. A fully equipped 60-ton per-day capacity mill is on the property which includes jaw crushers, ball mill, shaking tables, flotation cells, etc.

The principal vein at the Quail mine strikes N. 20-25° W. and dips 30-35° NE as it passes through the Juniper, Mammoth and Bonanza claims, but swings to a north strike as it crosses the Sunset claim. The width varies between 18 inches and 5 feet. According to Julihn and Horton (1940, p. 140) the vein fissure has been intruded over much of its length by and aplite dike about 2 feet wide which at some places follows the footwall side of the vein and at others the hanging wall. Vein matter consists of milky quartz with pyrite, galena, sphalerite, tetrahedrite and native gold. Wall rocks consist of mottled slate, sandy slate and associated schistose metasediments belonging to the Paleozoic Calaveras group. The slate cleavage in the upper adit strikes N. 40° W. and dips 75-80° NE.

The principal operating workings are two adits, one 90 feet vertically above the other. The lower adit is 700 feet long and the upper one 550 feet. These probably have not been reopened the full length of former workings. The upper tunnel is connected with the surface by a 135-foot raise 200 feet from the portal of the adit and the lower adit is connected to the surface by a 328-foot raise 300 feet from the adit portal.

These raises were in need of cleaning out in August. 1954. The adits had been newly re-timbered and were in good condition.

Queen Specimen Mine.   Location: Sec. 8, T. 4 S., R. 17 E., M. D., on Highway 49 a quarter of a mile north of the Pine Tree mine buildings, 3 miles north of Bear Valley and 1 mile south of Bagby on the Merced River. Ownership: Pacific Mining Company, 1022 Crocker Building, San Francisco, California, owns a large acreage of the former Las Mariposas Grant, which includes the Queen Specimen, Pine Tree and Josephine, French and Evans mines.

The Queen Specimen mine has a history more or less parallel to the Pine Tree and Josephine mine which it adjoins on the northwest (see Pine Tree and Josephine mine). It was discovered in the 1850s and worked in a small way prior to acquisition of Las Mariposas Grant in 1859 by Fremont. It was worked by the Mariposa Mining and Commercial Company in 1908 and in 1915, producing a total of 657.5 tons of ore yielding $6029.22 or a total of $9.16 per ton at the old price of gold (Logan, 1935, chart facing p. 188). During the period 1922-1924 the mine was operated under lease from the Mariposa Commercial and Mining Company. In this period 3000 tons of ore was milled which yielded $4 per ton and an undisclosed quantity of tailings which carried $2.50 per ton (Knopf, 1929, p. 84). Little has been done at the property since 1924 and the workings are caved and inaccessible.

The multiple vein on the Queen Specimen property is the northwestern extension of the Mother Lode as seen at the Pine Tree and Josephine mine. It has a general strike of N. 35° W., and dips 60-70° northeast and has an aggregate thickness of about 105 feet. Ore occurs almost entirely on the footwall side, varying between a few feet and 12 feet thick and yielding $4 to $12 per ton (Knopf, 1929, p. 84). Knopf describes the cross section along the main adit SW to NE as follows:

1. serpentine—130 feet
2. black slate—370 feet
3. quartz—3-12 feet
4. ankerite-quartz-mariposite rock netting with quartz veinlets—15 feet
5. talc schist (derived from serpentine)—30 feet
6. greenstone (locally called diorite, hydrothermally altered, pyritized and sheared—48 feet
7. black slate—extent not measured

The ore shoot was 1-1$\frac{1}{2}$ feet thick on the footwall side of the quartz member of the vein system. Primary minerals in the ore shoots are native gold, pyrite, tetrahedrite and chalcopyrite, probably with minor amounts of additional minerals similar to those at the Pine Tree and Josephine mine. Joints in the quartz member of the vein system were described by Knopf as being stained by azurite in some places.

Mine workings in 1924 consisted of: (1) A main 500-foot crosscut adit, driven northeast approximately perpendicular to the strike of the vein system, a connecting main drift 400 feet long and a raise with levels at 100 and 125 feet above the adit level. (2) An upper adit driven 200 feet (vertically) above the main adit connected with a 250-foot drift. (3) A lower crosscut adit, about 500 feet north of the upper adit, with minor drifts. (4) A 150-foot drift adit and stope about 400 feet northwest of the lower crosscut adit. None of these workings were accessible in September 1955.

Red Bank (Stevenson group) Mine. Location: Sec. 36, T. 3 S., R. 16 E., M.D., on the northeast bank of the Merced River 1½ miles northwest of Bagby. Accessible via half a mile of unimproved dirt road from Highway 49, the turnoff being 1½ miles from Bagby. Ownership: Percy L. Pettigrew, Box 639, Palo Alto, California, and Horace Meyer, Cathay, California, own the Daisy, Jubilee and Syndicate lode claims and three placer claims totaling about 165 acres. Most of this holding is patented.

The Red Bank mine was probably discovered in the late 1860s. Most of the early work done consisted of placering and ground sluicing. In 1881 hydraulic mining was started under Colonel Swadley (Min. and Sci. Press, vol. 43, no. 26, p. 429, 1881) but from the looks of the property no great yardage of material was handled in this way. The first extensive underground mining was done between 1894 and 1897 by the Redbanks Mining Company under Isaac Solhinger (Min. and Sci. Press, vol. 71, pp. 106, 138, 238). Two tunnels aggregating 400 feet were driven but the vein apparently was not reached. In 1897 the owners, Henry Bratnoble and A. Wartenweiler optioned the mine to Hamilton Smith and associates at an agreed purchase price of $100,000. This deal apparently was not consummated and the property was then bonded to the London Exploration Company. The Daisy tunnel was extended by this company to a total length of 900 feet, striking the vein about 1000 feet below its outcrop (Min. and Sci Press, vol. 75, pp. 30, 170, 434, 598). The vein was found to be 30 feet wide with ore assaying about $10 per ton. The mine closed down in 1898 before any considerable production was realized, apparently because of lease and ownership disputes. Some rehabilitation and development work was done on the mine in 1910, 1914, 1915, and 1917-18. E. C. Loftus of San Francisco had purchased the mine in 1906 (Eng. and Min. Jour., vol. 82, no. 11, p. 513, 1906), a French syndicate held the property in 1910 and ownership had passed to H. C. Callahan of San Francisco by 1914. A small tonnage of ore was milled between 1917 and 1918, ten tons of ore yielding 28 oz. of gold and 40 oz. of silver (U. S. Bureau of Mines records). A 5-stamp mill, separating tables and flotation were employed (Min. and Sci. Press, vol. 114, p. 104, 1917). By 1920 ownership had passed to Percy L. Pettigrew and H. C. Callahan, the mine being idle that year. By 1928 Percy L. Pettigrew had become the sole owner (Castello, 1921, p. 135; Laizure, 1928, p. 113). No mining was done in the 1920s but in 1935 the mine was leased by F. W. Draper. By 1937 a large tonnage of ore had been blocked and a 125-ton mill was being installed (Eng. and Min. Jour., vol. 140, no. 1, p. 70, 1937). During 1939 and 1940 a total of 12,796 tons of ore was milled which yielded 1551 oz. of gold and 6894 oz. of silver worth about $62,000—an average of about $4.74 per ton (U. S. Bureau of Mines records). The total production of the mine is well in excess of $100,000.

The Red Bank mine develops two well-defined veins belonging to the Mother Lode system, the Crown Peak and the Stevenson, as well as several smaller mineralized shear zones. These veins have a general strike of N. 40-45° W. and dip northeast at angles between 50 and 60 degrees. The Crown Peak vein over much of its length follows the fault contact between serpentine and Mariposa slate. The Anderson vein, on the footwall side of the Crown Peak vein, in most places is in slate. Vein matter in the Crown Peak vein commonly is 30 feet thick with ore-bearing parts averaging about 7 feet thick. Milky quartz, quartz-mariposite-ankerite rock and rusty silica-carbonate rock are the usual vein-rock types. The Anderson vein is similar to the Crown Peak but, in general, not so well defined or so thick. Ore minerals are pyrite, argentiferous galena and native gold with minor tetrahedrite, chalcopyrite and sphalerite. Ore milled in 1939-40 carried an unusually high percentage of silver for a Mother Lode mine.

The Red Bank mine has been developed chiefly by crosscut adits that explore the vein for more than 2400 feet of strike length and an inclined depth of over 1000 feet. Because of curvatures in the vein the roughly perpendicular crosscuts are not mutually parallel but trend north or northeast. The Daisy and Stevenson adits give access to the most extensive workings. The workings may be summarized as follows, west to east:

Name	Length of adit or shaft	Length of connecting workings	Elevation
Daisy adit	1821 feet	500 feet	900 feet
Unidentified adit	175 "	180 "	1000 "
Anderson adit	70 "	0 "	? "
Unidentified adit	30 "	0 "	? "
Croesus adit	180 "	40 "	1260 "
Ione adit	660 "	0 "	1005 "
Jubilee adit	370 "	500 "	967 "
Stevenson adit	550 "	500 "	860 "
Shaft on Crown Peak vein	50 "	0 "	1300 "

Red Cloud (Kate Kearney) Mine. Location: Secs. 22, 27, T. 2 S., R. 17 E., M.D., on Bear Creek $3\frac{1}{2}$ airline miles east of Greeley Hill or 10 miles east of Coulterville. Accessible by unimproved dirt road by way of the McDiermid guard station of the U. S. Forest Service. Ownership: Carl Harper, Coulterville, California, owns one claim of approximately 20 acres.

The Red Cloud mine was discovered prior to 1880 but the principal period of operation was 1885-1895 when most of the workings were driven and most of the production made. Castello (1921, p. 135) reports the estimated production to be about $1,500,000 but there are no records to substantiate this claim. In 1885 the mine was owned by Gaines and Carter of San Jose, California (Min. and Sci. Press, vol. 51, no. 7, p. 120, 1885). J. S. Carter was mine manager and R. B. Harper and John Guest were superintendents at various times under Carter. The shaft was down 360 feet late in 1885, was deepened in 1886 to 430 feet and again in 1893 to 700 feet. A 22-stamp mill was operating on the property in 1889 but fell idle soon thereafter, steam-powered. The operator in the middle 1890s was the Red Cloud Mining Company of Boston. Two short ore shoots were followed down to the 500 level during operation by Gaines and Carter, but these were lost between the 600 and 700 levels and the Red Mill Mining Company was reported to have spent $60,000 in fruitless exploration between the 600 and 700 levels (Min. and Sci. Press, vol. 80, no. 11, p. 265, 1900). In 1900 the property was optioned to A. P. Dron of Big Oak Flat but was not put into production. The owner in 1904 was Mrs. E. Whitman of Coulterville (Wilkinson, 1904, p. 13). By 1920 ownership of the mine had passed to Mrs. Emma McDiermid of Coulterville who retained possession of the mine throughout the 1920s. In 1935 the mine was leased for a short time by Fredericks and Hodge of Mariposa. The authors know of no active mining at the property since 1935.

The vein at the Red Cloud Mine strikes N. 40° E. and dips 60-80° north. Wall rocks are slaty metasediments of the Paleozoic Calaveras group. The vein averages about 5 feet wide, locally reaching a width of 15 feet. Vein matter is milky quartz with inclusions of brecciated wall rocks, locally banded or ribboned. Two ore shoots each about 400 feet long were worked in the upper levels to a depth of over 500 feet, ore being 3-5 feet thick. In the middle 1880s ore was running about $30 per ton mostly from native gold. The average content of sulfides was about 2½ percent, auriferous pyrite concentrates running between $100 and $200 per ton (Goodyear, 1888, p. 346). The vein is traceable for a distance of more than 4500 feet and the original property consisted of 3 claims oriented end to end on the main vein and a 4th claim near the east end of the vein at approximately right angles to the main vein. Apparently an intersecting or cutter vein was located in the workings which does not show at the surface.

Mine workings consist of a 8x10-foot inclined shaft 700 feet deep with levels at intervals of 100 feet. Most of the stoping has been done above the 500-foot level. Drifts aggregate more than 500 feet. The workings were all caved and inaccessible in October 1954.

Roma (Piedra de Goza) Mine. Location: Sec. 14, T. 4 S., R. 18 E., 1½ miles south and slightly east of Briceburg. Accessible by less than a quarter of a mile of trail from the Yosemite All Year Highway. Ownership: Frank and Nellie Harris, 595 Tunnel Avenue, San Francisco 24, California, own a patented property containing 25 acres.

The Roma mine was discovered in the 1850s or early 1860s by Spanish Californians and was known in the early days as the Piedra de Goza. Nothing concerning the early production of the mine has been recorded but the early workings were extensive and a substantial early production is probable. After lying idle for nearly 30 years the Roma mine was reopened in 1897 by the Holiday Mining Company directed by T. J. Parsons, G. W. Baker, J. Jacobs et al. with C. G. Rogers as superintendent (Eng. and Min. Jour., vol. 63, no. 7, p. 168). Several tunnels were opened. A 2½-foot vein was struck in the longest of these at a depth of 250 feet from the surface. The mine apparently was idle much of the time in 1898 and 1899 but in 1900 a long tunnel (adit) was started designed to strike the intersection of the Roma and Sierra Rica veins at a point 1400 from the adit portal (Eng. and Min. Jour. vol. 70, no. 6, p. 107, 1900). The Roma and Sierra Rica mines at this time were operated jointly by the same management (Storms, 1900, p. 146). By September 1, 1900, this adit had reached a length of 915 feet and ultimately reached 1000 feet. Its cross-sectional dimensions are 5 by 7 feet and it reached a depth of 600 feet below the outcrop of the Roma vein. Lewis Aubury superintended this operation. Very little work has been done on the mine since 1901. In 1904 the owner was listed as C. Kerrins of Mariposa (Wilkinson, 1904, p. 13). Later ownership passed to Matt Harris and has been held by various members of the Harris family up to the present time.

FIGURE 39.   The Ruth Pierce gold mine in the Hornitos district. Photo taken from beside the Hornitos-Mt. Bullion road with the observer facing south. The main Ruth Pierce vein extends across the hilltop from side to side of the photo and is well exposed where it crosses the road to the left of the photo. About $600,000 has been taken from this mine.

The main vein at the Roma mine averages 2 to 2½ feet wide locally thickening to as much as 5 feet. It strikes slightly north of east, dips 60-70° S. and is a continuation of the vein crossing the adjacent King Solomon claim. The wall rocks are massive pyroxene andesite greenstones of unknown age. Adjacent to the vein there has been rendered slaty to schistose by intense shearing.

Rutherford and Cranberry (Perrin and Craigue) Mine.   Location: Secs. 15, 22, T. 3 S., R. 19 E., M.D., just north of the Merced River and Highway 140, 4 miles by road west and slightly south of El Portal. Ownership: L. G. Goodrich, 1522 State St., Santa Barbara, California, owns a fractional claim, the Rutherford, containing 10.34 acres. Ownership of other claims of the Rutherford and Cranberry group were not determined.

The Rutherford and Cranberry mines are contiguous properties operated throughout much of their history under the same management. The Rutherford mine, closest to the Merced River, was discovered about 1863, probably by the Rutherford brothers. As early as 1865 there was a water-powered, 5-stamp mill on the Rutherford mill site (Min. and Sci. Press, vol. 10, no. 11, p. 163, 1865). According to Castello (1921, p. 112) the Cranberry mine, situated high on the hill north of the Rutherford mine, was discovered in 1863 and patented in 1873. The Rutherford mine was patented in 1874 (Laizure, 1928, p. 114). In 1867 a 50-foot shaft was sunk to connect with one of the tunnels. At that time ore ran $40 per ton as well as other miscellaneous workings. In 1868 the Rutherfords sold the properties to undisclosed individuals from San Francisco. By 1875 ownership was held by Perrin and Craigue with S. W. Craigue acting as mine superintendent. The mines were intermittently active until the summer of 1880 when the mine was closed down because of disputes among the owners. In 1879 ore was averaging about $10 per ton, but occasional masses of ore were found that ran as high as $4500 per ton. The principal owners in 1884 were Capt. A. H. Ward and S. W. Craigue. These individuals constructed 2 water-powered arrastras and considerable rich ore was taken from the Rutherford mine from a depth of about 70 feet. In 1885 a rich ore shoot was discovered in the Cranberry mine in a new crosscut driven about 500 feet southeast of the main shaft and on the opposite side of the ridge at a depth of 100 feet below the vein outcrop. At the time of the new discovery, the two arrastras each were milling 1 ton of rock per day which returned $34.60 per ton, mainly from ore taken from the old Cranberry shaft (Min. and Sci. Press, vol. 51, no. 14, p. 232, 1885). The vein encountered was described as 11 feet wide and very rich, but no statement as to the tenor of the ore was made.

Mining on the Rutherford property apparently was discontinued for a long period and effort was concentrated at the Cranberry mine. By 1888 the new ore-shoot had been developed for a length of 400 feet and a vertical depth of 320 feet by shaft and winze. The ore-shoot was described as 4 feet wide (Goodyear, 1888, p. 348; Storms, 1896, p. 217). Mining continued until about 1896 after which the property lay idle until 1927. A. H. Ward held ownership until the middle 1920s when the property passed into the hands of F. E. Bass of San Francisco. During that year Bass leased the property to R. A. Frederick of El Portal and considerable drifting was done on the Rutherford vein (Laizure, 1928, p. 114). There is no record of production for this period. About 1935 the property was leased by Yosemite National Gold Mines Company, John C. McGarry, manager (Laizure, 1935, p. 43). A new 200-foot shaft was sunk and considerable drifting done on the 200-level. In July, 1954, there was a good cabin on the Rutherford property, but no equipment, and the mine workings were inaccessible because of minor caving.

Although no production records are available for the Rutherford and Cranberry group of mines the total yield must have been impressive. The lowest grade of ore mined apparently ran about $10 per ton at the old price of gold and Laizure (1928, p. 114) states that several shoots averaged $20 per ton. Much ore ran considerably higher, particularly in the upper 100 feet of the veins where the ore was oxidized and enriched through removal of sulfides. Pyrite concentrates assayed $35 to $40 per ton in the late 1880s (Goodyear, 1889, p. 348). There is no indication that the veins were worked out and the total depth reached, about 320 feet, does not adequately explore the veins encountered on either the Rutherford or Cranberry properties.

There are two sets of veins at the Rutherford mine, one striking N. 10-15° E. and dipping 75-80° west and the other striking about N. 75-80° E., and nearly vertical. Some work has been done on at least four north-trending veins and one or more veins trending nearly east. Wall rocks are coarse-grained biotite granodiorite or quartz diorite cut by narrow dikes of fine-grained hornblendic quartz diorite. Vein matter is milky quartz with native gold, pyrite, arsenopyrite and galena. Pyrite makes up about 25 percent of the ore below the zone of oxidation (Goodyear, 1888, p. 349). The main vein is 1 foot to 5 feet wide averaging 3 to 4 feet. Others are thinner and less persistent along the strike.

Workings at the Rutherford property consist of a main drift adit with connecting shaft, a 200-foot shaft with workings on the 200-level and several less extensive adits. None of these were accessible to the authors in July, 1954. The shaft connecting with the main adit (located a short distance northwest of the cabin) has one level between the shaft collar and the haulage adit and both levels are connected to extensive stopes. The adit was 460 feet long in 1921 (Castello, p. 136). The 200-foot shaft sunk in 1935 could not be identified in July 1954 and presumably is caved at the collar.

At the Cranberry mine, about half a mile north and slightly west of the Rutherford mine workings, the main vein strikes roughly N. 50° W. and dips northeast at 40-50°. It averages 4 or 5 feet wide locally reaching widths of 11 feet. The vein cuts coarse-grained granitic rocks containing small pendants of dark slate, schist and hornfels. Topography is exceedingly rugged. Workings include 3 drift adits, all over 200 feet long and the longest at least 700 feet. The long tunnel connects with an 80-foot shaft (Castello, 1921, p. 136). Storms (1896, p. 217) also mentions a connecting 200-foot shaft 500 feet from the adit portal. There is a crosscut adit 500 feet southeast of the old shaft on the opposite (east) side of the ridge which cuts the vein at a depth of 100 feet below the outcrop (Min. and Sci. Press, vol. 51, no. 14, p. 232, 1885). This connects with extensive drifts and stopes. All of these workings are partially obscured by cave-ins and none had been entered recently in July 1954.

Schoolhouse Mine.   Location: Center sec. 10, T. 4 S., R. 16 E., M.D., in the central part of Hunter Valley about 7 airline miles northeast of Hornitos. Accessible via the paved Hornitos-Bear Valley and Hunter Valley roads. Ownership: Frank A. Maschio and John Maschio, Merced, California, and Joe Maschio, Snelling, California own one patented claim, a part of the patented Maschio Ranch.

The Schoolhouse mine is an old property which had extensive workings and dumps prior to 1912. During 1912 and 1913 the property was operated under bond by G. P. Gow of Berkeley with W. Taylor acting as superintendent (Eng. and Min. Jour., vol. 94, no. 21, p. 1002; vol. 95, no. 21, p. 1071). A 2-stamp mill was placed in operation crushing ore from old dumps. About 1916 some work was done by Milton Sutherland of Atwater, but soon closed because of disputes. The mine was reopened by Sutherland in 1923 and some very rich ore was taken out from 2 levels (Laizure, 1928, p. 115) between 1923 and 1925. A small production was recorded in 1931 (U. S. Bureau of Mines records) since when there has been little or no activity at the mine.

Mine workings are along the southeast extension of a prominent vein system locally known as the Blue Lead, which strikes N. 30-35° W. and dips about 60° E. The vein system cuts pyroxene andesite greenstone containing bands of red chert, these rocks belonging to the Upper Jurassic Amador group. Gold occurs in pockets in dark, bluish quartz. The principal working is a 60-foot inclined shaft serving two levels of undetermined length but the shaft is caved and the headframe has been removed. A blacksmith shop remains on the property. There had been no recent activity in June 1954.

Schroeder (Home) Group. Location: Sec. 16, T. 4 S., R. 18 E., M.D., on high ground west of Saxon Creek Canyon 3 airline miles northwest of Colorado School and 6½ airline miles north of Mariposa. Accessible by 2½ miles of unimproved dirt road from the graded Saxon Creek-Sherlock Creek road. Ownership: C. M. and R. E. Schroeder, Box 169, Mariposa, California own the patented Schroeder placer claim of 20 acres and 8 other unpatented claims including the Home, Rex, Missing Link, New Deal, Caldwell, Apex, Independence and Nelly Bly claims.

The Schroeder group of claims was discovered in the late 1870s or early 1880s by John Schroeder. The first sustained mining apparently began about January 1883 with hydraulicking and ground sluicing of soil and the upper weathered parts of the veins. Numerous sizeable nuggets and a great deal of fine gold was recovered, enough to finance construction of a ditch and pipeline as well as other miscellaneous equipment. In June 1883 the county was engaged in surveying for a pipeline designed to insure an adequate water supply for hydraulicking operations at the Schroeder and adjacent Schantz mines (Min. and Sci. Press, vol. 46, no. 22, p. 372, 1883). By May 1886 hydraulicking and ground sluicing operations had laid bare a blanket vein dipping 30-35° to the east (Min. and Sci. Press, vol. 52, no. 18, p. 288, 1886) as well as several other veins, and it became necessary to explore the property underground. During the middle 1890s the mine was sold to F. W. Keeney of San Francisco, J. A. Schroeder remaining as superintendent. A new mill was erected about this time. By April 1898 the mine had been bonded to W. Price who recovered a piece of ore weighing 13 pounds which yielded $2,000 (Min. and Sci. Press, vol. 76. no. 16, p. 422, April 16, 1898). Storms (1896, p. 223) describes the veins as pockety. Nevertheless 483.75 oz. of gold were produced from the mine during 1896 from an undisclosed quantity of ore (U. S. Bureau of Mines records). Between 1900 and 1906 approximately 133 tons of ore was mined which yielded 595.51 oz. of gold and 12 oz. of silver for an average of approximately $90 per ton.

In 1915 the mine was being operated by J. A. and C. J. Schroeder, P. W. Judkins and C. H. Weston (Lowell, 1916, p. 596). They installed a 5-foot, 20-ton Huntington mill. At that time the principal working was an open cut 200 feet long and 30 feet deep, but about 375 feet of drifts had been driven. By 1920 the open cut had been enlarged to 400 feet long and 60 feet deep. At present it can hardly be recognized from other gulches in the vicinity.

The Schroeders, Judkins and Weston retained ownership of the property through the 1920s but did very little work on it. In 1932 some production was recorded, apparently by the Schroeders and in January 1934 the mine was taken lease by Ashworth and Pehrson of Mariposa (Laizure, 1935, p. 44). Ownership has remained largely in the Schroeder family to the present time and there has been a substantial intermittent production of ore averaging slightly over 1 oz. of gold per ton. In the fall of 1954 intermittent mining was being carried on from a northwest-trending adit about 150 feet above and southwest of the mill but there was no one on the property at the time of the authors' visit.

Recorded production of the Schroeder group of mines since 1896 is $158,400 as calculated from U. S. Bureau of Mines records. Inasmuch as this does not include the very substantial production of the 1880s and early 1890s the total production of the group must be between $200,000 and $300,000.

By far the greatest production from the Schroeder group of mines has been from residual soil mantle and from the weathered, fragmental upper parts of the several veins which cut the property. A system of relatively narrow, more or less parallel, north-trending veins crosses the property. Most of these veins have a very steep east dip. Cutting this system is a somewhat discontinuous blanket-vein varying from a few inches to 3 feet wide. This dips east at angles varying from 15° to nearly 35°. Wall rocks are massive, schistose and slaty pyroxene andesite greenstones derived partly from tuff and partly from massive flows and dikes. The greenstones are of probable Paleozoic age. The schistosity of the greenstone, where well developed, strikes N. 60° W. and dips very steeply southwest.

In 1935 the principal working was an adit 215 feet long and a 15° inclined shaft 35 feet deep (Laizure, 1928, p. 44). In August 1954 there was a long adit driven southwest (N. 60° E.) from a point just west of the mill building and at the same level as the building. There are several adits of various lengths higher up the ridge from this lower adit that trend N. 30-40° W. The lower adit was not in active use, most of the current development work being done on the upper adit located nearest a small cabin or tool shed. This was not entered, in the absence of the owners, but was at least 150 feet long.

The mill includes a 5-foot diameter, 20-ton capacity Huntington crusher and appeared to be in running condition. There is also a serviceable compressor and compressor house and other miscellaneous mine buildings.

Silver Lead (St. Gabriel and Honeycomb, Silver Lead and Margaretta) Mine. Location: Sec. 28, T. 4 S., R. 16 E., M.D., 4½ miles north and slightly east of Hornitos. Accessible by 1½ miles of unimproved dirt road from the paved Hornitos-Bear Valley road. Ownership: E. G. Branson, et al., c/o Horace Meyer, Cathay, California owns two patented claims the Silver Lead and Margaretta totaling 39.84 acres. Property originally included the St. Gabriel and Honeycomb claims but these may have been re-named.

The Silver Lead mine was first discovered in the 1860s (Castello, 1921, p. 138) but was only active at wide intervals up to 1914. In 1880 some work was done at the mine by A.S.F. Company (Min. and Sci. Pres, vol. 40, no. 23, p. 357, 1880) but there is no record of production. In 1901 it was bonded for a short time to Richard O'Brien. In 1914 the property was acquired by the Mariposa Mining and Milling Company of Carson City, Nevada, with W. A. Bradley acting as mine superintendent (Lowell, 1916, p. 597). At that time Lowell reported the workings to consist of a 240-foot shaft with 422 feet of drifts. Bradley put a 5-stamp mill into operation and a moderate production of medium and low-grade ore was recorded. About 1919 the management was reorganized under the name Simpson Mining Company, W. A. Bradley remaining as superintendent, but this company ceased operations in 1922 (Laizure, 1928, p. 117). About 1935 the St. Gabriel and Honeycomb claims were taken under lease by the Sanchez Brothers of Hornitos (Laizure, 1935, p. 45) and there was some production by lessees up to 1947. There was no activity and no equipment at the property in July 1954.

Ore occurs in a narrow vein of glassy to milky quartz accompanied by a zone of quartz stringers 2-4 feet wide. The mineralized zone strikes about N. 30° W. and dips 40-45° northeast. Wall rocks are chiefly dense, massive, blue-black quartz-biotite hornfels and massive spotted slate. These apparently are cut at depth by green metavolcanic dikes and dark, quartz-rich granodiorite porphyry dikes. None of these crop out conspicuously at the surface. Both the metasediments and intrusives are of Upper Jurassic age. Ore minerals, in addition to finely divided native gold, include pyrite, chalcopyrite, galena, sphalerite, and arseno-pyrite. The silver content is no greater than most other gold ores of the Sierran foothill belt. Ore milled during the later periods of mining yielded $4-7 per ton. The principal working at the Silver Lead mine is an inclined shaft at least 250 feet deep serving 2 or 3 levels. This was flooded to within 50 feet of the collar in August 1954. In 1928 there were two levels aggregating about 500 feet (Laizure, 1928, p. 138). There are several short tunnels and shallow shafts and pits in a line extending northwest for about a quarter of a mile. The tunnel workings are both sides of a ravine which cuts across the trend of the vein system.

Spread Eagle Group (Farmers Hope, Miners Hope, Empire, Mohawk, Monarch, Fanny, Little Charlie, Tollgate, Bonanza). Location: Secs. 29, 31, 32, T. 4 S., R. 18 E., M.D., near the Whitlock Creek-Sherlock Creek road 4 airline miles north and slightly west of Mariposa. Accessible from State Highway 49 via 5 miles of graded dirt road and half a mile of unimproved dirt road. Ownership: Mack C. Lake, c/o Carlo S. Morbio, 58 Sutter St., San Francisco, California owns a property patented in 1934 which includes the Empire, Fanny, Little Charlie, Miners Hope (Farmers Hope), Mohawk, Monarch, Spread Eagle and Tollgate claims and the Empire millsite, a total of about 100 acres.

The Spread Eagle group of mines was discovered between 1850 and 1860 as an aftermath of placer-mining on Whitlock Creek (Julihn and Horton, 1940, p. 148). They are a consolidation of numerous claims separately owned in early days but gradually consolidated under one ownership. In 1896 the Farmers Hope (later called the Miners Hope) was owned by J. J. Ellingham and in 1903 by Ellingham and Grove—operated under bond to E. Waters and Sons who had a 5-stamp mill. The Farmers Hope was also extensively developed in 1905 by Stolder and Dussel, presumably under lease or bond, in conjunction with an adjoining claim called the Bonanza.

The Spread Eagle mine was operated in the late 1890s and early 1900s by Jacob Teats (also spelled Teets). Teats bought and installed a 2-stamp mill on the claim in the summer of 1904 (Eng. and Min. Jour., vol. 75, no. 3, p. 102, 1903). In August 1904 the Spread Eagle mine was under lease to Nevills and Hanna who took out 28 tons of ore which yielded $120 per ton (Eng. and Min. Jour., vol. 78, no. 5, p. 197, 1904).

By 1915 ownership of the Farmers Hope group of seven lode and two placer claims and one millsite had passed to G. L. Kennedy of Mariposa, and Nick Mullins of Whitlock held two claims under the name Spread Eagle mine (Lowell, 1916, p. 582, 597). At that time the Farmers Hope group was inactive and the Spread Eagle was active under lease. On January 1, 1916 the Farmers Hope claim was relocated by Nick Mullins and renamed the Miners Hope, assessment work having been allowed to lapse on the Farmers Hope group of claims (Castello, 1921, p. 126); Mullins held both the Miners Hope and Spread Eagle groups of claims until 1927 when they were reported sold to Belle McCord Roberts of Long Beach (Consolidated Gold Fields of Mariposa, Laizure, 1928, p. 117), but little or no mining was done under this ownership.

In the spring of 1934 the Spread Eagle—Miners Hope group of mines was reopened by Whitlock Mines Corporation under the management of Martin Tresidder and a 5-stamp mill was put into operation (Laizure, 1935, p. 38). The Whitlock company and later a leasing partnership of Martin Tresidder and R. C. Poor made a substantial production from the property between 1934 and 1939 from ore running slightly less than a quarter of an ounce of gold per ton (U. S. Bureau of Mines records). There has been no recorded production from the mines since 1939.

The following discussion on the geology and mine workings is drawn chiefly from Julihn and Horton (1940, pp. 148-150) the last authors to visit the mines while they were in operation. "The principal workings are on two nearly parallel veins that strike almost due north but diverge at depth. The Spread Eagle dips about 45° E., and the Miners Hope, which outcrops 600 to 700 feet to the west of it, dips about 60° E. . . : The average width of the Spread Eagle vein is about 20 inches but in stoped areas it has been considerably wider, up to a maximum of 7 feet. Both walls are greenstone. The ore consists of quartz carrying gold, pyrite, chalcopyrite and occasional galena, the sulfides composing about 1 percent of it. Above the adit level, which corresponds closely with the water level, the ore was almost wholly free-milling, while below it about one-fourth of the gold is contained in sulfides, the concentrates from which average about $250 per ton. The ore mined above the adit level averaged about 1 ounce of gold per ton, but the vein contained frequent pockets of high-grade, and 2- and 3-ounce ore was not exceptional. It is of interest that ore shoots were found at two places where the vein bends sharply to the east. . . ."

"The early workings of the Miners Hope included numerous deep trenches, shallow pits, and adits extending over 1000 feet along the vein but principally within 600 feet of the northern end of the claim. There the vein is said to have averaged over 4 feet in width for at least 300 feet, attaining a maximum width of 12 feet in one stope. South of this section, however, its width was 14 inches to 2 feet. The vein follows rather closely the contact of a schist (schistose greenstone) on the footwall and greenstone on the hanging wall. . . . The ore shoots are said to have been associated with small fissures coming through the schist to an abutment with the greenstone. Some of these fissures contained a fine-grained dike rock.

"In the Fanny claim west of Whitlock Creek is a strong quartz vein said to show no gold at the surface; but on the Milburn claim adjoining it on the south, good ore has been discovered . . . in the same vein.

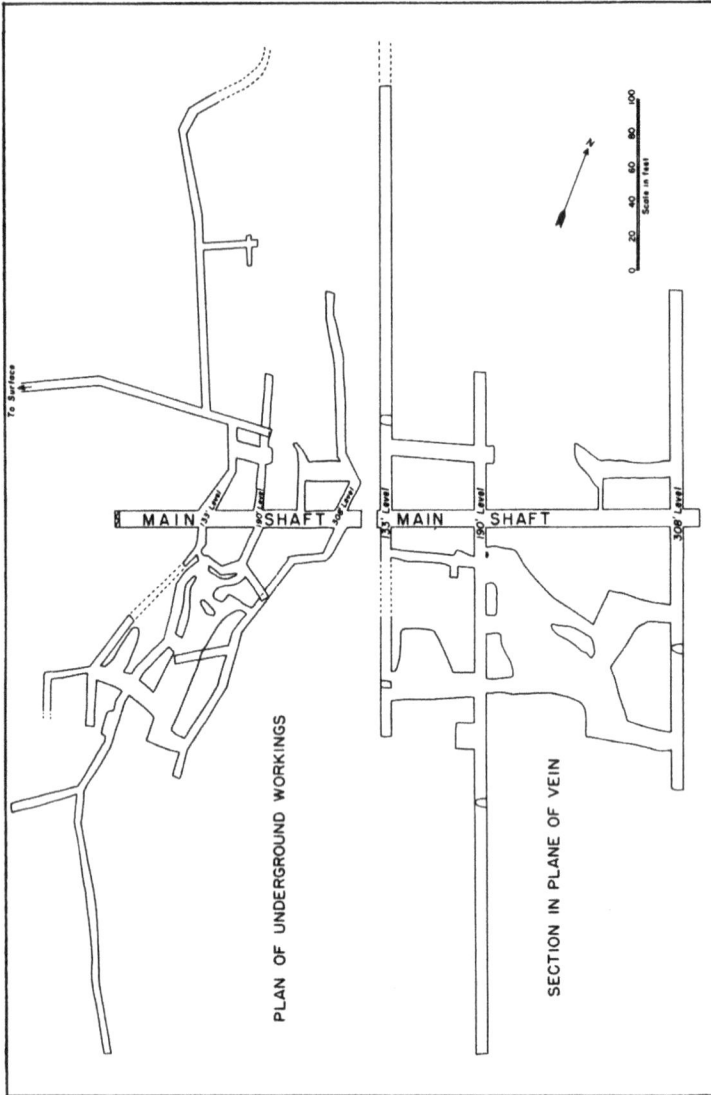

FIGURE 40. Longitudinal section of the workings of the Miners Hope mine in the Whitlock Creek district, one of the mines of the Spread Eagle group. The Spread Eagle group of mines, with a production of about $425,000, is noted for its pockets of high-grade ore. *Reproduced from a drawing by Bains and Tressider by courtesy of Thomas M. Bains III.*

In the Empire claim south of the Miners Hope, is a small vein called the Dolph, which is said to have produced $25,000. It lies about 100 feet east of the southern extension of the Miners Hope vein and dips 70° E. The Mohawk claim contains the northerly extension of the Whitlock vein, one of the famous early producers of the district."

The principal workings on the Spread Eagle vein are a 5- x 8-foot inclined shaft 345 feet deep, intersected by an adit at a depth of 125 feet (as measured along the dip of the vein). A drift driven at adit level both ways from the shaft is about 1000 feet long and has about 150 feet of auxiliary workings. There are about 1250 feet of drifts below the adit level. The upper part of the shaft is caved and most of the other workings were in need of cleaning out in August 1954.

On the Miners Hope vein the most recently used important working is a 2-compartment inclined shaft 328 feet deep as measured on the dip. It connects with an east-driven crosscut adit striking the vein at an inclined depth of 133 feet. These workings are about 800 feet from the north endline of the Miners Hope claim. The shaft has drift levels at 138, 190 and 308 feet (see accompanying plan and section). The 133-foot level extends more than 200 feet north and 200 feet south of the shaft but in 1939 was caved beyond these distances both north and south. Nearly all of the ore encountered above the 138-level has been mined. The 190-foot level extends north of the shaft 85 feet and 367 feet to the south. The 308 foot level consists of a 133-foot north drift and a 188-foot south drift. Neither the 190- nor 308-levels reached the region of stoping done from the 133-level north of the shaft, although the 308-level was within 100 feet of the stoped area in 1939 (Julihn and Horton, 1940, p. 150). According to these authors only about half of the ore between the 133- and 308-levels north of the shaft has been stoped. In 1937 an ore shoot 90 feet long was discovered on the 308-foot level and this was mined up to the 133-foot level. Its length decreased rapidly above the 190-foot level. Twenty-five hundred tons of ore was produced from this shoot which carried about $9.70 per ton, but only $7.20 was recovered by amalgamation. Seven-tenths of one percent of the ore from this shoot was auriferous pyrite sampling about 10 oz. of gold per ton of concentrate, on the average. Unfortunately the concentrates were lost in a storm.

Up to 1940 the best available information gathered by Julihn and Horton indicated the total production of the mines now included under the Spread Eagle group (including the Empire and Fanny mines) to be $425,000. No further production has been recorded since that time. Production reported to the U. S. Bureau of Mines for the period 1893 to 1955 aggregates approximately $100,200 in gold and silver. A small amount of copper was recovered in 1914 from the Spread Eagle mine. The ore milled before 1934 averaged a little over half an ounce of gold per ton and that mined in the 1930s ran approximately 0.26 oz. of gold per ton.

Sweetwater II Mine.   Location: Secs. 17, 20, T. 4 S., R. 19 E., M.D., on the west side of Sweetwater Creek 2 airline miles northwest of Jerseydale and half a mile northwest of the Early mine. Accessible from the paved Darrah-Jerseydale road via 3 miles of graded dirt road. Ownership: Hudson River Gold Mines, Ltd., c/o H. Vandel, P. O. Box 76, San Rafael, California, owns the Sweetwater, Crocker, Douty, Douty

Extension, Green, Sugar Pine 1 and 2, Gage, Gage Extension, Reddington, Berger and Wallace Amended lode claims and the N.I.R.A. placer claim.

The Sweetwater mine was discovered prior to 1884 but has been worked intermittently at widely spaced intervals. In 1903 the owners were E. J. and L. E. Hanchett with J. W. Jones as superintendent. A new 10-stamp mill was put into operation in 1905 but no production was recorded for the period 1903-1905. In 1914 the mine was reopened and developed by a man named McAllister. Two tons of selected ore mined by McAllister yielded 9 oz. of gold and 4 oz. of silver (U. S. Bureau of Mines records). In 1917 the mine was sold to E. C. Gamble and W. S. Stewart of Oakland who did some exploration but no sustained mining. By 1919 it had passed into the hands of the Dominican Mining Syndicate Inc., Coalinga, California (H. L. Schultz, Pres.) with M. Gamble acting as superintendent. The 10-stamp mill was overhauled (Castello, 1921, p. 118) but little or no production was realized from the mine. By 1928 the mine had been acquired by C. A. McCartney of Jerseydale who sold it that year to Frank Zuhosch and associates of San Francisco (Laizure, 1928, p. 118). Late in 1933 the property passed to the present owners. Between 1934 and 1937 approximately 300 tons of ore was milled which yielded 70 oz. of gold and 189 oz. of silver (U. S. Bureau of Mines records). Lee Rowland leased the mine in 1941 and 1942 and took out 32 oz. of gold and 6 oz. of silver from 9 tons of hand-picked ore. The mine was being operated under lease to Clyde Foster in the fall of 1954.

The Sweetwater vein averages close to 3 feet thick, strikes N. 55° W. and dips 13° to 27° near the surface. According to Castello (1921, p. 118) the vein dips 45° at depth. The thickness of the vein is very irregular and the ore is pockety. Vein matter is mainly glassy to milky quartz with native gold and pyrite. Concentrates produced in 1934-37 ran from $148 to $112 per ton, the tonnage of ore represented not being recorded (U. S. Bureau of Mines records). Wall rocks are coarse-grained, hornblende granodiorite or quartz diorite strongly weathered to depths as great as 50 feet. Much of the soil mantle above the vein is gold-bearing but the considerable clay content makes this ground difficult to handle. Some placer-mining has been done on the soil mantle. Pendants of slate probably belonging to the Paleozoic Calaveras group crop out northwest of the mine workings.

At present the mine is entered by a 600-foot crosscut adit which connects with a 350-foot southeast drift and an 800-foot northwest drift. There is an old inclined shaft 170 feet deep with a 123-foot level which was caved and not in use in August 1954. The drifts from the adit level had not been wholly cleaned out by the present operator in August 1954. There are also several extensive stopes which have not been recently explored. There is part of a 10-stamp milling installation on the property which might be rehabilitated.

The production record for the Sweetwater mine is fragmental and no reliable estimate can be made. Most of the ore appears to have been mined prior to 1903 but the ore mined since that time has been of considerably better than average grade.

Texas Hill (Carrie Todd, Texas) Mine. Location: Secs. 7, 8, T. 3 S., R. 18 E., M.D., high on the east wall of the North Fork of the Merced

River Canyon 2 miles southwest of Kinsley guard station of the U. S. Forest Service. Accessible from either Briceburg or Coulterville by way of the graded Kinsley road and 2½ miles of unimproved dirt road (Ponderosa Way). Ownership: G. Ross Frank and Mary E. Shell, c/o W. J. Beatty, Coulterville, California own the patented Carrie Todd claim of 20.66 acres and several adjoining unpatented claims.

The Texas Hill mine, first called the Texas mine and later the Carrie Todd was discovered about 1865 and produced intermittently in a small way up to the 1900s. In 1882 Francisco Bruschi bonded the property to Gilbert Douglass of Seth Cook and Company (Min. and Sci. Press, vol. 45, no. 26, p. 405, 1882) but no great amount of work was done at that time. It passed into the hands of Mrs. Hannah Douglass and remained under her ownership until the middle 1920s. A small production, presumably by leasers, was recorded in 1912-15. In 1919 the property was under bond to Milton Fraser of Kinsley and Mr. Sain of Coulterville who were in process of constructing a 10-stamp mill (Castello, 1921, p. 110). Estelle I. Fraser and George Frank of Coulterville acquired the mine in the middle 1920s. In 1927 the property was taken over for a short time by the Gentry Gulch Consolidated Mines Company of San Francisco, Thomas I. Box, president (Laizure, 1928, p. 82). In March 1934, the Texas Hill Mining Company of San Francisco, P. J. McLaughlin, president, leased the mine and operated it until about 1937. Leasers operating under ownership of G. D. Frank worked the mine from 1941 to 1943 since when the property has been inactive.

The known production of the mine is about $74,000 (McLean, W. D., personal communication, 1956). Ore produced in 1941-43 averaged about a third of an ounce of gold and a fifth of an ounce of silver per ton, plus a little copper and lead. Prior to that time the ore had run approximately $20.00 per ton based on the present price of gold.

The principal (Carrie Todd) vein at the Texas Hill mine averages 2-4 feet wide, strikes roughly east and dips 70-75° north. There are at least two minor parallel veins on the property developed by superficial workings. The veins cut dark slate and quartz-biotite hornfels probably belonging to the Paleozoic Calaveras group of metasediments. The Carrie Todd vein is developed by three drift adits each several hundred feet long. A winze connects the lower and middle tunnels (No. 2 and No. 3). The most recent work apparently was done from the No. 2 tunnel. Ore shoots are about 150 feet long and average about 2 feet thick.

Tyro (Rittershoffen) Mine. Location: Secs. 9, 10, T. 3 S., R. 16 E., M.D., 1¼ airline miles southwest of Coulterville. Accessible by 1½ miles of dirt road via the Malvina mine. Ownership: Louise A. Ward and Laurence Eaton, P. O. Box 223, Ross, Marin County, California own one patented claim of 20.15 acres.

The Tyro mine was probably discovered in the early 1850s at about the same time as the Malvina group. Although there were some workings prior to 1889, most of the development work and mining was done by the Tyro Mining and Milling Company under the management of Charles Sutherland and J. McLaughlin. The company was incorporated in 1889 (Eng. and Min. Jour., vol. 47, no. 20, p. 462, 1889). By 1895 the inclined shaft had reached the 600-level and drifts were being run on the 500-level (Eng. and Min. Jour., vol. 60, no. 6, p. 132). The shaft was deepened to the 700-foot level in 1896 and by the end of

1897 more than 1870 feet of drifts had been run. The company operated a 10-stamp mill. There were several changes of management during the late 1890s and the mine fell idle for several years. In 1902 and 1903 Thomas J. Brown operated the mine and rehabilitated the mill but no sustained mining was done (Eng. and Min. Jour., vol. 73, no. 15, p. 530; vol. 76, no. 15, p. 54). In 1909, McClure Gregory operated the mine for a short time but management again reverted to Joseph Buttgenback, a former president of the Tyro Mining and Milling Company, later that year. There has been little or no production from the mine since 1897. The workings were inaccessible and there was no equipment on the property in September 1954. The estimated production from 1893 to 1897 was $110,000 (Logan, 1935, p. 190).

Both the Tyro and the Malvina group of mines are on the west branch of the Mother Lode, the vein system at the Tyro mine being a continuation of the one that crosses the Malvina group of claims. It strikes about N. 20° W. on the Rittershoffen claim and dips from 60 to 70° northeast. Vein matter is largely milky quartz 2 to 7 feet wide. Ore shoots generally occur in a ribbon structure of quartz interleaved with thin sheets of wall rock (Fairbanks, 1890, p. 39). In addition to finely divided native gold there is considerable auriferous pyrite and chalcopyrite. Fairbanks also mentions the presence of covellite. The principal ore shoot was found between the 600- and 700-foot levels. This was at least 80 feet long and more than 100 feet deep. Most of the ore on the 700-foot level ran $20 per ton at the old price of gold and some ran as high as $500 per ton (Min. and Sci. Press, vol. 72, no. 13, p. 250; vol. 73, no. 11, p. 222).

The principal workings are an inclined shaft 700 feet deep with 6 levels and about 1870 feet of drifts (Logan, 1935, p. 190). Most of the ore has been stoped between the 600 and 700 levels in the main ore shoot and there has been considerable near-surface stoping, particularly north of the main shaft. In September 1954 this shaft was open to a depth of about 70 feet and there were several openings into the near-surface stopes. One of these may have been an old shaft. There has been no recent activity at the mine.

Virginia (Virginia-Belmont, White Gulch, Captain Aiken) Mine. Location: Secs. 13, 14, T. 3 S., R. 16 E., M. D., just west of Highway 49 about 4 miles south of Coulterville. Accessible from Highway 49 by a quarter of a mile of unimproved dirt road. Ownership: Sam E. Wells, 7533 Cecilia Ave., Downey, California, owns a patented property of 72 acres which includes the Virginia No. 2, Belmont, Claremont, Angus and Piedmont claims. Once included the Coe Gulch placer claim.

History: The Virginia mine was discovered by a group of miners in 1850 but was not extensively developed by them. Through the 1860s it was owned by Captain J. M. Aiken and was known as the Captain Aiken mine. Late in 1879 it was acquired by Senator Coe of New York. Coe and Lewis White sunk a shaft and drove a connecting adit during the winter of 1879-1880 and were reported to have taken out enough ore to pay for the mine (Min. and Sci. Press, vol. 39, no. 26, p. 413, 1879). During 35 days of mill operation during the spring of 1880 over $20,000 was recovered, and one pocket of specimen ore containing large flakes and stringers of gold yielded $10,000 (Min. and Sci. Press, vol. 40, no. 23, p. 357, 1880). The mine apparently remained under operation of

Coe until about 1884. Production under Coe is supposed to have been $100,000 (Min. and Sci. Press, vol. 75, no. 16, p. 366). The property fell idle between 1884 and 1888 but late in 1888 was bonded to Colonel Cook of Big Oak Flat, representing eastern interests. Cook unwatered and explored the mine but decided against reopening (Min. and Sci. Press, vol. 57, no. 18, p. 292, 1888). In 1894 the owner was listed as the Virginia Mining Company of Springfield, Ohio. Between 1894 and 1896 the property was described as in litigation (Storms, 1896, p. 224) but was leased, for a short time in 1895 to McKenzie, Hoxie and Donahoe. Late in 1895 the litigation apparently was settled and O'Toole and Arnold of San Diego emerged as owners—at a purchase price reported to be only $5,000 (Storms, 1896, p. 224; Min. and Sci. Press, vol. 75, no. 76, p. 366, 1897). In January 1896 a find of rich ore was reported. Later that year the mine was bonded to Paul Blackmar and the well known Clarence King, and was reactivated under the direction of J. J. Dolan in the summer of 1896 (Eng. and Min. Jour., vol. 62, no. 2, p. 36, 1899). This management was short lived and on January 1, 1897 the property was bonded by the California Exploration Company, Ltd. from O'Toole and placed under the superintendency of R. A. Parker. This company completed a 10-stamp mill and made considerable production before ceasing operations in the summer of 1898 because of a shortage of water. At that time the shaft was 470 feet deep and ore was milling $5.50 to $10 per ton (various issues of Min. and Sci. press).

In 1904 O'Toole operated the property under the name Virginia Gold Mining Company and produced 1000 tons of ore which returned about 0.74 oz. of gold per ton (U. S. Bureau of Mines records). In October 1907 the property was transferred to Garnett and Priest of Kansas City and T. W. McLean of Boise City, Idaho, McLean acting as mine manager. About 2000 tons of low grade ore were produced by the management but the operation again ended in litigation. S. E. Lewis of the Crown Lead mine leased the property for a time in 1912 but no production was recorded. In the spring of 1913 the mine was acquired by the White Gulch Mining Company of Oakland, a subsidiary of the Procter and Gamble Company (Min. and Sci. Press, vol. 95, no. 16, p. 826, 1913; Logan, 1935, p. 190). It was cleaned out and placed in operation under the direction of C. C. Powning of Oakland. Extensive development and production followed. By 1918 this company had deepened the shaft to more than 700 feet and water became a serious problem for the first time. Mining, however, was continued to a depth of 1000 feet and a large tonnage of ore was milled which averaged slightly more than 0.4 of an oz. of gold per ton. The White Gulch Mining Company ceased operations in 1920 and the Virginia Belmont Mining Company was organized in Nevada to continue operations on the deeper levels of the mine. W. V. Wilson was mine superintendent and B. C. Austin the consulting engineer for this company. This enterprise lasted until May, 1923, the shaft having been deepened to 1050 feet. Early in 1927 an unsuccessful attempt to reopen the mine was made by a group known as the Virginia Lode Mines Company, but in November of the same year the Nevada Hills Mining Company, Reorganized, of Reno, Nevada (H. G. Humphrey, President), took over the property and did considerable rehabilitation and development work under the direction of F. B. Hyder prior to the crash of 1929 (Min. and Sci. Press, vol. 127, no. 10, p. 413, 1929). The workings were deepened to 1300 feet during this period. In February 1932 the mine was reported sold to Fred Morris of San Francisco but the only work done on the mine prior to 1934 was by Grant Ewing. About 1035 tons of low grade ore was produced by leasers between 1934 and 1936. During the period 1952-53 Sam E. Wells and B. A. Miller drove a 450-foot exploratory tunnel on an old prospect southeast of the principal mine workings which is believed to have penetrated to a point only 80 feet from the vein. A compressor, 10-stamp mill and several mine buildings remain on the property (August 1954).

According to Logan (1935, p. 190) the total production of the Virginia mine between 1898 and 1931 was more than $660,000. Inasmuch as production under Coe (1979-1884) was about $100,000 and production since 1931 has been approximately $64,000 in gold and silver (U. S. Bureau of Mines records) it follows that the total production of the mine is at least $824,000 providing Logan's figures are correct. U. S. Bureau of Mines figures for the period 1898 to 1931 show a production of slightly less than half Logan's figure but some of the production may not have been reported.

The vein system at the Virginia mine is a segment of the east branch of the Mother Lode system which in this vicinity strikes N. 50-55° W. and dips 60° northeast, upon the average. Vein matter, which in places reaches a thickness of 150 feet (Fairbanks, 1890, p. 41), is predominantly ankerite-quartz mariposite rock, more or less banded and altered. The quartz sheets, in the vein near which the ore commonly is concentrated range from 8 inches to 4 feet wide. In addition to native gold, ore minerals include pyrite, chalcopyrite and galena. For the most part, the vein system follows the thrust-fault contact between serpentine, which forms the southwest or footwall side, and pyroxene andesite greenstone or mica schist on the northeast or hanging wall side. The serpentine is believed to be of Upper Jurassic age, the greenstone is either Jurassic or Paleozoic and the mica schist is part of the Paleozoic Calaveras group of metasediments. The principal ore shoot found northwest of the main shaft was mined down to 1200 feet and could not be located on the 1300-level. It varied in length from 180 to 300 feet above the 700-level, narrowed to 20 feet or less between the 700- and 1200-levels and apparently had a nearly vertical axis. Ore shoot width ranged from 8 inches to nearly 16 feet and averaged 3-4 feet. A second smaller and less persistent ore shoot was mined southeast of the main shaft.

Some bonanza ore was mined from the upper 80 feet of the vein, as previously mentioned. Much of the ore in the vicinity of the 400-level ran $15 per ton and $10 in the vicinity of the 300-level. On the 900-level much of the ore mined in 1922 averaged $15 per ton. In 1918 ore was reported to have averaged $17.50 per ton. In 1935, Logan stated that most of the production was from ore that ran more than $10 per ton. U. S. Bureau of Mines records indicate that the ore milled in 1904 averaged just under 0.74 oz. of gold per ton; in 1909 about 0.14 oz. of gold per ton; in 1914 approximately 0.45 oz. of gold per ton; and in the period 1919-1934, about 0.35 oz. of gold and 0.13 oz. of silver per ton. Several hundred pounds of copper was also recovered, as a by-product, in the early 1930s.

The principal working at the Virginia mine has been a 1300-foot inclined shaft, with levels at 100-foot intervals, and a connecting adit haulageway. The shaft is described as serving more than 10,000 feet of drifts (Laizure, 1928, p. 120). In addition to the shaft and adjoining workings there are three old adits south of but in the general vicinity of the main shaft, and a fourth crosscut adit (driven in 1952-53) 450 feet long which connects with an older working. The portal of last adit is 1700 feet southeast of the main haulageway adit portal, the mine buildings lying between these two localities. Most of the working could be entered in August, 1954 but undoubtedly are in need of rehabilitation.

Washington (Jenny Lind, Red Cloud) Mine.   Location: Secs. 4, 5, T. 5 S., R. 16 E., M.D., a quarter of a mile north of the paved Hornitos-Bear Valley road and 2 miles northeast of Hornitos. Ownership: Pacific Mining Company, Crocker Building, 620 Market St., San Francisco 4, California, owns the patented Washington claim, containing 13.77 acres, and adjoining patented agricultural land aggregating 509 acres.

The Washington mine was discovered in 1850 and prior to 1900 was probably the most productive mine among the many in the Hornitos district. As early as 1851 it was equipped with a 6-stamp mill which, for a time, produced 1000 oz. of gold per day! Except for a short period between 1864 and 1868 operation was continuous from 1850 to about 1882. In 1859 a 20-stamp mill was built on the Merced River to serve the Washington mine. Much of the early work was done by the Washington Mining Company in which George Webber and Moses L. Rodgers were the principal partners. A 10-stamp mill was erected near the Washington shaft in 1867 by this management. The mine was reported to be 1400 feet deep in 1881 and to have been deepened to 1600 feet during that year (several entries in vols. 42 and 43 of the Mining and Scientific Press), but was found to be a little less than 1500 feet deep when reopened by the Lind Mining Company in 1941. Late in 1881 the property was reported purchased by Marcus and J. W. Hulings of Pennsylvania, along with the Quartz Mountain, Number Nine and numerous other mines in the Hornitos district. The Huling venture proved unsuccessful and the Washington mine reverted to M. L. Rodgers. Considerable development work was done by Rodgers in 1884 but the stock company under which Rodgers operated ceased to function in 1885. In 1896 it was announced that Rodgers had sold the mine to S. W. Parker and associates of New York, the Hornitos Gold Mining Company, and that further sale to an English company was being negotiated. Nothing came of this transaction and the property remained idle until reopened in 1903 by Harmon and Stevens (Jenny Lind Mining Company), together with the Quartz Mountain mine. Operations were directed by W. F. Stevens. This operation was also short lived with little or no production realized. Several further attempts were made to promote the mine, notably in 1900, 1910 and 1917. In 1922 the owner was still listed as the Washington Mining Company. The Cora Belle Mining Company took out a little high-grade ore in 1928 (U. S. Bureau of Mines records). The Mariposa-Washington Mining Company of Hornitos built a new mill in 1939 and milled about 4000 tons of dump material between 1939 and 1940. By 1941 the property had been acquired by the Lind Mining Company, a subsidiary of the Pacific Mining Company, which took out a large tonnage of ore of moderate grade between 1941 and 1942. The company reopened and

FIGURE 41. Installations at the Washington or Jenny Lind mine in the Hornitos district. The attractive rolling pasture land in the background is typical of the landscape in western Mariposa County. The Washington mine, with a production of $2,377,000, is second in productivity among the mines of the Hornitos district. *Photo by Mary H. Rice.*

unwatered the caved Jenny Lind shaft, re-equipped the mine and did extensive exploration and development work on the 700-, 800-, 1000-, 1100- and 1500-levels. The Jenny Lind shaft was deepened 36 feet to a point slightly below the 1500-level. The 1000-level was cleaned out to the Washington shaft which was found to be open between the 900- and 1000-levels. A total of 7923 feet of workings were cleaned out or newly driven between 1941 and 1942.

Between 1943 and 1945 the mill and flotation plant were leased by Red Cloud Mines Company, a subsidiary of Hecla Mining Company of Wallace, Idaho, to process complex zinc ore from the nearby Blue Moon mine. Both mine and mill have been idle since 1945, reopening awaiting more favorable economic conditions.

Prior to 1900 the Washington mine had an estimated production of $2,247,000 and a recorded production of $1,099,000 (Bowen and Crippen, 1948, p. 38). Between 1939 and 1942 a total of 17,356 tons of ore was milled which yielded 3131 oz. of gold, 2436 oz. of silver, 1060 lbs. of copper and 111 lbs. of lead for an average of about $6.35 per ton (U. S. Bureau of Mines records). The total estimated production to 1954 is $2,377,000. Bonanza ore was found in the upper parts of the vein but most of the recent production has been from ore running between $6.50 and $7 per ton. According to Raymond (1875, p. 50), ore averaged about $25 per ton, at a price of only $13 an ounce, in the middle 1870s.

The principal vein at the Washington mine strikes N. 40° W., dips 70-75° S.W. and averages 6 to 8 feet wide. Vein matter is chiefly ankerite-quartz-mariposite rock in which large sheets and lenses of milky to glassy quartz several feet thick are locally developed. The character of the vein closely resembles typical Mother Lode veins seen 7 miles farther east in Mariposa County. Ore minerals are chiefly pyrite and chalcopyrite with native gold. In addition to the quartz vein-ore there are, in places, masses of wall rock impregnated with auriferous pyrite (gray ore). This material commonly runs 0.09 oz. of gold per ton.

Wall rocks are mainly micaceous and amphibole-bearing schists and dark quartz-biotite hornfelses of Upper Jurassic age. According to Storms (1896, p. 224) the vein is accompanied by a light-colored felsitic dike which adjoins the vein on the footwall side in the southeast workings and on the hanging wall side in the northwest workings. Small, irregular intrusions of hornblendic granitic rock penetrate to the surface west of the present mine buildings. Much of the metamorphic rock seen in the vicinity of the shaft has been derived from platy, tuffaceous metasediments.

The most recently used working entrance to the Washington mine is the Jenny Lind shaft inclined at an angle of 68° and reaching a depth of 1500 feet. It is surmounted by a lofty steel headframe and the mill buildings are closely adjacent. The Jenny Lind shaft originally had levels of 500, 600, 700, 1000, 1100, 1200, 1300 and 1400 feet but the ground has been stoped and the levels largely destroyed between depths of 500 and 700 feet. Approximately 670 feet northwest of the Jenny Lind shaft is the Washington shaft 1000 feet deep and inclined at about the same angle as the Jenny Lind. This is open from the surface to a depth of 1500 feet and is known to be open between 900- and 1000-levels. Workings connecting with the Washington and Jenny Lind shafts total nearly 2 miles.

About 200 feet north of the Washington shaft is the Franklin shaft, 200 feet deep, and 500 feet still farther north is a fourth old shaft 100-150 feet deep. There is a line of shallow workings between the two old northwesterly shafts.

There is (August 1954) crushing, hoisting and milling equipment on the property and there are several good mine buildings. A graded dirt road connects with the paved Hornitos-Bear Valley road.

White Porphyry Group. Location: Sec. 34, T. 3 S., R. 17 E., M.D., in the Cat Town district 3 miles southeast of Buckhorn Peak or 7½ airline miles southeast of Coulterville. Accessible from Highway 49 by way of the unimproved Buckhorn Peak-Solomon Gulch-Black Bart road via the Schilling Ranch and Rybergs. Ownership: several of the most northerly workings were being operated in September 1955 by F. E. Ryberg, Coulterville. Ownership of the other claims could not be determined.

The mile-long line of open stopes of the White-Porphyry group is a conspicuous landmark southeast of Schilling Ranch and Buckhorn Peak. It diagonally crosses section 34 on a gullied plateau. So far as the authors were able to discover, nothing has ever been written about these workings—extensive as they are. They undoubtedly date back to the 1870s or before in the heyday of the Black Bart group mines lying half a mile to the northeast.

Workings are arranged in two parallel lines along either wall of a white albitite dike 30 to 50 feet thick traceable at the surface for over

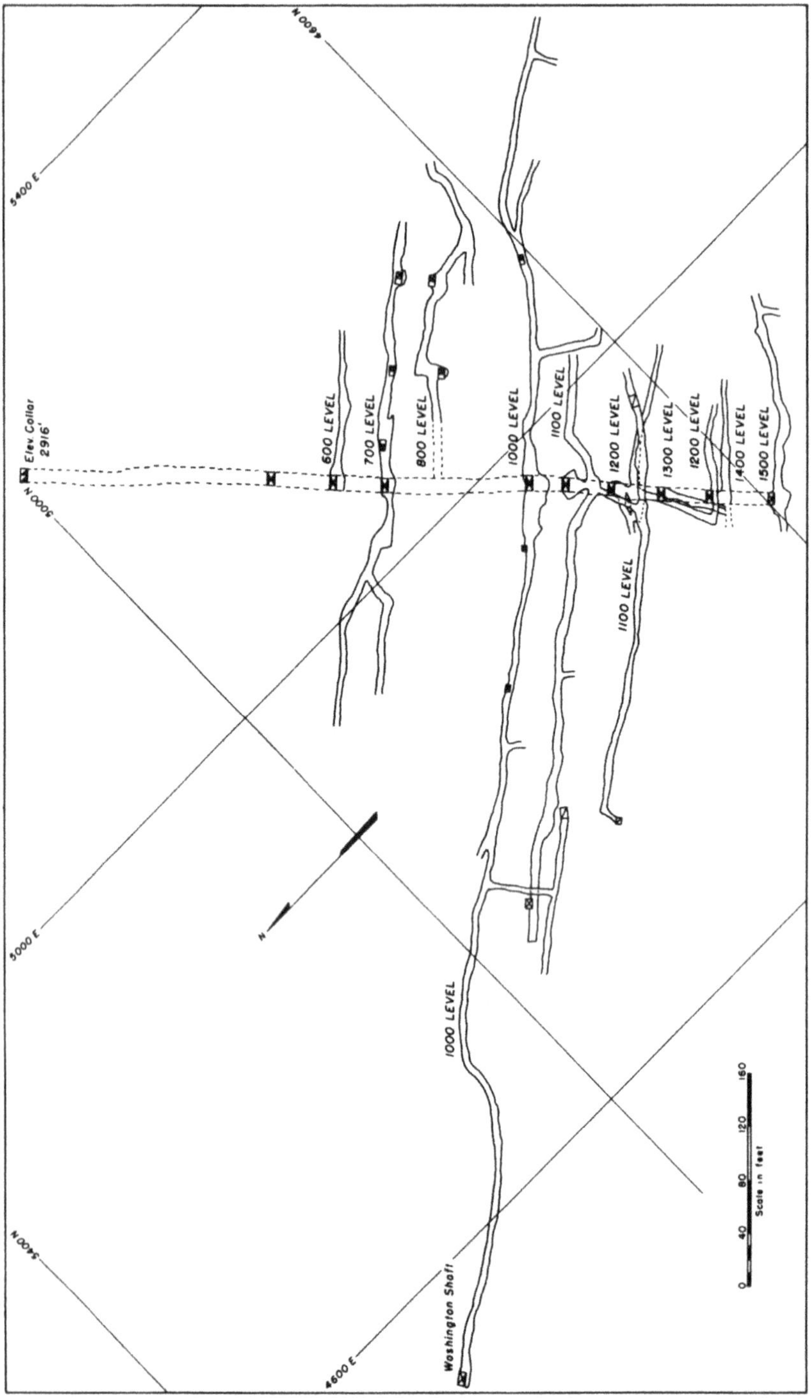

FIGURE 42. Sectional drawing through the workings of the Washington mine. *Reproduced through courtesy of the Pacific Mining Company.*

a mile. This dike strikes N. 40-45° W. and dips 75° northeast. Wall rocks are slate, schist and metachert of the Paleozoic Calaveras group. Gold tends to occur in rust colored pockets in the albite adjacent to quartz stringers and to the enclosing slaty wall rocks, or in the quartz stringers which cut both the albite and the wall rocks close to the contact. Ore apparently was confined to a narrow contact zone nowhere over 6 or 7 feet wide and commonly only 3 feet wide. The average depth of the open stopes is now only 30 to 40 feet but some caved workings probably extended down over 100 feet. Horses of schist, slate and metachert locally are caught in the albite and some appear to have been faborable places for concentration of ore.

No data are available on the production from this group of workings but from the length and persistence of the stopes and the known high grade of ore pockets in this district the yield must have been large. The relative shallowness of the workings indicates that the best ore was in the oxidized zone and may have been of considerably lower grade at depth.

Whitlock Group (Consolidated Whitlock and Alabama). Location: Sec. 32, T. 4 S., R. 18 E., M.D., east of the Whitlock Creek road 3½ airline miles northwest of Mariposa and 4½ miles by graded dirt road from Mt. Bullion on Highway 49. Ownership: Dr. Frank E. Gallison, P. O. Box 491, Ventura, California, and E. J. Freethy, 1432 Kearney St., El Cerrito, California, own 3 patented claims, the Westward, Alabama and Whitlock, aggregating 47.68 acres.

The Whitlock group of mines was discovered prior to 1870, probably in the 1850's or 1860's as an aftermath of placer-mining on Whitlock Creek. In 1871 the property was transferred under option for $20,000, to Stephens and Lamber (Min. and Sci. Press, vol. 22, no. 9, 1871). Some production was made by this partnership that year and there was intermittent activity at the mine by others, probably leasers, through the 1870s. After a period of inactivity of several years' duration in the early 1880s, the property was reopened or at least re-explored by Captain Diltz of the Diltz mine. At that time the ore was reported to average about $20 per ton and the vein to be 5 feet thick (Min. and Sci. Press, vol. 46, no. 21, p. 256, 1883). During the early 1890s the mine was owned and operated by the Ward Brothers. About 1895 the mine was sold by Wards to the Sierra Buttes Gold Mining Company of London, England, and most of the development work and sustained mining at the property was done in the ensuing 5 years (Storms, 1896, p. 24; Min. and Sci. Press, several issues 1892-1895). The Sierra Buttes company erected a 20-stamp mill, replacing an earlier, smaller mill, and maintained a steady production from 1895 to 1900. About $482,000 was recovered during this period. Mining ceased with "considerable ore still in sight" because of the high operating costs (Castello, 1921, p. 142). Jacob Teets was listed as the owner in 1904 (Williamson, p. 13) but by 1914 the property had passed into the Gallison family and Lizzie Sain. A small production by Gallison and Sain was recorded between 1940 and 1942 when 942 tons of ore milled yielded only 133 oz. of gold and 33 oz. of silver. Little or no work has been done on the property since 1942. The total production of the mine is estimated to be about $500,000.

FIGURE 43. Panorama across the rugged country in the Cat Town district, south of Buckhorn Peak and east of Highway 49, observer facing slightly east of south. Slightly to the right of the center of the photo 2 parallel lines of dumps and open stopes may be seen which are on the White Porphyry group of claims. Gold occurs there in pockets along either wall of a 40-foot-wide albitite dike. The pockets are in places found in quartz stringers cutting the albitite or the adjacent wall rocks or else are in rusty spots in the albitite dike itself. *Photo by Mary H. Rice.*

There are three veins on the Whitlock group of claims, the Whitlock, Alabama and Westward. The Whitlock vein strikes north to N. 10° E., dips 75° east to nearly vertical, and ranges from 4 to 10 feet wide. Vein matter is chiefly milky quartz. Ore minerals are chiefly pyrite and minor chalcopyrite with native gold, but locally argentiferous galena is prominent—particularly in the ore mined in 1900. From the stopes described in old accounts there apparently were two main ore-shoots, one about 150 feet long and one about 240 feet long. Wall rocks are massive pyroxene andesite greenstone of unknown age. The Alabama and Westward veins are roughly parallel with the Whitlock vein but diverge slightly toward it to the north. They are less well defined and apparently contained less ore than the Whitlock vein. In any event the Westward vein was very superficially developed and workings on the Alabama vein are much less extensive than those on the Whitlock vein.

Principal workings on the Whitlock vein are a 900-foot, steeply inclined shaft, with 4 levels, and a 600-foot crosscut adit with connecting drifts. Drifts connecting with the main shaft aggregate about 3000 feet (Castello, 1921, p. 142) and those connecting with the crosscut total several hundred feet. There are several other crosscut adits of unknown length. None of the workings have been recently used and must be cleaned out (August, 1954). The Alabama vein has been developed by an inclined shaft and connecting drifts of unknown extent. These workings are also inaccessible.

Williams Brothers (Gibbs) Mine. Location: Secs. 31, 32, T. 3 S., R. 20 E., M.D., on the southeast slope of Brown Peak 400 feet below the divide at Canty Meadow. It is 3 miles south and 1 mile west of El Portal and is accessible by about 9 miles of steep mountain road via Cold Canyon.

Ownership: Earl, Ray, Fred and G. H. Williams and Ralph Zelers, address El Portal, California or Rt. 1, Box 1061 E., Modesto, California, hold four unpatented claims the Williams Brothers and the 2, 4 and 5 extensions, totaling about 80 acres.

The Williams Brothers mine, known in the early days as the Gibbs mine, was probably first worked in the 1860s about the time of discovery of the nearby Hite mine and the adjacent Mexican or El Carmen mine. It was developed by 3 drift adits which are now being rehabilitated. Driving of these workings was done partly in 1875 (Min. and Sci. Press, vol. 30, p. 364). Because of its relative inaccessibility little or no work had been done for several decades until the present owners acquired the property in the late 1940s. In 1949 a 5-ton Gibson mill was packed to the mine by horseback for mill testing and some high-grade ore found on the dump was processed at this time. In 1950 a 9-mile road into the property from Highway 140 was started by the Williams partnership. Pacific Placers Engineers Company entered into a joint contract with the Williams group for construction of the road to the Williams Brothers and Kaderitas mines. The road was completed to the Williams Brothers mine in December 1952. A 5-stamp mill was completed in 1953 and for some time ran on ore taken from old dumps. The dump material yielded about $10 per ton. In 1954 work was started on the lowermost of three old adits. Some ore mined that year ran $27.22 per ton in gold. Development was still progressing in September, 1955.

The principal vein on the property strikes about N. 40° W. and dips northeast at angles varying from 25° to nearly vertical. Vein matter is chiefly milky to glassy quartz containing wall-rock fragments. Vein matter is 6 to 20 feet thick averaging perhaps 6 or 8 feet. Gold occurs in small shoots and pockets in the quartz commonly associated with fractured, friable pyrite. Flecks of other sulfides such as sphalerite, galena and arsenopyrite may be seen in some of the ore but are not common. The vein matter characteristically is full of vugs or small cavities lined with masses of spongy to friable pyrite. The trend of the ore shoots had not been determined at the time of the author's visit to the mine in September, 1954.

Workings consist of 3 adits 150, 75 and 200 feet long, respectively, upper to lower. These were in process of being cleaned out in September, 1954 and some raising and stoping was being done above the lower adit level. There apparently had been considerable stoping from the upper levels in early days.

The 5-stamp mill has a capacity of 1 ton per hour. Ore is crushed in an 8-inch jaw crusher and the 5 stamps to pass a 45-mesh screen. It is

FIGURE 44. Main adit at the Williams Brothers mine on the northeast slope of Browns Peak in the Hite Cove-El Portal district. Ore is dropped down the incline to road level and trucked to the mill.

FIGURE 45. Mill buildings at the Williams Brothers mine. Known in the early days as the Gibbs mine, this old working was reactivated in 1949 and had operated most of the time up to 1956, producing small tonnages of good-grade ore during development work.

then passed over amalgamating plates and a concentrating table. Sulfide concentrates will be saved for future smelter shipment.

Yellowstone (Yellowstone-Sirocco, Sirocco) Mine. Location: Sec. 19, T. 4 S., R. 17 E., M.D., just north of the paved Bear Valley-Hornitos road 1½ miles west of Bear Valley. Ownership: Not determined. Last operator was Glenn Coburn of Mariposa (1941).

The Yellowstone mine was discovered sometime prior to 1893. During that year it was operated by Burt and Bach of Bear Valley (Storms, 1894, p. 175) under the name Sirocco mine and had extensive underground workings and steam hoisting equipment. In 1896 the operators were listed as the Yellowstone Mining Company of San Francisco, C. H. Burt of Bear Valley acting as superintendent (Storms, 1896, p. 224). The mine was then called the Yellowstone-Sirocco mine. About 1897 the property was taken over under bond by a Chicago group with G. Seymour acting as superintendent (Min. and Sci. Press, vol. 74, no. 8, p. 154, 1897). At that time the main shaft was 420 feet deep and a second shaft at the west end of the property was down 70 feet. Good ore was reported to be coming from both shafts (Eng. and Min. Jour., vol. 63, no. 10, 240, 1897). About $22,000 in gold was recovered between 1896 and 1897 (U.S. Bureau of Mines records). The mine apparently lay idle until about 1935 but about 650 tons of low-grade ore was produced between 1935 and 1941, most of this by Glenn Coburn of Mariposa (U. S. Bureau of Mines records). There has been no recent activity at the property except for utilization of the extensive dumps for road base and fill.

The Yellowstone mine workings are chiefly on a narrow shear zone striking N. 40-45° E. and dipping 75° northwest. Quartz occurs in pods and larger lenticular masses in sheared greenstone (pyroxene andesite). Storms (1896, p. 224) describe calcitic masses in the vein but none of these were observed at the surface. Gold occurs in quartz and in adjacent schistose wall rock, according to Storms (1896, p. 175). No ore minerals were found on the dump by the authors.

Workings originally consisted of a 420-foot and a 70-foot inclined shaft and at least 2 crosscut adits. One of these trended about N. 45° W. and the other nearly due north. None of the workings are accessible owing to caving. Judging from the size of the dumps the underground workings must be extensive. Ore apparently was very pockety.

Placer Mines

Placer deposits in Mariposa County were worked as early as 1848, and possibly earlier, by Californians of Spanish descent. These operations apparently began on Agua Fria Creek, rapidly spread to the Hornitos district and before 1870 had spread to almost every creek and gulch west of a cross-county line joining Wawona and El Portal. There are few if any gold-quartz veins east of this line. During the first two decades mining was carried on almost exclusively by hand methods. In the late 1860s and during the 1870s some hydraulicking and ground sluicing were done, notably along the Merced River west of Bagby and at the Schroeder placer ground west of Sherlock Creek. Very little drift mining for buried placers has been done in Mariposa County as such deposits are rare. Some drift mining was carried on

FIGURE 46. A typical scene on the Weston placer ground on Sherlock Creek half a mile southeast of the Diltz gold mine. In places the productive part of the old Pleistocene (?) channel gravels were followed underground in tunnels. In others the entire pile of gravel was washed for gold. *Photo by Mary H. Rice.*

FIGURE 47. Dragline dredge tailings along Burns Creek south of Hornitos. As gravels may be handled very cheaply by this method of mining the gravels in this vicinity, which ranged from 12 to 20 cents per cubic yard in gold, could be profitably handled during the depression years of the 1930's. *Photo by Mary H. Rice.*

at the Weston bench-gravel ground southeast of the Diltz mine on Sherlock Creek and at the Schroeder property previously mentioned.

Only 3 patches of Tertiary gravels have been found in the county, one about 4 miles northwest of Forty Nine Gap in the Blanchard district of northwestern Mariposa County, a second (the largest) on the county line south of Jawbone Ridge, Tuolumne County, and the third high on the ridge lying between Moore and Jordan Creeks 4 to 5 miles northwest of Bower Cave. These produced considerable gold in early days but have long since been worked out.

Bench gravels of Quaternary, possibly Pleistocene age, have been worked on the old Weston property east of Sherlock Creek and southeast of the Diltz mine and there probably were others of this type in the county. Unfortunately, these deposits have been so thoroughly disturbed that their original character and distribution cannot be easily ascertained.

Much of the placering, hydraulicking and ground sluicing done in the county has been on the weathered, uppermost parts of the quartz veins themselves or the soil mantle covering them. Deposits of this sort commonly were very rich in coarse gold and more than a few important underground mines were financed by the easily recovered gold found near the vein outcrop. Notable amounts of gold were recovered from such material at the Diltz, Schroeder, and Mt. Gaines mines and at many of the mines in the Colorado district such as the Landrum, Artru and Champion.

All of the other placer deposits in the county are Quaternary stream-bed deposits or shallow bench gravels lying only a few feet above the present high water levels of the contemporary streams. Although there is a steady dribble of placer gold coming from small patches of ground overlooked in the early days or from material reconcentrated during times of high water by present streams, most of the production during the last 2 decades has been from dredges processing material too low in grade to have attracted the early-day placer miner. These only became profitable to work when heavy equipment was designed to handle huge yardages at very low cost. Dredges operated within the past 25 years have, for the most part, processed ground that yielded from 13 to 25 cents per cubic yard. A little ground was worked that ran as low as 7 cents per cubic yard. Most of the recent dredging has been done along Mariposa, Burns, Eldorado, Cotton and Temperance Creeks.

Julihn and Horton (1940, p. 159) have estimated that the total placer gold production up to 1870 probably did not exceed $10,000,000. If this figure is accepted then the total production of placer gold in Mariposa County to 1955 is about $12,000,000.

Inasmuch as most of the placer ground has been pretty well worked out and as a majority of the properties are no longer classed as mines, none are discussed here in detail. Operations within the past 25 years have been described by Julihn and Horton (1940, pp. 158-162) and Averill (1946, pp. 261-262). Some data on most of the placer mines will be found in the accompanying tabulated list.

Lead

Lead has been produced in substantial quantities in Mariposa County from complex zinc ore and as a by-product in the milling of gold ore.

The total recorded production of lead up to 1954 has been about 709,508 pounds. The principal producer has been the Blue Moon zinc mine with a production of 533,753 pounds. Gold mines which have yielded by-product lead in substantial amounts are the Bondurant (85,000 pounds), Malvina group (about 25,000 pounds), Pine Tree and Josephine, Mount Gaines and Washington or Jenny Lind. The complex zinc ores have been worked almost entirely in wartime. Production of by-product lead follows fairly closely the production of gold.

The principal primary lead ore-mineral in both the zinc and gold ores is galena, although minor amounts of lead carbonate and sulfate probably are present in oxidized parts of the veins. The complex ores of the Blue Moon mine are confined to shear zones in hydrothermally altered greenstone, where they form banded deposits parallel to the schistosity. The greenstone is commonly converted to sericite schist adjacent to the ore bodies. Some galena is generally present in all types of gold-bearing veins.

The various lead-producing mines, which number about 20, are discussed elsewhere under the headings zinc and gold. With the exception of those listed above, these have produced less that 3000 pounds of lead each.

Manganese

Manganese has been mined in a small way in Mariposa County during wartime periods of high price and government subsidy. Only a few hundreds of tons of ore have been marketed. Most of the known occurrences fall in two districts, the Granite Springs-Jasper Point district 5 airline miles southwest of Coulterville, and the Sweetwater district 2 miles northwest of Jerseydale and 10 miles northeast of Mariposa.

Manganese occurs in sedimentary deposits associated with chert or metamorphosed sedimentary deposits, chiefly quartzite derived from chert. Volcanic rocks are commonly associated with the cherts, and the manganese is generally believed to have originated in mineral water introduced onto the sea floor by volcanic activity. None of the deposits are in the form of hydrothermally deposited veins. The manganiferous cherts southwest of Coulterville are found in a northwest-trending belt in the Hunter Valley Chert formation of the Upper Jurassic Amador group (Taliaferro, 1943, p. 283). Associated with the chert, which occurs in lenticular masses up to several hundred feet thick and 4 miles long, are tuffaceous sediments and submarine lava flows. Manganiferous parts of the chert masses generally are only a few feet wide and less than a hundred feet long, although a manganiferous zone over 750 feet long has been found at the Caldwell mine. In the Sweetwater district manganese ore occurs in small lenticular and irregular masses within quartzite. The quartzite is a laminated rock derived from chert and is associated with schist and phyllite derived from shale. These rocks belong to the Paleozoic Calaveras group and occur in the Jerseydale district principally as pendants in granitic rocks.

Most of the manganese occurrences have been explored very superficially and no large reserves have been proved thus far. All of the known deposits in California were studied in considerable detail by Federal agencies during the wartime periods of World Wars I and II. The results of these studies have been published as Bulletins 125 and

152 of the State Division of Mines. The following mine description and the tabulated list data have been abstracted from pages 106-112 of Bulletin 152 written by Ivan F. Wilson and N. L. Taliaferro. The deposits have been inactive for more than 10 years and were not visited by the authors during this investigation.

Caldwell (Daly) Mine. Location: NE¼ sec. 14, T. 4 S., R. 15 E., M. D., on the Caldwell Ranch 1.5 miles south of Granite Springs School and 7 miles west of Coulterville via State Highway 132. Ownership: The mine is on patented agricultural land, ownership of which was not determined.

The Caldwell mine is the most extensively developed manganese property in the county and the only one having recorded production. First worked during World War I as the Daly mine, it produced between 100 and 200 tons of ore. About 1937 the mine was reactivated and produced some ore between 1937 and 1943. The total production has been about 265 tons of ore averaging nearly 45 percent manganese.

The principal (north) group of workings lies along a single bed or connected series of lenses 2 to 5 feet thick enclosed in chert of the Upper Jurassic Hunter Valley formation. The ore is associated with thick-bedded red chert, thin-bedded buff-to-gray chert and thick-bedded white chert. These cherts are enclosed in gray and green metatuff and andesite greenstone. The manganiferous zone is traceable for more than 1000 feet in the northern group of workings. Ore minerals are chiefly black psilomelane and pyrolusite but there is considerable carbonate ore consisting of brown, gray and pink manganocalcite. Manganocalcite ore runs 20-30 percent manganese, which in past years has been too low grade to mine. Over much of its length the ore zone strikes N. 5° E. and dips 50-80° east, but is markedly sinuous over part of its length and the dips locally change to steeply west.

There is an inclined shaft 57 feet deep along a 45° incline with a 26-foot drift at the bottom. A second shaft 10 feet deep is 500 feet south of the main shaft and a third shaft 15 feet deep lies 800 feet south of the main shaft. There are also about 30 open cuts scattered along the ore-zone for a distance of more than 1000 feet.

A quarter of a mile south of the principal or northern group of workings and apparently on the same chert belt is a second group of shallow workings covering a distance of 170 feet. There the chert is thin-bedded and has thin interlaminations of shale. These beds strike N. 45° W. and dip 55° N.E. A few inches to 2 feet of soft, shaly, black oxide ore is exposed at the southern end of the ore-zone and the dump contained about 13 tons of ore in 1942.

The 1918 production of the mine (about 100 tons) averaged 45 percent manganese, 6 percent iron and 13 percent silica. Nine tons of ore shipped in 1942 ran 49.3 percent manganese, 4.1 percent aluminum, 2.1 percent iron, 0.038 percent phosphorus and 11.3 percent silica. Ore running at least 20 percent manganese remained in the mine in 1942.

Nickel

Nickel occurs in Mariposa County in minor amounts in Mother Lode gold-quartz veins, notably at the Pine Tree and Josephine mine just south of Bagby. It is found in the primary minerals niccolite (nickel arsenide) and millerite (nickel sulfide) associated with pyrite, arsenopyrite, chalcopyrite, sphalerite, galena and native gold. The nickel content of the sulfide concentrates from the Pine Tree and Josephine mine ranged from 0.35 to 1.35 percent; during 1936 the concentrates averaged 0.77 percent (Julihn and Horton, 1940, p. 110). No attempt has been made thus far to recover the nickel.

Nickel is also found in trace amounts (0.2-0.8%) in almost all of the serpentines of the foothill gold belt and may possibly be found within the county concentrated in small garnierite-rich lateritic deposits on ancient, deeply weathered surfaces worn on serpentine.

Platinum

Finds of platinum have been reported in Mariposa County from time to time; the Platinum King group in sec. 31, T. 3 S., R. 17 E., and the Clark or Devils Gulch claims in secs. 1 and 12, T. 4 S., R. 19 E., M.D., have allegedly contained the greatest amount. No platinum has been marketed from any of these vein deposits and its presence has never been authenticated. On the Devils Gulch claim platinum was reported to occur in 2 zones each a few inches wide on both sides of barren quartz vein matter associated with gold and with minor cobalt, nickel and tin. Such an association is improbable and it is possible that the metals were misidentified.

Platinum-group metals occur in minor amounts in placer gold deposits of the county but so far the authors have been able to determine only an ounce or two has been recovered—from dredge ground in the Hornitos district.

Quicksilver

Cinnabar in small quantities has been found in the Crystal mine on the southwest slope of Horseshoe Bend Mountain in secs. 15, 16, T. 3 S., R. 16 E., M.D. Two specimens of high-grade ore from this locality are on display in the Division of Mines museum in the Ferry Building, San Francisco. The vertical Crystal vein worked principally for gold, strikes N. 60° E., intersecting the Cabinet and Lookout veins at right angles. Cinnabar occurs as drusy crystal groups and blebs in calcite vein-matter and in hydrothermally altered greenstone close to the vein. The quantity of ore uncovered apparently was very small and no quicksilver was ever recovered from the cinnabar ore.

Silver

Mesothermal quartz veins carrying native gold and silver and the silver sulfo-salts proustite and pyrargyrite are found in the Bootjack district of southern Mariposa County, notably at the Silver Bar and Silver Lane mines. This type of occurrence is unique among the mines of the county, most of the silver having been recovered from ores containing gold-silver alloys or from argentiferous galena in gold mines. The total silver recovered from this type of deposit has been small, however, and the very substantial silver production of Mariposa County, amounting to about $345,000, has come principally from mines worked primarily for gold.

Silver Bar (Bryan, Silver) Mine. Location: Secs. 8, 9, T. 6 S., R. 19 E., M.D., on the Wallman ranch 3 miles south and 1 mile west of Bootjack and 3½ airline miles southeast of Mormon Bar. Accessible by

FIGURE 48. Headframe and idle mining equipment at the Silver Bar silver mine in the Bootjack district 3 airline miles southeast of Mormon Bar. The Silver Bar mine is one of a small group of quartz vein mines in the Bootjack district where the principal ore mineral is pyrargyrite.

3½ miles of dirt road from the graded Mormon Bar-Moore Hill road. Ownership: Jeffery Investment Co., 7041 Thornhill Dr., Oakland 11, California, holds four claims.

The Silver Bar mine known also as the Bryan or Silver mine, was worked in a small way prior to 1927. A small production was recorded for the years 1909, 1920 and 1921. During 1927 the property was operated under lease by A. D. Lane, et al., of Mariposa. This management did considerable development work and built a small mill, and a small production was realized in 1930. During the late 1930's the mine was operated by Richard E. Jeffery of Mariposa. Most of the silver production was made in 1930. In other years gold has been the principal metal recovered, from the point of view of value.

The Silver Bar vein strikes N. 60-65° W., dips 45° southwest and ranges from a few inches up to 5 feet wide. Near the surface the width seldom exceeds 1 foot. Ore exposed on the dump showed galena, pyrite, arsenopyrite and argentite. Proustite, pyrargyrite, native silver and gold are also reported in the ore from the mine (Julihn and Horton, 1940, p. 163). Wall rocks are hornblende biotite granodiorite or quartz diorite of Upper Jurassic or early Cretaceous age. In 1939 ore was said to average 15 to 20 oz. of silver and 0.15 oz. of gold per ton (Julihn and Horton, 1940, p. 163). The tonnage of ore mined or blocked out at that time was about 4000 tons.

Workings (September, 1954) consist of a 257-foot shaft, inclined 45° southwest, surmounted by an ore bin and crusher housing, and a partly caved drift adit striking N. 65-70° W. The headframe of the shaft has been demolished and the shaft is partly caved and inaccessible. The tunnel is in need of cleaning out but can be entered. The shaft serves levels at 100 and 200 feet, the 100-level having a 235-foot east drift and an 85-foot west drift, and the 200-level having a 263-foot east drift and a 245-foot west drift.

Milling equipment includes a 5-foot Huntington mill, small tube mill, two small ball mills, a Union compressor, and two Kraut flotation cells. One small cabin is the only building on the property. The mine makes about 750 gallons of water a day and water shortage has limited past operations.

Silver Lane (Goldenrod, Wallman Ranch) Mine. Location: Secs. 8, 9, 16, 17, T. 6 S., R. 19 E., M.D., 4 airline miles south and 1 mile west of Bootjack and half a mile south of the Silver Bar mine. Accessible by 3½ miles of dirt road from the graded Mormon Bar-Moore Hill road. Ownership: Not determined.

The early history of the Silver Lane mine has not been recorded although much of the area has been ground-sluiced and presumably started as a placer area. The principal period of operation was 1933 to 1938 by the Silver Lane Gold Mining Company, A. D. Lane, et al.,

FIGURE 49.   Part of an old smelter at the Silver Lane silver mine in the Bootjack district southeast of Mormon Bar. Sulpho-salts of silver, mainly pyrargyrite, were roasted in wood- or charcoal-fired furnaces to free the silver in the ore.

then of Ben Hur, California. Equipment constructed to employ combined roasting and chlorination, included a 50-ton crusher, rod mill, roasting furnace and leaching tank (Laizure, 1935, pp. 44-45). The hoist and compressor were steam-powered as steam was to be utilized in the extraction process. Ore reserves believed to contain $60,000 were developed but only a small part of the ore was ever processed. The operation was greatly hampered by insufficient water. All equipment including the headframe, ore bin and buildings has been removed and the property was idle in September, 1954.

At the surface a quartz vein 8 inches to 14 inches wide strikes N. 70° W. and dips 35-45° southwest. The vein is reported to average 4 feet wide at depth (Laizure, 1935, p. 44) and to contain the ore minerals argentite, proustite, pyrargyrite, cerargyrite, native silver, pyrite, galena and a little gold (Julihn and Horton, 1940, p. 164). Ore remaining on the dump in 1954 contained pyrite, chalcopyrite, pyrargyrite, galena and argentite (?). Gangue material is chiefly milky and glassy quartz with vugs and crude comb structures. Wall rocks are hornblende biotite granodiorite or quartz diorite. Several tons of ore shipped to Selby smelter in 1938 yielded $38.50 in silver and $5.50 in gold (Julihn and Horton, 1940, p. 164).

Workings consist of a 240-foot inclined shaft with 2 levels and a second shaft of unknown extent. There is a 90-foot east drift on the 100-level and a 72-foot east and 21-foot west drift on the 200-level. Both shafts were caved and inaccessible in September 1954.

Tungsten

Tungsten occurs in Mariposa County in mesothermal gold-bearing veins and as replacement disseminations in metamorphic rocks adjacent to granitic intrusive contacts. In both of these types of occurrences the principal ore mineral is scheelite, calcium tungstate. Wolframite (iron-manganese tungstate) has been found in small amounts in quartz near the Buchanan mine 6 miles southwest of Raymond, Madera County, and might be expected in adjacent southwestern Mariposa County, also. Contact metamorphic rocks (tactites and skarn rocks) containing scheelite most readily develop at the borders of limestone and dolomite pendants but are also found in aluminous rocks such as mica schists and spotted slates. At the Big Grizzly group of claims in Ned Gulch 1½ miles northwest of Clearinghouse scheelite occurs in gold-bearing quartz veins (Tucker and Sampson, 1941, p. 580). Scheelite also is known to occur disseminated in quartz veins at the Early mine in the Sweetwater district 2 airline miles northwest of Jerseydale.

Most of the recent activity in tungsten has centered around El Portal where the Incline Mining Company of San Francisco has been developing several groups of tungsten claims. A new tungsten-ore-processing mill was built by the company and placed in operation by John P. Jones. It is close to the north bank of the Merced River half a mile by road west of the El Portal barite-processing plant on Rancheria Flat. Storms, which washed out roads to the tungsten mines, caused a temporary shutdown of the mill in February, 1956. Future activity in tungsten in Mariposa County will depend largely upon the continuation of the government stockpiling program and upon the tungsten price supports as there are no known high-grade deposits.

FIGURE 50. One of the adits and a small ore-pile at the Blue
Dipper group of tungsten claims operated by Incline Mining Company. The mines are in El Portal district 6 airline miles southwest
of Crane Flat on Highway 120.

Blue Dipper Group.   Location: Secs. 2, 11, T. 3 S., R. 20 E., M.D.,
on steep terrain on the east side of Dry Gulch 3½ airline miles northwest of El Portal. Approached by about 12 miles of dirt road from
Crane Flat on Highway 120. Ownership: L. W. and W. J. Barnett,
Box 33, El Portal, California, John Milanovitch estate, Box A., El Portal, California, and Otto Mayer, Box 33, El Portal, California 1954),
hold several claims. Under lease to Incline Mining Company, Lee Cuneo,
L. Garribotti, et al., of San Francisco.

The Blue Dipper mine is a recent discovery on which there has been
considerable capital outlay in road building and other development
work. Some ore was produced and milled at the Incline Mining Company mill on the Merced River during 1955 before the November-
December floods washed out the road to the mine. Operations were
scheduled to continue later in the spring of 1956.

Scheelite occurs in tactite in a small pendant of limestone cropping
out over an acre or two of very steep terrain. The general strike of the
tactite bearing limestone is N. 30° W. and the dip is 70° northeast.
The limestone pendant is bordered on the east by horblende granodiorite or quartz diorite. The tactite is crudely banded through increase
in the garnet and epidote content. The chief mineral in the tactite is
quartz with subordinate calcite, garnet, epidote, pyrrhotite, scheelite,
tremolite and wollastonite. Some scheelite occurs in the granodiorite
which has intruded the pendant in small patches and to some extent in
the bordering granitic massif. The ore averages about 0.75 percent

FIGURE 51.   Ore pile and adit portal at the Blue Dipper group of tungsten mines in El Portal district. Scheelite ore is disseminated in garnet-epidote tactite near a granite contact with limestone.

tungsten. In August 1954 the principal workings were an upper tunnel, with several galleries, striking N. 55° E. and about 100 feet long. Scheelite ore was almost continuously exposed in this tunnel. A second tunnel has been driven southeast about 65 feet below the main tunnel. This was about 50 feet long in August 1954.

Zinc

Complex ores containing substantial amounts of zinc, copper, and lead and minor amounts of gold and silver have been found in Mariposa County in a narrow, northwest-trending belt astride the Merced River between Hornitos and Webb Station. The principal mines in this belt are the Blue Moon and the American Eagle. Nearly all of the zinc production for Mariposa County was made during the World War II years, notably 1944 and 1945, when the Blue Moon mine produced 14,687,920 pounds (Eric and Cox, 1948, p. 145).

Minor amounts of zinc are present in most gold ores of the county but not in quantities sufficiently large to attempt recovery. Zinc is an undesirable constituent in gold ores and concentrates that are to be smelted, and if present in more than trace amounts many commercial smelters assess a penalty on shippers proportional to the amount of zinc present. Consequently, zinc is seldom recovered as a by-product of gold mining.

The principal zinc-bearing mineral is the sulfide sphalerite, both in the complex zinc-copper-lead ores and in the gold-quartz veins. In the zinc belt ore commonly occurs in steeply dipping tabular masses in shear zones in hydrothermally altered greenstone. In the gold-quartz veins sphalerite generally is sparsely disseminated through the ore shoots along with other sulfides such as pyrite, galena, tetrahedrite. arsenopyrite and galena.

Akoz (B.A.B., Radium, Asposozien) Mine. Location: Secs. 9, 10. T. 4 S., R. 15 E., M.D., near Webb Station on the Merced Falls-Granite Springs road and 5¼ airline miles north of Merced Falls. Accessible by three quarters of a mile of dirt road from Webb Station. It is 7.8 miles from Merced Falls. Ownership: Carlin Estate, c/o Mrs. P. Erickson. La Grange, California.

The Akoz mine apparently was prospected prior to 1900 and may have produced a little gold-silver ore from the surface oxidized zone, but there is no early record of production. In the early 1900's the un- usual triboluminescence of the sphalerite was mistaken for the glow given off by radium, and a small quantity of ore was mined, pulverized and marketed for medicinal purposes by the Asposozien Company. At that time the mine was variously known as the Radium, Akoz and Asposozien, Akoz being the trade name of the pharmaceutical product. A 90-foot shaft was sunk during this operation.

About 1941 the mine was leased by a partnership of L. M. Brady, E. M. Aldrich and Menard Brihten (Wiebelt, 1947, p. 3) and the mine was renamed the B.A.B. Under this management considerable stripping was done and the shaft was reopened to the 60-level. In 1944 about 81 tons of ore was shipped that yielded 1.41 oz. of gold, 52 oz. of silver, 1068 pounds of lead and 3623 pounds of copper. Zinc was not recovered from any of this ore and there was no attempt to recover lead or copper from part of it. The yield was between $5 and $6 per ton (Wiebelt, 1947, p. 3).

Further development work was done between 1944 and 1946, the latest by F. H. A. Williams of New York. This culminated in the sam- pling and diamond-drilling of the deposit by the U. S. Bureau of Mines in 1946. Mining was discontinued in 1947 and no further work has been done.

The following discussion is derived chiefly from the U. S. Bureau of Mines Report of Investigations No. 4144 by Frank J. Wiebelt, with minor additions and corrections by the authors.

Ore occurs in irregular replacement masses in micaceous schist de- rived from greenstone and cut by greenstone dikes. The mineralized zone is roughly 400 by 800 feet in plan and is elongated roughly north- west. Among the primary sulfide minerals sphalerite predominates; other species include pyrite, chalcopyrite and argentiferous galena. Barite and quartz are the common gangue minerals, quartz being pre- sent most conspicuously as stringers and pods. The ore has been oxi- dized almost completely to a depth of 35 feet and partly oxidized 15 feet deeper. The best precious-metal ore has come from the oxidized zone but the best zinc-ore lies below the oxidized zone. A sample taken by the U. S. Bureau of Mines across a width of 3 feet of massive sulfide ore at a depth of 55 feet in the main shaft tested 0.12 oz. of gold and 10.9 oz. of silver per ton and 12.83 percent zinc, 0.79 percent lead and 0.87 percent copper. None of the other samples taken underground contained over 3.5 percent zinc and 3.2 oz. of silver per ton and none of the surface samples contained over 3.5 percent zinc. The principal values in the near surface samples were in silver and gold, several of the samples being in the range of 9 to 21 oz. of silver per ton. Complete sampling and diamond drill data are listed by Wiebelt (pp. 4-6).

The principal working at the Akoz mine is a 90-foot shaft from which short drifts were run from the 35, 60 and 90-foot levels. This is on the main ore zone which is also developed by numerous surface cuts. About 800 feet east of the main shaft is a second shaft 15 feet deep. Promising ore apparently was not found there as there are few additional workings.

American Eagle (Bullion Hill, Blue Bell and Bonanza) Mine. Location: N. ½, Sec. 30, T. 4 S., R. 16 E., M.D., 3 miles by road northwest of Hornitos. Adjoins the Blue Moon mine on the south. Ownership: Jack I. and Irene Kopp (½), 650 Liberty St., Santa Clara, California, Edward C. Morrison (¼), Korbel, California, and Valberde Brothers, Box 568, Portola, California.

The American Eagle mine is an old property worked in the 1890s and early 1900s for gold and silver. A substantial amount of gold was produced in 1899 and the 300-foot-long American Eagle adit was driven. Owners in 1910 were listed as L. Valberde and John Morrison and the mine was being operated by the Bullion Hill Mining Company. (Min. and Sci. Press, vol. 90, 91, several entries). Considerable ore was shipped. In 1911 the mine was sold to J. M. and A. L. Richardson of Minneapolis and a Huntington mill was installed. The mine fell idle about 1912 and little was done with it until about 1942 when F. H. A. Williams of New York did some development work and shipped some ore (Eric and Co., 1948, p. 143). The property has been inactive since 1942.

The American Eagle and Blue Moon mines were intensively explored and studied by the U. S. Bureau of Mines and the U. S. Geological Survey during World War II. The following discussion is abstracted largely from the report on the American Eagle and Blue Moon mines by U. S. Geological Survey geologists J. H. Eric and M. W. Cox published in California State Division of Mines Bulletin 144 (1948, pp. 133-150).

There are six overlapping, en echelon mineralized zones on the American Eagle property strung out in northwesterly direction over a length of over 1000 feet. Only one, zone A, the most westerly, penetrated by the main crosscut adit, gives promise of bearing ore of economic grade and quantity. This has been explored for a strike length of about 200 feet and varies from a few inches to 4 feet wide. Sixteen channel samples cut from the bottom and sides of a 40-foot winze, sunk on zone A from the adit level at a point about 60 feet north of the crosscut yielded assays within the following ranges:

copper	1.53 to 11.01 percent
zinc	2.59 to 8.29 percent
lead	0.71 to 1.02 percent
gold	0.01 to 0.22 oz. per ton
silver	0.99 to 2.58 oz. per ton

The ore body where the samples were taken is about 4 feet wide and consists of partly oxidized sulfides. Ore minerals include sphalerite, pyrite, tetrahedrite, galena and chalcopyrite. Gangue minerals are barite, glossy and milky quartz, sericite and calcite. Above the adit level the ore is largely oxidized showing streaks of azurite and malachite. Most of the high-grade gold-silver ore taken out in early days came from near-surface, oxidized ground.

Ore bodies tend to be tabular and are elongated parallel to the schistosity of the enclosing schist. The principal mineralized zone is found in sheared felsite breccia or tuff near its contact with massive felsitic volcanic rock. These metavolcanics are cut by greenstone dikes. The ore-enclosing schist has formed by shearing and hydrothermal alteration of the metavolcanic rocks.

Most of the workings of the American Eagle mine connect with the main crosscut adit-level which penetrates about 120 feet below the top of the ridge. This adit is somewhat sinuous and is driven slightly north of east for about 300 feet, intersecting both ore zone A and a secondary zone (named B). From the adit level is a 100-foot southeast drift driven on ore zone B and a 200-foot sinuous drift driven slightly west of north from the crosscut, largely on ore zone A. About 60 feet north of the crosscut adit is the winze previously mentioned and just north of the winze is a 100-foot raise to the surface. South of the winze is a short stope about 25 feet high. The surface croppings of the various mineralized zones have also been explored by about 50 shallow trenches, pits and shafts.

Although zinc and copper are the principal constituents of the complex ores at the American Eagle mine, the principal production, in regard to value, has been in gold and silver. A satisfactory total production figure is not available for this mine.

Blue Moon (Blue Cloud, Red Cloud, Porcupine) Mine. Location: Secs. 19, 30, T. 4 S., R. 16 E., M.D., 4 airline miles northwest of Hornitos. Accessible by 5 miles of dirt road from the paved Hornitos-Bear Valley road, the turnoff being 1.2 miles north of the Merced Falls-Hornitos-Mt. Bullion road. Ownership: Not determined. Last operator (1945) was Hecla Mining Company of Wallace, Idaho. The property includes the Porcupine, Porcupine Fraction and Blue Moon group of claims.

Although the Blue Moon deposits were probably discovered and superficially worked for gold and silver at various times during the last half of the nineteenth century, all but a minor fraction of the production was made between 1943 and 1946 under the impetus of the wartime demand for base-metals. A little gold, silver and copper production was recorded during the middle 1930s from near-surface oxidized ore (U.S. Bureau of Mines records). From 1940 to 1943 the Red Cloud Mines, Inc. explored the mineralized zones by shallow underground workings and by diamond drill-holes aggregating 5000 feet. Two zones containing promising deposits were outlined as well as an additional silicified ore zone bordering the west ore zone on the north. In 1943 the property was taken under active exploitation by Hecla Mining Company of Wallace, Idaho. The mine was equipped to handle 200 tons of ore daily and flotation concentrate was made at the mill of the Washington mine, 4 miles distant to the southeast. Mining ceased in November 1945 because of caving in the mine and because of the uncertainty of postwar conditions. The property has been idle since 1945.

The total production of the Blue Moon mine has been in excess of $2,054,000, nearly all of the total having been between 1943 and 1946. During this period the 55,656 tons of complex ore mined yielded nearly $37.00 per ton—$4.78 per ton in precious metals and $32.07 per ton in base metals, predominantly zinc. Consequently the Blue Moon mine ranks among the 11 highest producing mines in the country and is by far the highest producing base-metal mine in the county.

The following discussion on the geology and workings of the Blue Moon mine is drawn largely from Eric and Cox (1948, pp. 145-148). The property was studied in detail as part of the wartime investigation of the foothill copper-zinc belt by the U. S. Geological Survey.

Ore occurs in 3 outcropping, tabular, mineralized zones in altered sericite schist derived from sheared, fragmental, andesitic volcanic rocks by hydrothermal alteration. The No. 1 ore body is a lenticular mass striking north to N. 15° E. and dipping west at angles varying between 48° and 87°. It is an irregular, lenticular mass reaching a maximum width of 38 feet and a strike length of 180 feet, and has been worked to a depth of 450 feet. The No. 2 ore body, separated from the No. 1 mass by 10 to 20 feet of sheared tuff-breccia roughly parallels the main body. It has an average width of 5 feet and a strike length of about 80 feet at the 165-foot level, but is believed to pinch out both above and below the 165-level. It apparently converges and joins the No. 1 ore body at its south end. This ore body has not been exploited. A third mineralized zone in the vicinity of the Number 2 shaft is believed to be a southern extension of the main ore zone. About 40 feet of unmineralized material separates the No. 1 and No. 3 ore bodies. Its stope length varies from 50 to 80 feet and its width from 4 to 9 feet. Ore extends well below the 165-foot level but has not been exploited beyond that depth. The ore body strikes N. 5° E. and dips about 80° west.

About 160 feet west of the No. 1 shaft is a silicified zone not mineralized at the surface but containing sulfide ore at depth as demonstrated by several diamond drill holes. This zone has not been exploited but is believed to contain several lenticular, south-plunging ore bodies carrying 6 to 10 percent zinc and having a higher lead content than the main ore zone. Detailed data concerning the geology, and disposition of ore bodies and workings of both the main and west ore zones are contained in the paper by Eric and Cox, previously cited.

The mineralized zones are found in metavolcanic rocks of the Peñon Blanco formation of the Upper Jurassic Amador group (Taliaferro, 1943, pp. 283-284). This belt of rocks trends about N. 10-20° W. and dips steeply east. Felsitic greenstone and greenstone tuff-breccia are the principal rock types in which mineralization took place and shear zones controlled the influx of mineralizing solutions.

The principal working at the Blue Moon mine is a 500-foot shaft, the No. 2 inclined 80° W. with levels at 125, 165, 275, 385 and 450 feet. There are about 200 feet of drifts and subsidiary workings on the 125-level, 600 feet on the 165-level, 250 feet on the 275-level, 225 feet on the 385-level and 150 feet on the 450-level. Two hundred and fifty feet south and slightly east of the main or No. 2 shaft is the No. 1 shaft originaly 35 feet deep but later connected to the 165-level of the main working by a raise. In addition to these underground workings there are several open cuts, 2 over 100 feet long. The main shaft and adjoining workings are not accessible because of caving.

Primary ore generally consists of sphalerite and pyrite with minor amounts of tetrahedrite, galena and chalcopyrite in a gangue of sericite, barite, clear quartz and calcite. This ore is massive in some places and banded in others, banding being due to sphalerite-rich layers and elongated sphalerite grains. Massive siliceous pyritic rock is found along the foot-wall of the northern ore zone and ore locally consists of massive sphalerite-barite rock. Ore in the main or No. 1 ore body averaged 13.1 percent zinc, 0.36 percent copper, 0.47 percent lead and 0.061 oz. of gold and 3.75 oz. of silver per ton.

Non-Metallic Minerals

Andalusite

Andalusite, an aluminum silicate mineral commonly used in the manufacture of electrical porcelain insulators, occurs rather abundantly in western Mariposa County in black slate, phyllite and schist which have undergone contact metamorphism. Andalusite-bearing slates and schists most commonly are found within a few hundred yards of granitic intrusions but occasionally occupy still larger zones adjacent to large intrusions. In many of the occurrences the andalusite is in the form of the variety chiastolite, which contains some organic matter and is commonly altered in some degree to pinnite, a hydro-mica. Chiastolite and pinnitized chiastolite have little economic value, but unaltered andalusite, when present in high concentrations near to transportation, has potential economic importance. The authors have examined one such occurrence described below, and there undoubtedly are others along the foothill belt, particularly in Indian Gulch and Raynor Creek quadrangles.

Southwest of Three Buttes Deposit. Location: SW¼ sec. 17, T. 6 S., R. 16 E., M.D., astride the Indian Gulch-Planada ranch road 1¼ miles southwest of Tres Cerritos or Three Buttes. Accessible from Indian Gulch by 4¼ miles of unimproved dirt road or from Highway 140 via the old Merced Falls dirt road, a distance of 6½ miles.

Ownership: not determined, probably Ed. P. Waltz, Reno, Nevada. On patented agricultural land.

The Southwest of Three Buttes deposit is undeveloped. Andalusite occurs in colorless-to-pink, glassy crystals averaging 3x6 mm in section and 20 mm long. Only a small percentage of the crystals contain organic matter and none appear to be pinnitized. Andalusite-bearing slate crops out over an area of at least 1 acre on rolling terrain, the concentration of andalusite in the slate varying from 15 to 25 percent.

Asbestos

There are numerous occurrences of both cross-fibered chrysotile asbestos and slip-fibered tremolite asbestos in the serpentine masses of Mariposa County. Thus far no concentrations of economic importance have been found. Serpentines along Highway 49 between Bagby and Coulterville contain veinlets of cross-fibered chrysotile from a fraction of a millimeter to 3 mm wide but the chrysotile does not make up more than a few percent of the rock. It is possible that a systematic search

FIGURE 52.   Outcroppings of chiastolite-mica schist in the Three Buttes district 3½ miles southwest of Indian Gulch. Such rock develops from slate by heating of the rock near a granitic intrusion. The chiastolite variety of andalusite has little economic value except for semi-precious gems, but fresh, glassy andalusite when found in large concentrations has potential value as a source of aluminum silicate for porcelain insulators. *Photo by Mary H. Rice.*

FIGURE 53.   A closeup view of large chiastolite crystals in mica schist from the same series of outcrops shown in figure 52. Some of the crystals are several inches long and half an inch across. *Photo by Mary H. Rice.*

of the serpentine masses of the county would reveal locations worthy of development.

Barite

Large deposits of barite have been found in the El Portal district in the Pigeon Gulch-Rancheria Flat vicinity, astride State Highway 140, and on a ridge lying between Devils Gulch and Granite Creek 2 miles northwest of Chowchilla Mountain. The two districts are about 6 airline miles apart. In the El Portal district massive, medium-to-fine-granular barite occurs in vein-like masses in metasediments of the Calaveras group. Associated rocks are quartzite, limestone, slate, phyllite and schist. In the vicinity of Chowchilla Mountain barite also occurs in the Calaveras group of rocks in massive quartzite and micaceous quartzite without apparent association with limestone.

Barite also occurs in large quantities as a gangue mineral in the zinc-lead-copper ores of the Blue Moon and American Eagle mines northwest of Hornitos. Thus far no attempt has been made to recover barite from these ores.

Almost all of the barite production of Mariposa County has come from El Portal mine, most of this production being between the years 1910 and 1948. Within this period 398,613 tons of barite was marketed, valued at $2,760,493. This amounts to about 73 percent of the total California production of barite and compares favorably with the production of many of the best gold mines. Mariposa County remains one of the largest potential producers of barite in California, but present market requirements are such that ore must be either very selectively mined or else ordinary mine-run material must be beneficiated (upgraded). Mines in Nevada currently supply most of California's barite requirements.

Egenhoff (Devils Gulch) Deposit. Location: Secs. 17, 21, T. 4 S., R. 20 E., M.D., near the top of the main ridge that lies between Granite Creek Canyon and Devils Gulch, 2 miles northwest of Chowchilla Mountain and 4½ airline miles east of Jerseydale. Accessible by 1½ miles of trail from the end of the U. S. Forest Service road on Chowchilla Mountain. Ownership: California Barite Corporation, c/o Edwin Earl, 1102 Rowan Building, Los Angeles 13, California owns a patented property of 44 acres which includes the Barium No. 2, Barium No. 3 and North Barium Fractional claims. The company also holds 3 unpatented claims, the Barium No. 3, South Fractional and Camp Lode claims.

This deposit was discovered in 1917 by W. D. Egenhoff who did extensive exploration and development work on the property. Sold to the California Barite Corporation in 1929, about $50,000 was spent by this company in further development work (Laizure, 1935, p. 46) during the early 1930s. The only substantial production recorded from the mine was in 1937 when about 800 tons was marketed.

The authors are indebted to Charles J. Kundert, formerly barite specialist for the Division of Mines, for most of the following data. Mr. Kundert visited the property in October, 1955.

Barite occurs in metasedimentary rocks of the Paleozoic Calaveras group. The barite-bearing beds are nearly everywhere enclosed in massive, blue-gray quartzite which strike N. 25° W. and dip 65° northeast.

Within a foot of the contact of quartzite with the barite rock the quartzite is micaceous. The barite rock is white, granular and contains some silica. Part of the 1937 production was reported to be witherite, indicating that the character of the ore at the Egenhoff and El Portal deposits is similar, but no witherite was observed on the property at the time of Mr. Kunderts' visit in October 1955. The barite rock is foliated parallel to the borders of the ore masses, the foliation planes being about 2 inches apart and most conspicuous near the contact with the enclosing micaceous quartzite. A brown carbonate mineral, probably calcite, occurs along the foliation planes and as blotches throughout the barite rock. No data are available as to the barite content of the ore.

The barite horizon is readily traced for 50 feet southeast of the main adit portal and for a hundred feet to the northwest. According to Laizure (1928, p. 144) the horizon has been traced for a total distance of 4500 feet. In 1942 the owners reported that 2000 tons of ore remained on the dumps and that 25,000 tons had been blocked out in the mine.

Workings consist of a drift adit at least 100 feet long, which develops extensive backs, and numerous surface cuts. The adit is driven N. 25° W. from a gulch which indents the southwest side of the main N.-NW. trending ridge. More than 50 feet of backs are developed toward the rear of the adit.

El Portal Barite Mine.   Location: Secs. 18, 19, T. 3 S., R. 20 E., M.D., astride Highway 140 and the Merced River 1½ miles west of El Portal. Ownership: El Portal Mining Company (a subsidiary of Baroid Division of National Lead Company), c/o R. J. Groves, Trust Department, American Trust Company, 464 California St., San Francisco 20, California, owns a patented property of 81.42 acres which includes the Barium 1 and 2, North Barium and South Barium claims.

According to Young (1930, p. 7) the North mine of the El Portal Barite property was discovered and first worked in the 1880s; the South mine was discovered later. First production was recorded in 1910 by the El Portal Mining Company of San Francisco (a different group than the present owners), which operated the mine intermittently until about 1914. Barbour Chemical Company of Alameda operated the property under lease for a short period in 1914-15 (Lowell, 1916, p. 571), and the Western Rock Products Company had it for some time in the early 1920s. In 1927 the mine was acquired by Yosemite Barium Company of El Portal. This company opened up the series of lenses south of the Merced River called the South Mine (Laizure, 1928, p. 144). In 1929 or 1930 the National Pigments Company succeeded the El Portal Mining Company as operators. This company and its subsequent reorganizations, the National Pigments and Chemical Division and Baroid Division of National Lead Company, operated the mine more or less continuously until 1948. The property has been idle since 1948 because more satisfactory ore can be procured from the company's mines in Nevada at less cost than can be met at the El Portal plant. Loss of rail facilities when service was discontinued on the Yosemite Valley Railroad was the main factor behind discontinuance of operations of the El Portal mine.

The barite masses occur as apparent replacements in metasediments of the Paleozoic Calaveras group. In the South Mine some limestone is found adjacent to the barite-bearing zone and the barite masses have

FIGURE 54. Mill installation and mine workings at El Portal barite mines astride Highway 140 between Briceburg and El Portal. Barite-witherite rock was mined chiefly for use in weighting oil-well drilling-mud. Workings of the North mine, shown at the right of the photo, are in two parallel veins of grayish-white rock which strike directly away from the observer and dip very steeply toward the right (east). The lenticular masses of rock range from 3 or 4 feet in width to more than 20 feet in width and have been mined to a depth of over 200 feet with no lessening of ore at depth. To keep producing a uniform product having a specific gravity of 4.2 it was necessary to beneficiate or upgrade the ore by flotation in the mill seen at the left of the photo. The mines closed down about the time that service was discontinued on the Yosemite Valley Railroad in 1944. *Photo by Mary H. Rice.*

long been considered to be replacements of limestone. Whatever the nature of the parent rock, the barite occurs in apparently homoclinal sequences of beds associated with quartzite, slate, schist and phyllite.

In the North Mine the main barite-bearing zone ranges from 4 to 20 feet wide, dips 70° west and, although slightly sinuous because of local thickening and thinning, strikes generally N. 12° E. Within the barite zone the rock is principally a light-gray to white granular, crystalline rock. In most places it is about 85-90 percent barite and in other places barite is mixed with considerable calcite and quartz. Clean, white barite rock commonly occurs in lenticular masses flanked by several feet of more impure, sheared material. Contorted structures formed by plastic flow of the barite rock are common within the main ore body. The barite rock is most commonly a dense, equigranular aggregate of grains averaging less than half a millimeter in longest dimension. It is commonly banded parallel to the walls of the tabular ore bodies, the darker colored bands containing the most impurities. Impurities include grains of quartz, pyrite, sphene, tourmaline, magnetite, chlorite, actinolite and a green pyroxene (Fitch, 1931, p. 462). Witherite locally is abundant and in places forms distinct ore bodies. In other places calcite takes the place of the normally more abundant barite and the rock approaches a crystalline limestone in composition and appearances. Ore normally contains about 85 percent barite but was beneficiated to about 94 percent $BaSO_4$ (Julihn and Horton, 1940, p. 169; Young, 1930, p. 71).

A second roughly parallel barite bearing-zone has been developed 75 to 100 feet west of the main zone. This zone, although from 1 to 3 feet narrower than the main zone, is similar in other respects. South and east of the North mine across the Merced River is another southeast-trending group of tabular barite masses which have had considerable development work done upon them. Wall rocks are slate and metachert having cleavage and bedding in most places parallel to the planarity of the barite masses. Locally, however, the barite zone bevels the edges of the beds. The barite zones are sharply defined between walls of metachert and slate. There are no limestone strata in the vicinity of the North mine except the calcitic matter found in the barite zones themselves. There is a strong suggestion that the barite zones are actually metamorphosed fissure-vein deposits and not replacements of a pre-existing bed of limestone. Detailed field and petrographic work would be necessary to determine whether or not this suggestion is correct.

The main barite zone is developed by a long adit which begins as a northeast-trending crosscut and then veers north and follows the vein. The line of overhand and underhand stopes developed from this adit has almost obliterated the haulageway in some places. The stopes are open to the surface for long distances and the zone has been mined for nearly 1000 feet north from the adit entrance. The second productive zone has been developed by several crosscut adits which connect with a long line of stopes that are open to the surface. These aggregate about 500 feet of strike length.

Geologic conditions at the South mine are similar to those at the North mine, although the strike of the ore zone is more to the northwest. Varying from a width of 6 inches to more than 20 feet the lenticular masses strike N. 45° W., upon the average, and dip 70° northeast (Kundert, C. J., field report dated in August, 1953). The higher grade shoots vary in length from 50 to 300 feet. Ore varies from white, uniform, medium-grained material to dark-colored, intricately banded rock having abundant pyrite in very fine particles (Julihn and Horton, 1940, p. 169). It commonly contains 85 percent barite, 2-3 percent witherite, 0.5 percent pyrite and about 12 percent quartz, but considerable barite rock falls below 85 percent barite and has not been considered good enough to process.

The South mine is developed by 2 crosscut adits at the same elevation and approximately 150 feet apart driven southwest into the west side of Pigeon Gulch. These connect with a drift level about 1500 feet long. This level is connected by a 90-foot raise to an upper level about 1000 feet long. Ore has been stoped to the surface for several hundred feet above the upper level at the northern end of the mine. None of the workings reach a depth of over 200 feet below the surface (Julihn and Horton, 1940, p. 169) and ore continues below the adit levels for an unknown distance. Reserves of high and medium grade ore in both the North and South mines are believed to be very large.

Good accounts of the processing of the barite ore may be found in Young (1930, p. 71) and Julihn and Horton (1940, p. 169-170). Most of the ore mined was used to add weight to oil well drilling fluids. Ore was transported from the South Mine to the mill on Rancheria Flat near the North Mine by aerial tram. The mill and loading facilities remain intact but have been inactive for several years.

Clay

Clay has not been produced in quantity in Mariposa County thus far, although a little local material may have been burned into common brick prior to 1870. Alluvial clay and clay mantle suitable for common brick and large tile-bodies probably occur at numerous places within the county boundaries. So far as the authors know there has been no systematic search for ceramic materials in the county, due mainly to the distances to market that are involved. Presence of patches of the Eocene Ione formation along the western borders of the county suggest the possibility of the presence of refractory, white clays similar to those mined in Amador and Placer Counties. White clay of unknown extent was encountered in a drill-hole in the Huelsdonk Dredging Company placer ground in sections 5 and 6, T. 4 S., R. 15 E., M.D., on or adjoining the old Prouty ranch 7 miles northeast of Merced Falls (L. L. Huelsdonk, personal communication, 1954).

Limestone and Dolomite

There are immense deposits of carbonate rock in Mariposa County in the Jenkins Hill, Hites Cove, Bower Cave and Exchequer Lake districts. Because the deposits are relatively inaccessible or are a long distance from potential markets, there has been little incentive to develop any of the deposits in recent years. With the exception of the deposits on Jenkins Hill, little has been done on the chemical and physical properties of any of the carbonate rocks in order to determine their best utilization. In general, the carbonate rocks of the Bull Creek-Bower Cave district are either magnesian limestones or dolomites; those of Jenkins Hill tend to be graphitic and slaty (aluminous) ; and those of the Hite Cove district are at least in part magnesian. Most of the crystalline limestone on the Hite Cove district is a fine-grained gray and white marble that takes a good polish. Some of the rock in the Bower Cove district is similar. The limestone at Jenkins Hill is nearly black because of the presence of about 1½ percent of finely divided graphite (Laizure, 1928, p. 148).

Between 1927 and 1944 a total of 2,430,843 tons of limestone was quarried in Mariposa County, almost entirely from the Emory deposit on Jenkins Hill. Although this was captive tonnage utilized in manufacture of portland cement, it would have been worth between $2,000,000 and $3,000,000 on the open market. Only a small fraction of the reserves were utilized.

Bower Cave Deposits. Location: Secs. 19, 20, 29, 30, T. 2 S., R. 18 E., M.D., astride the Briceburg-Kinsley-Coulterville road 19 to 20 miles by road from Briceburg. Ownership: not determined. Portions of the limestone-bearing land are owned by several individuals.

Lenticular masses of dove-gray carbonate rock crop out along the east side of the Kinsley-Coulterville road from a point about half a mile northwest of Bower Cave southeast for a distance of over 3 miles.

These are part of the metasedimentary portion of the Paleozoic Calaveras group.

There is some high-calcium limestone in the vicinity of Bower Cave, the extent of which is not known to the authors, but most of the rock exposures examined consist mainly of dolomite or a mixture of dolomite and calcite too intimately mixed to mine separately. Much of the rock is dense and fine to medium-grained and will take a high polish, but the colors are not particularly attractive for use as marble in building facings etc., and the deposits are rather remotely situated from transportation and markets. None of the deposits have been developed or adequately sampled so far as the authors could find out.

Cotton Creek Deposit.   Location: Sec. 18, T. 4 S., R. 16 E., M.D., just north of the Cotton Creek arm of Exchequer Lake 3 airline miles southeast of Granite Springs School. Accessible only by trail from a ranch road leading southeast from Granite Springs School. The old road from Hornitos has been inundated by Exchequer Lake. Ownership: not determined.

A small tonnage of limestone was quarried and burned into lime at the deposit prior to 1910. No recent work has been done. The limestone occurs in a narrow, northwest-striking lens enveloped in slaty and schistose marine metasediments of Upper Jurassic age. The limestone is dense, fine-grained, gray rock apparently high in calcium and low in magnesium and other impurities. No chemical analytical data are available on the deposit. The mass is several hundred feet long but does not exceed a hundred feet wide, and in most places is much thinner than that. Except for a few small pits the mass is undeveloped.

Jenkins Hill (Emory, Richardson, Timertone, Yosemite Cement) Deposit.   Location: Secs. 7, 18, T. 3 S., R. 19 E., M.D., high on the west side of Miller Gulch 1 mile north of the Merced River and Highway 140 and 8 miles northeast of Briceburg. Ownership: Johnny Richardson, 1124 South Fifth Avenue, Arcadia, California owns several claims covering most of the limestone-bearing land.

The Jenkins Hill deposit was put into production in 1927 when the first rock was shipped to Merced for use in manufacture of portland cement. Between 1927 and 1944 approximately 2,430,843 tons of limestone was quarried. Production ceased in 1944 when service on the Yosemite Valley Railroad was discontinued. Limestone was quarried from long benches situated about 800 feet above the railroad grade, put through a 30-inch gyratory crusher and a heavy-duty hammer mill, delivered to the upper storage bin by belt conveyor, and transferred to the railroad-side storage bin by electrically driven, cable-drawn incline cars. Each car carried 20 tons of minus-one-inch rock and could travel at a rate of 500 feet per minute. All equipment and supplies were delivered to the quarry area by the inclined cableway track which was nearly 2000 feet long. The limestone was broken by massive blasts which brought down as much as a million tons of rock.

Limestone is found in a north-trending lenticular mass nearly half a mile long and several hundred feet wide at the thickest part of the lens. The strata dip about 80° east. The mass crosses Miller Gulch and continues south as far as the Merced River, but it has narrowed to less than 50 feet wide at that point and is quite slaty. The limestone is ex-

FIGURE 55. Trestle, incline and rock-storage bin at the former Yosemite Cement Company property near Jenkins Hill north of Highway 140 between Briceburg and El Portal. The limestone, a black, rather slaty rock, came from quarries located off of the photo to the right rear. More than 2 million tons of rock was quarried during the 17 years of operation.

posed by erosion for a depth of more than 800 feet and continues below the bottom of Miller Gulch for an unknown depth. The limestone is dark-gray to black and rather thinly bedded. The magnesium oxide content averages considerably less than 3 percent. Little else regarding its chemistry has been recorded, but some of the rock is slaty and there-fore aluminous. In the manufacture of cement the limestone was blended with alluvium taken from pits near Merced. Only a small per-centage of the reserves were depleted but the deposit is a mile north of Highway 140 on rugged terrain and nearly 60 miles to the nearest rail-shipping point at Planada.

Kinsley (Jenkins Ranch) Deposit.   Location: Secs. 4, 9, T. 3 S., R. 18 E., M.D., on the Jenkins ranch half a mile west and slightly south of the Kinsley guard station of the U. S. Forest Service and about 11 miles by graded dirt road from Briceburg. Ownership: on patented agricultural land, mostly part of the Jenkins ranch near Kinsley.

Light gray to grayish white limestone crops out over several acres about half a mile southwest of the Jenkins ranch house. The rock is a fine-grained dolomitic marble but no chemical or physical data on it are available. The deposits are undeveloped and the reserves are un-known. Distance from market, lack of heavy-duty transportation fa-cilities and the magnesian character of the rock have hindered develop-ment of the deposit.

Marble Point (Bondshu, Hite Cove) Deposits.   Location: Sec. 2, T. 4 S., 19 E., M.D., on the south side of the South Fork of the Merced River 1½ miles south and slightly east of Hite Cove. Accessible by 7 miles of jeep trail from Jerseydale.

The Marble Point deposit is a large, undeveloped mass of finely crystalline carbonate rock much of which is reported to be good quality commercial marble. The type rock is white with dark veining and is dolomitic. The carbonate mass reaches a width of 3000 feet, is more than half a mile long and has been eroded to a depth of 600 feet. It is slightly elongated north and dips very steeply east. Little or no work has been done on the deposit to determine its physical and chemical properties and it has remained undeveloped because of its inaccessibility.

O'Brien Deposit.   Location: N¼, NE¼, sec. 11, T. 4 S., R. 18 E., M.D., near the crest of a ridge about 1 mile southeast of Briceburg. Accessible by 5 miles of graded dirt road, via Trabucco Creek and the Feliciana mine, and 2 miles of truck trail. Ownership: Mrs. Ethel R. O'Brien, et al., 1534 Clay St., San Francisco owns a patented property of 80 acres which includes much of the limestone land.

A northwest-striking, nearly vertical lens of limestone several hundred feet thick can be seen cutting across the ridge east of the Briceburg grade. There has been no production and reserves have not been proven beyond the estimate of several million tons (Logan, 1947, p. 253). The best quarry sites are about 800 feet above the bend of the Merced River. Although the property is 7 miles by road from Highway 140 near Timber Lodge, it would be possible to drop the limestone down to highway level at Briceburg over a distance of only 1 mile—at the expense of losing 800 feet of altitude. Three samples analyzed by Santa Cruz Portland Cement Company (Logan, 1947, p. 253) gave the following results:

Oxide	Sample 1 (white)	Sample 2 (black)	Sample 3 (mottled)
SiO₂	0.28	1.18	0.20
R₂O₃ (R = Al & Fe)	0.60	1.02	0.36
CaO	54.10	53.40	54.60
MgO	not done	not done	not done
CaCO₃	96.54	95.29	97.44
Ign. loss	44.22	43.92	44.48

Logan does not state how the sampling was done or from what parts of the deposit the samples were taken, but two-thirds of the deposit contains rock resembling sample No. 3.

Miscellaneous Occurrences.   There is a small mass of limestone close to the north side of Highway 49 one mile northeast of Mt. Bullion, but it is not sufficiently large or of sufficient purity to be of economic importance. Another deposit, reported to be 3000 feet long and from 25 to 100 feet thick, is about 1½ miles east of Bagby and about 1 mile north of the Merced River. The deposit, sometimes called the Welch and Farney deposit (Logan, 1947, p. 253), is undeveloped and nothing is known of its chemical or physical characteristics. Small pendants of limestone in granitic rock are reported in the Ben Hur district of southwestern Mariposa County but these are undeveloped.

FIGURE 56. One of several muscovite schist quarries along Brushy Canyon in the White Rock district. The rock was ground and used as a lubricant in manufacture of rubber tires. The schist is derived from rhyolite tuff of the Cosumnes member of the Upper Jurassic age.

Magnesite

There is one known occurrence of magnesite in the vicinity of Mormon Bar. There, several narrow magnesite veins are found in weathered serpentine associated with siliceous hot spring deposits (Laizure, 1928, p. 148). No magnesite has been produced in the county and the occurrence is not believed to be sufficiently large to be of economic importance.

Mica

There are large deposits of muscovite-quartz schist in the Brushy Canyon vicinity of the White Rock district 2 miles north and slightly west of White Rock School. The schist has formed from metamorphism or rhyolite tuff of the lowest member of the Upper Jurassic Mariposa group (Taliaferro, N. L., personal communication, 1938). It occurs in at least two northwest trending (N. 55-65° W.) belts that dip steeply northeast. The width of the relatively clean, pearly white schist ranges from 40 to over 100 feet and, in the main zone has a strike length of at least 1½ miles. The main zone is repeated at least once by folding, so that there are two or more subparallel zones separated by several hundred feet of schistose rocks of much lower muscovite content.

Numerous pits and several small quarries expose a considerable tonnage of uniform rock in which quartz is the only major mineral other than muscovite. Several million tons of material could be developed.

FIGURE 57.   Detail of the quarry face in the muscovite schist quarry shown in figure 56. The lightest bands are quartz-rich rock which in places is a troublesome impurity.

The quality of the material is dependent upon the proportion of quartz present, for most uses, and insufficient testing has been done to establish the amounts of rock of various grades that are present. There was a small, intermittent production of the schist in the 1930's and 1940's principally for use as a lubricant in manufacture of automobile tires. So far as the authors know, there has been no production since 1946.

There is one quarry on the main schist zone 15-20 feet wide, 85 feet long and 20-30 feet deep.

A series of open cuts extends north from the main quarry over 400 feet. There are also several pits northeast of the main workings about a quarter of a mile away.

Ornamental Stone

The brightly colored, green variegated mariposite-ankerite-quartz rock from Mother Lode and adjacent vein systems has been finding favor, in recent years, as an ornamental stone in gardens, patios and the like. It has been quarried recently along Highway 49 between Bagby and Coulterville near the Specimen group of gold mines and near the Virginia and Mary Harrison mines. Production has been intermittent because of vagaries of demand and because the select, unstained mariposite rock tends to occur in small masses. The chief detrimental impurity is discoloring iron oxide derived from original sulfide minerals. Select rock brings 20 to 30 dollars a ton on the retail market but selective mining and hand sorting make the initial cost of production high.

Rock, Sand and Gravel for Aggregate

An intermittent production of rock, sand and gravel has been recorded from Mariposa County since 1907. During the period of operation of the Yosemite Valley Railroad there was a substantial production of chert and greenstone for road metal, road base and railroad ballast from quarries near the present location of Exchequer Lake. All activity there ceased in 1944. Sand and gravel deposits in Chowchilla Creek, Mariposa Creek and the Merced River between El Portal and Briceburg have been exploited intermittently in a small way for many years. The rock particles making up these deposits consist predominantly of ancient, dense, volcanic porphyries, quartzite, vein-quartz and numerous varieties of crystalline metamorphic rocks, but in the Ben Hur-Chowchilla Creek area the deposits are predominantly of granitic rock-debris.

The extensive dumps of many of the mines along Highway 49 have been utilized in quantity, from time to time, for fill, road base and macadam. Such rock has also been used sparingly in concrete although not generally as satisfactory as stream gravel. Extensive deposits of quartz sand, the product of the mills operated for gold ore, have yielded substantial tonnages of material for concrete and plaster. The dumps of the old Princeton and Mariposa mills have been nearly depleted of such material. There is a huge dump of crushed quartz south of the Merced River at Bagby but it is not readily accessible and is a considerable distance from potential markets. Inasmuch as the population of Mariposa County is only about 5000 and inasmuch as the mileage of state highways and other surfaced roads is relatively small, there has been little incentive to develop the vast reserves of hard, durable rock found within the county. These include the granites of the Cathay, El Portal, Mormon Bar, Ben Hur, and Schultz Mountain districts; the cherts and greenstones of the Hornitos Bear Valley and Coulterville districts; and the crystalline limestones and dolomites of the Briceburg, Hite Cove and Bower Cave districts.

Mariposa Sand and Gravel Company. This company, a partnership of George P., J. G. and E. C. Greenamayer, operates out of Mariposa and in recent years has operated a crushing and screening plant on Mariposa Creek opposite the fair grounds 1½ miles south of Mariposa. Much of the material processed consists of old dredge tailings. A steady, substantial production has been maintained over the past 5 years or more. The company also produces a little by-product gold and silver.

Other Producers in the Vicinity of Mariposa. W. J. Saye of Mariposa has produced sand and gravel in recent years from shallow deposits on Humbug Creek 2½ miles south of Bootjack. These deposits are chiefly weathered granitic rock that has been moved only a short distance by flood waters of Humbug Creek. In past years the Duke Brothers have produced sand and gravel from Chowchilla Creek about 7 miles south of Mariposa. Paving contractors such as George E. France, Ralph Maxwell and the Piambo Construction Company produce rock intermit-

FIGURE 58.   The sand, gravel and crushed rock operation of Mariposa Sand and Gravel Company on Mariposa Creek between Mariposa and Mormon Bar. A little placer gold is produced as a by-product. *Photo by Mary H. Rice.*

tently using portable crushers that can be moved to deposits adjacent to their various jobs.

Roofing Granules and Terrazzo Chips

There has been considerable demand, in recent years, for colored, durable rocks for use on built-up, asphalt-base roofs. The red and yellow jasperoid chert of the Jasper Point district would undoubtedly be suitable for this use if satisfactory roads were constructed to the vicinity. No work has been done on them since discontinuance of service on the Yosemite Valley Railroad. The massive greenstones of the Sierra foothills have been utilized in other counties for granules but so far none of the Mariposa County deposits have been activated.

For the past year or two, dark green serpentine has been quarried by the Sonora Marble Aggregates Company from a roadside quarry on Highway 49 approximately 2½ airline miles northwest of Bagby. The quarry is in the NW¼, sec. 19, T. 3 S., R. 17 E., M.D. About once a month the company trucks several loads of serpentine to their crushing and sizing plant at Shaws Flat near Sonora, Tuolumne County. The serpentine is dense and tenacious and has about the same hardness as marble. Production of terrazzo chips in California is pretty well limited to limestone, marble and serpentine because of the way in which the chips are used. Terrazzo walls and floors are made by setting the chips in light-colored, fine-grained concrete and then polishing the surface with carborundum wheels. Materials harder or less tenacious than marble are unsuitable inasmuch as grinding of hard materials is excessively costly and use of materials of mixed hardness tends to yield an uneven polished surface.

FIGURE 59.   LeGrand (White Rock) silica quarry in southwestern
Mariposa County just north of the Ganns Creek (LeGrand-Mari-
posa) road. Between 1942 and 1952 Kaiser Aluminum and Chemical
Company produced approximately 148,000 tons of quartz for use in
manufacture of ferrosilicon.

Silica

Vein-quartz has been quarried from massive deposits in western
Mariposa County for the manufacture of ferrosilicon and, to a minor
extent, for chemical and metallurgical use. Between 1942 and 1952 the
Kaiser enterprises quarried 147,964 tons of quartz from their White
Rock (LeGrand) deposit in sec. 14, T. S., R. 17 E., M.D., for use at
their Permanente ferrosilicon plant. This operation was discontinued
as the wartime government demand for ferrosilicon fell off. Prior to
1942 there had been a small intermittent production of vein quartz for
miscellaneous uses, chiefly in the late 1920's.

There are large deposits of vein-quartz cutting the slate belt that
lies between Mariposa and Planada. Unlike the vein matter in the
Mother Lode, most of this material is low in sulfides and carbonates
and would be a suitable source of high-grade silica. The deposits, however, tend to be of very irregular shape and often require expensive
selective quarrying. The potential reserves of silica in this part of Mariposa County run into many millions of tons.

Slate

In common with several other Sierran foothill counties, Mariposa
County has immense reserves of black slate suitable for roofing sheets,
flagstone and roofing granules. Good quarry sites can be found in all
of the three main metasedimentary units in the county, the Paleozoic

FIGURE 60. Rock crusher, storage bins and loading facilities at Kaiser Aluminum and Chemical Company's LeGrand (White Rock) silica quarry.

FIGURE 61. Detail of chiastolite-mica schist forming the south wall of LeGrand silica quarry. Rock was derived by contact alteration of black slate near a granitic intrusion.

Calaveras group, the Upper Jurassic Mariposa slate and in the slaty, tuffaceous parts of the Upper Jurassic Amador group. All of these rock units have broad distribution in the county. At least 20 percent of the land area of Mariposa County is underlain by slate and there are innumerable places where sound, cleavable material can be quarried. Because of the high cost of construction of buildings that are to be roofed with slate there has been no demand for roofing slate in California in recent years. Crushed slate granules for built-up, asphalt-base roofs are used to some extent in California but have not been produced in Mariposa County within the last decade. The most probable outlet for slate at the present time is for flagstone for California's flourishing construction industry. The use of flagstone for building facings, patios and ornamental walls and walks is ever-increasing and although black rock is not in as great demand as the light-colored or strongly colored sandstones and volcanic rocks of Arizona, Nevada and other parts of California, the consumption of slate is sure to increase. Up to the present time there has been no vigorous attempt to put slate flagstone onto the California market. Inasmuch as slate is softer and has a greater tendency to spall than many other kinds of flagstone it is necessary to select the soundest material available in order to successfully compete with these other materials.

During the early part of the present century sheets of good roofing slate were produced by the Cunningham Corporation of Planada from a property covering sections 6, 7, 8 and 17, T. 7 S., R. 17 E., M.D., astride Mariposa Creek 6 airline miles south of Cathay. At about the same time roofing-slate sheets were also produced by Pacific Slate Company of Merced from a property in sec. 16, T. 6 S., R. 16 E., M.D., about 3 miles southwest of Indian Gulch (Laizure, 1928, p. 151). There is a series of small quarries on the northwest side of Highway 140, the Yosemite All Year Highway in sections 2 and 3, T. 7 S., R. 16 E. and sec. 35, T. 6 S., R. 16 E., M.D. These quarries are less than 2 miles from the intersection of the highway with the Mariposa-Merced County line and 7 miles from the railroad at Planada.

The only slate property that has produced flagstone in recent years is in sec. 21, T. 5 S., R. 18 E., M.D., at the north end of Nigger Hill close to Agua Fria Creek and Highway 140 approximately 9 miles by road from Cathay. There has been a small intermittent production of flagstone from this place sold locally and called for by the customer. Formerly part of Las Mariposas Grant the property is now separately owned.

Talc, Pyrophyllite and Soapstone

Talc occurs at numerous places in Mariposa County as a hydrothermal alteration product of serpentine. It is found in and adjacent to many of the quartz veins that cut serpentine but is also found along shear zones in serpentine not associated with quartz veins. Because most of the talc formed by alteration of serpentine contains considerable iron it cannot be used for most ceramic purposes, and its usefulness is limited to such products as lubricants, insecticides and, when sufficiently pure and white, cosmetics. Up to the present deposits that are larger and nearer the marketing centers have supplied most of the requirements of the California market in talc of the lower grades, but some of the Mariposa County occurrences may be used in the future. There has been no systematic exploration or testing of any of the Mariposa County deposits to determine the extent and quality of the material available. Most of the occurrences discovered thus far in the county are in the Flyaway Gulch vicinity of the Bagby district.

Talc rock or soapstone was used all along the Mother Lode in the early days for building facings as it looked good and was easy to shape. Talc for this purpose was produced in the hills east of Mariposa and on Greeley Hill east of Coulterville.

Pyrophyllite in beautifully crystallized rosettes occurs with quartz on Three Buttes (Tres Cerritos) 2 miles southwest of Indian Gulch. It has developed by hydrothermal alteration of greenstone of the Upper Jurassic Amador group along a northwest-trending shear zone. The pyrophyllite is too intimately mixed with quartz to allow selective mining and the total pyrophyllite content of the quartz mass is low. Presence of iron oxide from altering pyrite also detracts from the economic value of the deposit. The pyrophyllite-bearing zone is at least half a mile long and 20 to 40 feet wide.

REFERENCES

Averill, C. V., et al., 1946, Placer mining for gold in California: Calif. Div. Mines Bull. 135, pp. 261-262.

Aubury, L. E., et al., 1905, The copper resources of California: Calif. Min. Bur. Bull. 23, pp. 203-216.

Aubury, L. E., et al., 1908, The copper resources of California: Calif. Min. Bur. Bull. 50, pp. 251-269.

Aubury, L. E., et al., 1906, The structural and industrial materials of California: Calif. Min. Bur. Bull. 38, pp. 100, 150-152.

Boalich, E. S., 1922, Mariposa County: Calif. Min. Bur. Rept. 18, pp. 363-366.

Bowen, O. E. Jr. and Crippen, R. A. Jr., 1948, Geologic maps and notes along Highway 49, California Div. Mines Bull. 141, pp. 35-86.

Bradley, P. R., Jr., 1954, A brief mining history of Mariposa County: Mariposa Centennials, Mariposa, pp. 21, 32.

Bradley, W. W., 1930, Barite in California: California Div. Mines Rept. 26, pp. 45-63.

Braun, L. T., 1950, Barite: California Div. Mines Bull. 156, pp. 130-132.

Browne, J. Ross, 1868, Mineral resources of the states and territories west of the Rocky Mountains: U. S. Government, pp. 28-35.

Browne, J. R. and Taylor, J. W., 1867, Mineral resources of the states and territories west of the Rocky Mountains: U. S. Government, pp. 40-43, 150-152.

Calkins, F. C., 1930, The granitic rocks of the Yosemite region: U. S. Geol. Survey Prof. Paper 160, pp. 120-129, 1 map.

Castello, W. O., 1921, Mariposa County: Calif. Min. Bur. Rept. 17, pp. 86-143.

Cater, F. W., Jr., 1948, Chromite deposits of Tuolumne and Mariposa Counties, California: Calif. Div. Mines Bull. 134, pt. 3, ch. 1, pp. 14-15.

Cox, M. W. and Wyant, D. G., 1948, La Victoria copper mine, Mariposa County, California: Calif. Div. Mines Bull. 144, pp. 127-132.

Engineering and Mining Journal, Numerous entries, vols. 1-155, 1872-1954.

Eric, J. H., 1948, Tabulation of the copper deposits of California: Calif. Div. Mines Bull. 144, pp. 199-357.

Eric, J. H., and Cox, M. W., Zinc deposits of the American Eagle-Blue Moon area, Mariposa County, California: Calif. Div. Mines Bull. 144, pp. 133-150.

Fairbanks, H. W., 1890, Geology of the Mother Lode region: Calif. Min. Bur. Rept. 10, pp. 23-90.

Fitch, A. A., 1931, Barite and witherite from near El Portal, California: Am. Mineralogist, vol. 16, no. 10, pp. 461-468.

Forstner, William, 1908, Copper deposits in the western foothills of the Sierra Nevada: Min. and Sci. Press, vol. 96, pp. 743-748. Mariposa County section on p. 747.

Gillice, J. O., The Iron Duke mine: unpublished private report.

Goodyear, W. A., 1888, Mariposa County: Calif. Min. Bur. Rept. 8, pp. 343-349.

Heizer, R. F. and Fenenga, Franklin, 1948, Survey of building structures of the Sierran gold belt: Calif. Div. Mines Bull. 141, pp. 91-164.

Heyl, G. R., 1948, Foothill copper-zinc belt of the Sierra Nevada, California: Calif. Div. Mines Bull. 144, pp. 11-29.

Hudson, F. S., 1955, Measurement of the deformation of the Sierra Nevada, California, since middle Eocene: Geol. Soc. America Bull., vol. 66, pp. 835-870.

Julihn, C. E. and Horton, F. W., 1940, Mines of the southern Mother Lode region, pt. 2, U. S. Bur. Mines Bull. 424, pp. 94-173.

Knopf, Adolph, 1929, The Mother Lode system of California: U. S. Geol. Survey Prof. Paper 157, 85 pp.

Laizure, C. M., 1922, Mining activities in Mariposa County: Calif. Min. Bur. Rept. 18, pp. 9, 144.

Laizure, C. M., 1928, Mariposa County: Calif. Div. Mines and Mining, Rept. 24, pp. 72-153.

Laizure, C. M., 1935, Mariposa County: Calif. Div. Mines Rept. 31, pp. 27-46.

Lang, Herbert, 1907, The copper belt of California: Eng. and Min. Jour., vol. 84, pp. 963-966.

Logan, C. A., 1947, Limestone in California: Calif. Jour. Mines and Geol., vol. 43, no. 3, pp. 252-253.

Logan, C. A., 1935, The Mother Lode gold belt of California: Calif. Div. Mines Bull. 108, 240 p.

Lowell, F. L., 1916, Mariposa County: Calif. Min. Bur. Rept. 14, pp. 569-604.

Mathes, F., 1930, Geologic history of Yosemite Valley: U. S. Geol. Survey Prof. Paper 160, 120 p.

Mining and Scientific Press, Numerous entries, vols. 1-125, 1860-1921.

Odgers, H. H., 1955, News notes on Mariposa County mines: Calif. Min. Jour., vol. 25, No. 3, p. 27.

Preston, E. B., 1890, Mariposa County: Calif. Min. Bur. Rept. 10, pp. 300-310.

Raymond, R. W., 1869, Mineral resources of the states and territories west of the Rocky Mountains: U. S. Government, pp. 11-15.

Raymond, R. W., 1870, Mineral resources of the states and territories west of the Rocky Mountains, pp. 23-24.

Raymond, R. W., 1871, Mineral deposits of the states and territories west of the Rocky Mountains, U. S. Government: pp. 29-33.

Raymond, R. W., 1872, Mineral deposits of the states and territories west of the Rocky Mountains: U. S. Government, pp. 57-59.

Raymond, R. W., 1874, Mineral deposits of the states and territories west of the Rocky Mountains: U. S. Government, pp. 65-68.

Raymond, R. W., 1875, Mineral deposits of the states and territories west of the Rocky Mountains: U. S. Government, pp. 50-55.

Raymond, R. W., 1876, Mineral deposits of the states and territories west of the Rocky Mountains: U. S. Government, pp. 36-40.

Reid, J. A., 1908, The foothill copper belt of the Sierra Nevada: Min. and Sci. Press, vol. 96, pp. 388-393, November 21.

Stevenson, W. C., 1927, New plant of the Yosemite Portland Cement Corporation at Merced, California: Rock Products, vol. 30, no. 12, pp. 79-84.

Storms, W. H., 1894, Geology of a portion of Madera and Mariposa Counties: Calif. Min. Bur. Rept. 12, pp. 165-176.

Storms, W. H., 1896, Mariposa County: Calif. Min. Bur. Rept. 13, pp. 216-225.

Storms, W. H., 1899, Mariposa County, in California Mines and Minerals, California Miners Association, San Francisco, pp. 360-369.

Storms, W. H., 1900, The Mother Lode region of California: Calif. Min. Bur. Bull. 18, pp. 142-147.

Taliaferro, N. L., 1943, Manganese deposits of the Sierra Nevada, their genesis and metamorphism: Calif. Div. Mines Bull. 125, pp. 277-331.

Trask, P. D., et al., 1950, Manganese in California: Calif. Div. Mines Bull. 152, pp. 106-112.

Tucker, W. B. and Sampson, R. J., 1941, Recent developments in the tungsten resources of Californai: Calif. Div. Mines Rept. 37, pp. 565-588. Mariposa County mines pp. 580-581.

Turner, H. W. and Ransome, F. L., 1897, U. S. Geol. Survey Atlas, Sonora folio (No. 41), 9 pp. 4 maps.

Van Norden, R. W., 1917, A narrow gauge alternating current mine locomotive: Eng. and Min. Jour., vol. 103, no. 16, pp. 698-702.

Wiebelt, Frank, J., 1947, The Akoz mine; Mariposa County, California: U. S. Bur. Mines Rept. Inv. 4144, 6 p.

Wilkinson, E. M., 1904, Mines register, Mariposa County, 16 pp., 1 map.

Young, Geo. J., 1929, Making a 30-ton California gold mine pay: Eng. and Min. Jour., vol. 127, pp. 45-48.

Young, Geo. J., 1930, Mining and milling barite: Eng. and Min. Jour., vol. 130, pp. 70-71.

TABULATION OF MARIPOSA COUNTY MINERAL DEPOSITS

The following table lists Mariposa County mineral deposits in alphabetical order by commodity. The number in the first column refers to the location on the county map, plate 4, in pocket.

The references given in the *Remarks* column refer to the bibliography accompanying this report. Only the last name of the author is given. The first number following the author's name is the abbreviated date of publication as given in the bibliography; the second number, that following the colon, is the page reference.

FIGURE 62. Ruins of one of the first copper smelters built in California. In the Green Mountain district near the Raymond Mariposa road. *Photo by W. O. Castello, October 1919.*

CHROMIUM

MAP NO.	CLAIM, MINE, OR GROUP	OWNER NAME, ADDRESS	SEC.	T.	R.	B & M	REMARKS
1	Fossow property	Oro Rico Mines Co., c/o A. D. Vencile, Rm. 14, 1584 Washington Boulevard, Los Angeles	NE¼ 29	2S	16E	MD	Near Penon Blanco, 3 miles northwest of Coulterville (Cater 48:14; herein).
	Purcell-Griffin						See Riverside chrome.
2	Reed property	Walter J. Lautenschlager, 626 South Catalina Street, Los Angeles 5	SW¼ 27	2S	16E	MD	On Blacks Creek 1 mile southeast of Fossow property (Cater 48:14; herein).
	Riverside chrome (Purcell-Griffin)	Alfred W. Stickney, 435 Hillcrest Road, San Mateo	NE¼ 22	3S	16E	MD	Southeast of Coulterville 3½ miles (Cater 48:14; herein).

COPPER

MAP NO.	CLAIM, MINE, OR GROUP	OWNER NAME, ADDRESS	LOCATION				REMARKS
			SEC.	T.	R.	B & M	
3	A. C. Smith		3	7S	17E	MD	Discovered about 1883. Intermittent production 1893-1908. Adjoins Great Northern and has similar geology. Had a shaft 80 feet deep in 1908. One shipment of ore was reported to run 15 percent copper (Aubury 08:262). One mile west of Pocahontas mine.
	Amador						See Green Mountain group.
	American						See Green Mountain group.
	American Eagle						See under zinc.
4	Barrett (Baretta, Berette, Barrette, Beaudry, New Year, Wildcat)	Angelo Beaudry and Inez Bouvier, 1431 Moraga Street, San Francisco	29 30 32	3S	16E	MD	Hunter Valley district (Aubury 05:215; Lang 07:966; Aubury 08:267; Castello 21:107; Eric 48:265; herein). Includes New Year, Wildcat, and Mountaineer patented claims, and Berettes Enclosure, a patented property which includes the old Berette and Beaudry claims.
	Barretta						See Barrett.
	Barrette						See Barrett.
5	Beaudry						See Barrett.
	Berette						See Barrett.
	Bouvier						See Carson under Gold.
	Bruschi	Not determined	SE¼ 30?	3S	16E	MD	Workings in 1902 consisted of several shafts, cuts and tunnels. Located northwest of Castagnetto II mine. (Aubury 05:215).
	Buena Vista						See Green Mountain group.

COPPER (CONT.)

| MAP NO. | CLAIM, MINE, OR GROUP | OWNER NAME, ADDRESS | LOCATION | | | | REMARKS |
			SEC.	T.	R.	B & M	
	Carson						See under gold.
6	Castagnetto II	Not determined	SW¼ 30	3S	16E	MD	A quarter of a mile northwest of New Year claim of the Barrett mine in the Hunter Valley district 10¼ airline miles north of Hornitos. Geology and occurrence of ore similar to Barrett mine. Ore contains both gold and copper. Best copper ore was in the oxidized zone and was carbonate type (Aubury 05:215; 08:267).
	Cavan						See Good View.
	Cavan - San Jose						See Good View.
7	Chemisal (Castagnetto III)	Not determined	NE¼ 31	3S	16E	MD	Close to Last Chance gold mine and Barrett copper mine. Ore contains both gold and copper, best copper ore oxidized. In 1902 there were two shallow shafts 15 and 20 feet deep, respectively (Aubury 05:213).
	Copper Chief						See Green Mountain group.
	Copper Hill	Not determined	14	6S	16E	MD	Patented mine with shaft 60 feet deep according to Wilkinson 04:15.
	Copper King I						See White Rock.
	Copper King II						See Green Mountain group.
	Copper Mountain						See Green Mountain group.
	Copper Peak						See Green Mountain group.
	Copper Queen I	Not determined	19	5S	19E	MD	Produced a small amount of gold in 1890 and 5750 pounds of copper in 1912; also a small amount of gold and silver, according to Housemann (in Aubury 08:268). The

COPPER (CONT.)

MAP NO.	CLAIM, MINE, OR GROUP	OWNER NAME, ADDRESS	SEC.	T.	R.	B. & M.	REMARKS
	Copper Queen I (continued)						vein strikes northwest between schist walls. Upper part of vein is carbonate ore with native copper, and lower part is blue sulfide. Width of vein, 4 feet. Workings in 1908 consisted of a shaft 40 feet deep and another incline 15 feet deep. (Aubury 05:216; 08:268; Laizure 28:125.)
	Copper Queen II						See Green Mountain group.
	Cornett	Not determined	30	6S	17E	MD	Cathay district. Wall rocks schistose greenstone. Vein is 34 inches wide. Near granodiorite contact. Sulfide ore; 160 sacks yielded 20 percent copper, $2.26 to $4.60 in gold per ton. (Aubury 08:264; Eric 48:266.)
	Crown Point						See Good View
	Discovery						See Green Mountain group.
	Ferrari	Not determined	30	3S	16E	MD	Near Castagnetto mine. In 1902 had shaft 50 feet deep and several open cuts; $5000 pocket of gold found in mine about 1902. Copper ore reported by Aubury 05:215 to be of fair grade.
							See Green Mountain group.
8	Good View (Cavan-San Jose)	Mine-Metal Properties, Inc. c/o C. C. Kellogg, 50 - 26th Street, Merced	4,5 8,9 32	8S 7S	18E 18E	MD MD	Patented. 161.3 acres; 3 miles west of Green Mountain School (Aubury 05:206-209; 08:257; Lowell 16:573; Castello 21:103; Laizure 28:77; herein). Property includes Rothschild. Sunset, S.A., Crown Point, Little Giant, San Jose, Stonewall Jackson, and Copper King claims. Property formerly called Cavan and Cavan-San Jose mine.

COPPER (CONT.)

MAP NO.	CLAIM, MINE, OR GROUP	OWNER NAME, ADDRESS	SEC.	T.	R.	B & M	REMARKS
9	Great Northern	Not determined	2,3 10 11	7S	17E	MD	Vein 1 to 8 feet in schistose greenstone. Workings once included three shafts 25, 70 and 110 feet deep, and 40 feet of crosscuts from deepest shaft. Ores consist of chalcopyrite, pyrite, bornite, and green carbonates. (Aubury 05:209; Lang 07:1010; Aubury 08:260; Laizure 28:77). One mile west of Pocahontas.
	Green						See Johnnie Green.
10	Green Mountain group	Dr. Felix A. Smith, 507 Medical Bldg., Oakland 12	3 10	8S	18E	MD	Not patented, (Aubury 05:206; Eng. and Min. Jour. 07, vol. 84, p. 964; herein), includes Amador, Copper Chief, Copper King, Copper Mountain, Copper Peak, Discovery, Francisco, Green Mountain, Juliet and Last Chance claims. Once included the Lone Tree, Hecla, Ironclad, Verde, Iron Mask, Calumet, Monte, Nevada, Montana, American, Copper Mount, and Copper Queen claims. Some of these were later renamed and some have passed into different ownership. (Laizure 28:77,125).
	Heiser	Not determined	19	5S	19E	MD	Located 3 miles east of Mariposa; adjoins Copper Queen I mine. About 1908 had a shaft 30 feet deep. Ore reported to run 19 to 37 percent copper. (Aubury 08:268.)
11	Indian Peak group	Not determined	19 20	6S	20E	MD	Indian Peak claim located about 1897; others in 1913. Carbonate and sulfide ores in limestone gangue. Wall rocks limestone and granodiorite. In 1914 workings consisted of two inclined shafts each 35 feet deep, both sunk in ore; also four tunnels 120, 20, 50 and 20 feet long respectively. (Lovell 16: 574.)
	John Diaz						See Owl Copper.

COPPER (CONT.)

MAP NO.	CLAIM, MINE, OR GROUP	OWNER NAME, ADDRESS	SEC.	T.	R.	B & M	REMARKS
12	Johnnie Green (Green, Johnny Green Jr.)	Not determined	31	7S	18E	MD	(Lang 07:964; Eric 48:267; herein.)
	Johnny Green						See Johnnie Green.
	Johnny Green Jr.						See Johnnie Green.
	Juliet						See Green Mountain group.
13	Klondike and Last Chance	Not determined. Probably Joe A. Marino (½), Rt. 2, Box 239, Merced; and John V. Johnson (½), Hornitos	NW¼ 16 / NE¼ 17	7S / 7S	18E / 18E	MD / MD	Small production of relatively low-grade ore in 1912-13 by R. S.Calhoun of Lewis, California (Eric 48:268). Ore contains some gold.
14	La Victoria (La Victoire, Victoire, Tandem)	Ralph E. and Libbie N. Dailey, 1165 West 22d Street, Merced	4,9 / 10	4S	16E	MD	(Browne 68:207-219; Aubury 05:213; Lang 07:1010; Aubury 08:265; Castello 20:103-104; Laizure 28:79; Julihn and Horton 40:167-168; Cox and Wyant 48:127-132; herein) 10 acres, patented.
	Last Chance						See Green Mountain group.
	Little Giant						See Good View.
15	Lone Tree	Not determined	4 / 33	8S / 7S	18E / 18E	MD / MD	Produced 12,451 pounds of copper in 1905. Well-defined mineralized shear-zone strikes N. 30° - 40° W., 300 feet wide. Within this zone are two well-defined veins between which the wall rocks are more or less mineralized. Mine produced carbonate ore to depth of 150 feet; chiefly sulfides below that level. In 1908 there were several shafts between 25 and 100 feet deep and a main

COPPER (CONT.)

MAP NO.	CLAIM, MINE, OR GROUP	OWNER NAME, ADDRESS	LOCATION				REMARKS
			SEC.	T.	R.	B B M	
	Lone Tree (continued)						shaft 200 feet deep with drifts and stopes on the 150-foot level. (Aubury 08:257; Laizure 28:77.) Operated as part of the Green Mountain mine in early 1900's.
	Lookout	Not determined	NE¼ SE¼ 32	7S	18E	MD	A mile and a quarter northwest of Green Mountain Forest Service Lookout. Oxidized copper ore, chiefly malachite, occurs in a northeast-trending shear zone in phyllite and amphibolite schist of Upper Jurassic age. Wall rocks adjacent to the shear zone are porous and thoroughly leached by hydrothermal solutions. Workings consist of two old shafts 60 feet and 30 feet deep, respectively, and numerous shallow shafts and cuts. The most recent working is a shallow shaft inclined 80° E. supported by 8-inch stalls and entered by a chain ladder. It is equipped with a lightweight hoist. Exploration work was being carried on intermittently in the summer of 1956.
16	Mammoth group	Not determined	8,9	8S	18E	MD	Adjoins Good View claims on south. Property originally consisted of 14 claims. Three steeply dipping, northwest-trending veins within a zone 600 feet wide. In 1908 the east vein had a shaft 80 feet deep, the middle vein had shafts 20 and 25 feet deep, and the west vein one shaft 50 feet deep; west shaft exposed 3 feet of copper ore containing some gold. Some sulfide ore shipped ran 20-22 percent copper (Aubury 08:260). Country rocks in this vicinity are slate and chiastolite-mica schist.
	Mormon Bar group	Not determined	30	5S	19E	MD	In vicinity of Mariposa district fair grounds; copper showing in quartz porphyry country rock (Lowell 16:574).
	Mountaineer						See Barrett.

COPPER (CONT.)

MAP NO.	CLAIM, MINE, OR GROUP	OWNER NAME, ADDRESS	LOCATION SEC.	T.	R.	B & M	REMARKS
	New Year						See Barrett.
	Owl Copper (John Diaz)	Joe C. Souza and Tony Diaz, 542 R Street, Merced	12	6S	16E	MD	Vein 3 feet wide strikes northwest; schistose greenstone walls. Ore minerals include cuprite, azurite, chalcopyrite, and chrysocolla; had shaft 24 feet deep in 1902 (Aubury 05:213; 08:265).
17	Pocahontas	Estate of Eben N. Briggs, c/o L. M. Olds, Room 402, 57 Post Street, San Francisco	1,12 14	7S	17E	MD	Most productive copper mine in Mariposa County. (Aubury 05:209-210; Lang 07:1010; Aubury 08:260-262; Forstner 08:747; Lowell 16:573; Castello 20:103; Julihn and Horton 40:166; herein).
	Rihn Ranch	Not determined	13	3S	15E	MD	Two shafts 60 and 40 feet deep expose chalcopyrite and oxidized copper ore (Aubury 05:215).
	Robinson	Georgia Robinson, address not determined	N½ 21	6S	18E	MD	On the north bank of Mariposa Creek about 100 yards west of the White Rock fire-control road. Quartz vein in schist. Ore is chiefly chalcopyrite and pyrite in quartz and although the precious metal content is promising the copper content makes the ore unacceptable at some smelters. Main working is a shaft with several short laterals. Worked intermittently by lessees.
	Rothschild						See Good View.
	San Jose						See Good View.
	Stonewall Jackson						See Good View.
	Sunset						See Good View.
	Tandem						See La Victoria.

COPPER (CONT.)

MAP NO.	CLAIM, MINE, OR GROUP	OWNER NAME, ADDRESS	LOCATION				REMARKS
			SEC.	T.	R.	B & M	
18	Toad	Not determined	31	7S	18E	MD	Small masses of high-grade sulfide ore occur in schist. Main shaft was down 200 feet in March 1904. One carload of ore was shipped early in 1904. (Eng. and Min. Jour. 04, vol. 77, no. 9 p. 375; 04, vol. 77, no. 11, p. 575; Lang 07:964; Eric 48:269).
	Victoire						See La Victoria.
19	Ward	Not determined	34	6S	19E	MD	Adjoins Indian Peak group and geology is similar; ores contain some native copper.
	White Knob						See White Rock.
20	White Rock (Copper King, White Knob)	James Helm Le Grande (1943)	NW¼ 14	7S	17E	MD	A quarter of a mile south of White Rock silica mine. Discovered in 1900. Ore occurs in a shear zone in schist, mineralized rock being 25 or more feet wide. Oxide and carbonate ore reaches a depth of 100 feet, sulfides below that depth. In 1908 the principal working was a 150-foot shaft from which 175 feet of drifts had been driven; also numerous minor workings. Several carloads of ore were shipped 1900-02 that yielded 35 percent copper; one carload ran 40 percent copper. Oxidized ore reported to carry $1.50 to $2.50 in gold per ton at the old price of gold, and 3½ ounces of silver. Produced 17,000 pounds of copper in 1916. (Aubury 05: 210-212; 08:262-264; Reid 08:49; Laizure 26:211; Julihn and Horton 40:211; Eric 48:270.) Reported to be a patented mine (Eric 48:270) but was not on County Assessor's list in 1954.
	Wild Cat						See Barrett.

LODE GOLD

MAP NO.	CLAIM, MINE, OR GROUP	OWNER NAME, ADDRESS	SEC.	T.	R.	B & M	REMARKS
21	A-J (Burkhart)	B. F. and Ruth L. Burkhart, Bear Valley (1954)	26	4S	16E	MD	Located 3½ airline miles west of Bear Valley. Discussed herein.
22	Adelaide and Anderson	Alfred W. Stickney, 435 Hillcrest Road, San Mateo	NE¼ 22, NW¼ 23	3S	16E	MD	Four miles south of Coulterville and 1 mile north of Merced River in Horseshoe Bend Mtns. (Fairbanks 90:57; Storms 95:217; Eng. and Min. Jour. 05, vol. 79, p. 1292, 1910; Lowell 16:575; Laizure 20:104; herein.)
	Aden						See Menlo Consolidated.
23	Agua Fria Canyon	Not determined	NW¼ 20	5S	18E	MD	Four-inch vein on west side of canyon half a mile north of confluence of Carson Creek. Two small lots of ore mined in 1900 and 1908 totaled 12.65 tons and yielded $131.07 or $10.37 per ton. (Logan 35: chart facing p. 188.)
	Akoz (B.A.B. Radium, Asposozien)	Carlin estate, c/o Mrs. P. Erickson, La Grange	9,10	4S	18E	MD	See under zinc. Small gold production 1941 and 1944-47.
	Alabama						See Whitlock group.
	Aladin	Not determined. Under lease to H. H. Odgers	NW¼ 30	4S	18E	MD	In the Whitlock Creek district. Adjoins the Nutmeg mine on the northwest. Property being developed by H. H. Odgers (1956).
	Alarid prospect	Frank Casaccia, Mariposa	8	5S	18E	MD	Located 2½ miles northwest of Mariposa. Quartz vein about 2½ feet wide strikes east and dips 60° S. Small tonnage of ore taken out in 1917-18 that averaged $15.00 per ton at the prevailing price of gold. At that time workings consisted of a shaft 20 feet deep and a drift 100 feet long.

LODE GOLD (CONT.)

MAP NO.	CLAIM, MINE, OR GROUP	OWNER NAME, ADDRESS	LOCATION SEC.	T.	R.	B & M	REMARKS
	Albert Austin group						See Banner group.
24	Alice (Alice Quartz I)	Horace Meyer, Mariposa	NW¼ 16, NW¼ 17	5S	17E	MD	Located 4½ miles west of Mt. Bullion just south of the Mt. Bullion-Hornitos road and about 1200 feet east of the Long Mary mine. Equipped (Sept. 1954) with a wooden headframe but no hoisting or milling machinery. Property consists of two claims located in the 1850's. Early history not recorded. Active in 1901, 1902, 1903, 1906, and 1908, producing 1284 tons of ore which yielded $13,500 or about $10.51 per ton at old price of gold. Mined for a short time in 1936 producing 268 tons of ore which yielded approximately $7.54 per ton at present price of gold. Quartz vein averages about 3 feet wide, strikes northwest and is nearly vertical. Slate wall rocks. In 1914, workings consisted of a 200-foot shaft, three levels with 360 feet of drifts, one 27-foot winze and a 38-foot raise. Some additional work was done in 1936 (Lowell 16:575; Castello 20:105; Logan 35:188).
	Alice Quartz II	Chris Mills, % Bagby Store, Coulterville	36	3S	16E	MD	Adjoins claims of the Red Bank mine 2½ miles northwest of Bagby. Pocket mine in greenstone. Small production during last 3 years. Worked intermittently.
25	Alta	Lena Giusto, Box 66, Angeles Camp	19	3S	17E	MD	Located 6 miles southeast of Coulterville near head of Flyaway Gulch. Part of the Flyaway group of mines. Lies between Talc and Silver Right claims. Property consists of 15.25 acres. A pocket mine with small intermittent production. Quartz vein in greenstone and serpentine walls strikes northwest and dips steeply northeast. Superficial workings. Idle.
	Alvina						See Quail.
	Amelia						See Ruth Pierce.

LODE GOLD (CONT.)

MAP NO.	CLAIM, MINE, OR GROUP	OWNER NAME, ADDRESS	SEC.	T.	R.	B & M	REMARKS
	American Eagle						See under Zinc.
	Anderson I						See Clearinghouse.
	Anderson II		NE¼ 22	3S	16E	MD	See Adelaide and Anderson.
	Anderson and Aul						See Aul and Anderson.
26	Anita	Not determined	6,7	2S	16E	MD	On Willow Creek close to State Highway 132. Quartz vein in greenstone wall rocks.
	Ann	Not determined	22	7S	17E	MD	White rock district near Ganns Creek. Produced a small tonnage of low-grade ore in 1927.
27	Annabelle prospect	Mr. and Mrs. Sherman B. Pickard and Bryan A. Miller, Coulterville	NW¼ 19	3S	17E	MD	Located 50 feet north of Highway 49 and 50 feet northwest of Scotch Gulch, 4 airline miles southeast of Coulterville. Unpatented claim owned and operated by Pickard and Miller (discussed herein).
	Apolinaris	Not determined	22	5S	18E	MD	Former Mariposa Grant mine. Land has reverted to agricultural status. Last worked in 1908 when 9 tons of ore were mined which yielded $126.29 or an average of $14.05 per ton at the old price of gold. (Logan 35: 186.)
	Apollo	Not determined	13	5S	17E	MD	One of Las Mariposas Grant mines. Land has reverted to agricultural status. Near Princeton mine. Active 1936-38 producing 99 tons of ore which yielded an average of 32.69 oz. of gold and 8 oz. of silver, an average of $11.70 per ton (Logan 35:188).
	Arcturus (B.V.D.)	Not determined	30	6S	18E	MD	In the Guadeloupe Mountains 4 airline miles southeast of Cathay School. Small tonnage of low-grade ore was

LODE GOLD (CONT.)

MAP NO.	CLAIM, MINE, OR GROUP	OWNER NAME, ADDRESS	SEC.	T.	R.	B & M	REMARKS
	Arcturas (B.V.D.) (continued)						produced in 1908 (U.S. Bur. of Mines records). Principal development work 1921-23 by B.V.D. Mining Co. which had six unpatented claims, the Rich-Luckett, Arcturas, Day Break, Moore Hill, Arcturas No. 9, and Arcturas No. 7. Northwest-trending northeast-dipping vein up to 9 feet wide cuts granodiorite. Ore is high in sulfides. Mine is half a mile southwest of the Rich mine and west of the Francis mine. Idle (Boalich 22: 364; Laizure 28:82).
28	Argo (Pioneer)	Walter D. McLean, Coulterville	NW¼ 15, NE¼ 16	2S	17E	MD	Located 7 airline miles northeast of Coulterville in the Greeley Hill district. Two claims (Eng. and Min. Jour. 31, vol.132, no. 2, p. 84; Julihn and Horton 40: 135; herein).
	Argonaut						See Quartzburg group.
	Arkansas Flat	Not determined	21	5S	18E	MD	Located two miles west of Mariposa on a ridge between Carson Creek and McBrides Gulch. Pocket mine in slate and greenstone (?). Active 1906-08 and 1911. In that period 8.75 tons of ore were milled that yielded $953.91 or an average of $109.02 per ton. A Las Mariposas Grant mine. Has reverted to agricultural land status (Logan 35:188).
	Arkell group						See Hite.
29	Artru (Dorothy)	C. W. and Velma Worley, Mariposa	NW¼ 27	4S	18E	MD	Patented property consisting of 6.45 acres in the Colorado district near the junction of Long Canyon and Saxon Creek, 5 airline miles north of Mariposa. Worked by H. Artru for a 15-year period ending in 1912. No material amount of work done since that time. Pockets of high-grade ore, including crystallized gold, occur in an altered porphyry dike cutting crinkling slate (Lovell 16:576).

LODE GOLD (CONT.)

MAP NO.	CLAIM, MINE, OR GROUP	OWNER NAME, ADDRESS	SEC.	T.	R.	B & M	REMARKS
	Artru (Dorothy) (continued)						Dike strikes roughly NW and dips 70° SW (Castello 20:106). In 1914, workings consisted of a 75-foot shaft, 450 feet of drifts, four crosscut tunnels, each about 140 feet long, and several raises (Lowell 16:576).
	Aul and Anderson	Not determined	30	5S	18E	MD	Close to Highway 140, 5 miles west of Mariposa in the vicinity of the Sorrel (Sarle), Turner, and Ortega mines. Vein system strikes N. 10° W. to N. 20° W., approximately vertical. Wall rocks are granodiorite, granodiorite porphyry and aplite. During 1900, a return of $76.74 came from 4.18 tons of ore for an average of $18.35 per ton (Logan 35:188).
	Austin group						See Golden Key.
30	B and M (Fournier)	Not determined	3	6S	18E	MD	On Buckeye Creek 4 airline miles south of Mariposa or 2 miles SW of Mormon Bar. Last known period of activity was 1934-39 when 55 tons of ore were shipped which yielded $5000.00 in gold and silver. May be the same property from which 8.5 tons of pocket ore yielded $7523.04 between 1911 and 1914 (Logan 35:188). Two approximately parallel veins strike N. 55° W. and dip southwest. Wall rocks are biotite hornblende quartz diorite. East vein has an apparent dip at the surface of about 70°. It is equipped with a vertical shaft between 65 and 110 feet deep (largely filled with debris) with a northwest drift about 50 feet long. West shaft was originally 110 feet deep with 185 feet of drifts (Laizure 35:28). Ore carries only 3 percent of sulfides. Average milling ran $25.00 to $30.00 per ton with high-grade running as high as $643.00 per ton.

LODE GOLD (CONT.)

MAP NO.	CLAIM, MINE, OR GROUP	OWNER NAME, ADDRESS	SEC.	LOCATION T.	R.	B & M	REMARKS
31	Badger (Prescott)	Mrs. Charles B. Cavagnaro, Hornitos	2	5S	16E	MD	In the Quartzburg district about 3½ airline miles northeast of Hornitos. Adjoins the Number 5 mine on the north (Laizure 28:80; 35:29; herein).
	Baker prospect	Not determined	18	5S	18E	MD	Located 2000 feet southwest of Princeton vein and mine. Produced 18 tons of ore in 1912 from which $248.24 was realized or an average of $13.81 per ton (Logan 35: 188). Northwest-trending vein, slate walls.
32	Balance	Walter J. Lautenschlager, 626 South Catalina Street, Los Angeles 5	NE¼ 3	3S	16E	MD	On east branch of Mother Lode half a mile south of Coulterville. Patented property consisting of 2.34 acres. Adjoins Louisa claim on south and Venture claim on north. Quartz-mariposite-ankerite vein strikes N. 45° W. and dips 50° NE. Ore minerals are free gold, auriferous pyrite, and galena. Slate and greenstone wallrock. Production and extent of former workings not recorded. Idle for more than 50 years. See Mary Harrison group.
	Bald Eagle						See Sultana group.
	Bandaretta						See Bandarita.
33	Bandarita (Bandaretta, Eclipse)		NW¼ 7	3S	18E	MD	On the North Fork of the Merced River at its confluence with Gentry Gulch. (Storms 94:167; 96:216; Laizure 28:89; Julihn and Horton 40:144; herein.)
	Bank of California	Not determined	27	4S	18E	MD	Mine not identifiable from available descriptions. Could be one of two workings in E½ sec. 27. Probably one of mines later gathered together under the name Mountain Belle group. Owner (1914) G. E. Dunbar of Kalamazoo, Michigan. Quartz vein in porphyry dike cutting slate. Workings once consisted of a 50-foot and a 75-foot shaft, both caved. (Lowell 16:576.)

LODE GOLD (CONT.)

MAP NO.	CLAIM, MINE, OR GROUP	OWNER NAME, ADDRESS	SEC.	T.	R.	B & M	REMARKS
34	Banner group (Albert Austin group, London)	Not determined	NE¼ 31, SE¼ 30	4S	18E	MD	Several claims lying between the Geary and Nutmeg mines 4 miles northwest of Mariposa via the Whitlock Creek road. Principal vein strikes N. 20° W., dips nearly vertically, and from the disposition of the workings it appears that there were several cutter veins. Schistose greenstone wall rocks. Main shaft, vertical, is partly caved and a line of caved stopes extends north of the shaft for several hundred feet. Several east-trending open cuts apparently are sunk on cutter veins.
	Barfield						See Mt. Gaines.
	Barley Field						See Landrum.
	Barney Kane (Kane Quartz, Mayflower)	Not determined	2	5S	19E	MD	East slope of Buckingham Mountain 8 airline miles northeast of Mariposa and 1 mine northwest of Buckingham Mountain School. Owned in 1927 by C. N. Kane, Mariposa; in 1932 by Mrs. Barney Kane of Merced and Roy Green, Usona. Quartz vein about 4 feet wide strikes northeast and dips 55° southeast in granitic wall rocks. Vein reputed to sample 1.31 oz. of gold. Ore is native gold in quartz. Workings consist of two tunnels 60 feet and 140 feet long, a 35-foot vertical shaft, and a connecting 18-foot drift (Laizure 28:119; 35:29).
	Barrel Springs	Not determined; probably Horace Meyer, Cathay	SE¼ 23	5S	17E	MD	Last active in 1905 when 47½ tons of ore yielded $708.68 or an average of $14.89 per ton (Logan 35:188). Land has probably reverted to agricultural status. One of the mines of Las Mariposas grant.
	Bart						See Black Bart group.
	B. B.						See Bogan and Baitelle.

LODE GOLD (CONT.)

MAP NO.	CLAIM, MINE, OR GROUP	OWNER NAME, ADDRESS	LOCATION				REMARKS
			SEC.	T.	R.	B & M	
35	Bean Creek (Epperson) group (Kohler, Walsh, Caroline, and Blowout claims)	Mrs. Gladys Schutte, Coulterville	21,28	2S	17E	MD	Patented property consisting of 66.6 acres on Bean Creek 2 airline miles east of Greeley Hill School. Vein system strikes N. 85° W. with an average dip of 65° N. Wall rocks are slate, schist, and hornfels of the Paleozoic Calaveras group. Idle.
	Bear Creek						See Malone.
	Bearfield						See Mount Gaines.
	Belces	Not determined	19	3S	17E	MD	Mariposa County Register of Mines for 1903 indicates that mine is close to the Buckhorn Peak road (southwest side) at the extreme eastern edge of the section. Same source lists the vein as 3 to 5 feet wide with a northeast strike and 45° SE dip. Ore reported to contain native gold and auriferous pyrite. Workings in 1903 reported to consist of a 55-foot inclined shaft and a 50-foot open cut. (Wilkinson 04:10.) Land has probably reverted to agricultural status.
36	Big Break, Big Break Ext. No. 1, Contact lode	Bryan A. Miller, Box 36, Coulterville (1954)	14	3S	16E	MD	Coulterville district, about 2½ airline miles southeast of Coulterville; adjoins Virginia mine on the northwest. Three unpatented contiguous claims. Country rock is greenstone, slate, and serpentine cut by basic dikes. Quartz vein strikes northwest and dips northeast, developed by 40-foot shaft with some connecting drifts. Development work is done intermittently; no recorded production.
	Big Buck						See Yellow-Metal group.

LODE GOLD (CONT.)

MAP NO.	CLAIM, MINE, OR GROUP	OWNER NAME, ADDRESS	SEC.	T.	R.	B & M	REMARKS
37	Big Lode	(2/3) James K. and Bernard Lindsey, 5301 Harter Lane, La Canada (12 acres) (1/3) John Lindsey, Coulterville (6 acres)	29,30	3S	17E	MD	Located 7 miles southeast of Coulterville near head of Flyaway Gulch. One of the Flyaway group of mines discovered prior to the 1890's. Vein occurs along a series of multiple dikes of albitite, norite, and diorite. The dike series divides serpentine from sheared greenstone. Vein-dike series strikes N. 50° W. and dips 55-60° NE. In addition to pockets of gold which occur in both albitite and quartz there are masses of mineralized serpentine which contain some gold. During 1913, 117 tons of ore mined by G. A. Helander of Coulterville yielded 68.02 oz. of gold and 13 oz. of silver. Shaft and adjacent workings inaccessible in September 1954. Idle.
	Big Oak	Not determined	NE¼	5S	17E	MD	Las Mariposas Grant mine active 1902, 1904, 1906 and 1909. During this period 98.78 tons of ore was mined which yielded $1564.71 or an average of $15.85 per ton. (Logan 35:188.)
	Billings						See Granite King.
	Bill Jones						See Number Nine.
	Bill Jones and McCall						See Number Nine.
38	Black I	Robert Gemblin, Coulterville	12	3S	17E	MD	Quartz vein in slate probably on same vein system as Sunshine claims. No recorded production. Had a shaft 170 feet deep in 1868. Wall rocks described as limestone and yellow slate. Vein strikes west and dips south, according to Browne (68:31).
	Black II						See Confidence.

LODE GOLD (CONT.)

MAP NO.	CLAIM, MINE, OR GROUP	OWNER NAME, ADDRESS	LOCATION				REMARKS
			SEC.	T.	R.	B & M	
39	Black Bart group (Cal-Penn-Tex group, Bart)	Franz E. Ryburg, Coulterville	27	3S	17E	MD	East side of Solomon Gulch above junction with Black Bart Gulch, 11 miles southeast of Coulterville or 3 miles southeast of Buckhorn Peak via Highway 49 and the Buckhorn Peak road. Property consists of three unpatented claims, the Black Bart, Worms Turn, and Cal-Penn-Tex. The ground surface in the vicinity has been placered. Ore occurs in a quartz vein which parallels and in places is interlaminated with a light-gray albitite dike. The dike-vein system strikes N. 65° W. and dips 25-30° NE. Gold also occurs in rust-colored hydrothermal alteration pockets in albitite associated with chlorite (Storms 94:169).
	Black Cat	Etta Barnard and S. W. Wilmert, Hornitos	21	4S	16E	MD	Hornitos district, 2½ airline miles north of Quartzburg School and half a mile north of the Silver Lead mine. Occurrence similar to Silver Lead mine. Undergoing development work 1953-55.
40	Black Hill (Pumpkin)	Louise A. Boyd, 265 California Street San Francisco 11	33	2S	16E	MD	Adjoins Margaret claim on the north, 1/2 mile northwest of Coulterville. Patented property consisting of 40 acres. (Storms 96:218; 00:147; Eng. and Min. Jour. 00: vol. 76, p. 167, 647; herein.)
	Black Jack	Not determined	1	3S	16E	MD	Quartz vein in greenstone wall rocks. Produced 406.35 oz. gold in 1900-01 from an unknown tonnage of ore, and 1043.35 oz. gold and 14 oz. silver in 1909-10, also from an unknown tonnage of ore(U.S. Bur. Mines records).
	Black Log	Not determined; probably Catherine B. Trabucco and Elsie R. Davidio, Midpines	13	4S	18E	MD	Close to Buffalo or San Domingo mine. Quartz vein 6 inches to 1 foot wide in prophyry wall rocks. Chief working in 1914 was a shaft 70 feet deep. Working caved and inaccessible in August 1954 (Lowell 16:576).

LODE GOLD (CONT.)

MAP NO.	CLAIM, MINE, OR GROUP	OWNER NAME, ADDRESS	SEC.	T.	R.	B & M	REMARKS
	Black Oak	Not determined	8	5S	20E	MD	Three miles east of Darrah. Last period of activity was 1933-36 during which time 153 tons of ore was milled which yielded 144.33 oz. of gold, 37 oz. of silver and 1507 pounds of copper (U.S. Bur. Mines records).
41	Black Spider (Gray Eagle, Blue Jay)	Walter McLean, Coulterville	19	3S	17E	MD	Located 6 miles southeast of Coulterville. Adjoins Flyaway group of mines on northwest. Quartz vein strikes N. 60° W. and dips 50° NE. Vein is at contact of greenstone on northeast and serpentine on southwest. Gold occurs in pockets and in sulfide-impregnated serpentine. Workings consist of several tunnels driven N. 40° E. and several caved shafts. Worked intermittently by hand methods. During 1935, about 480 tons of ore yielded 85.48 oz. of gold. In 1914 workings consisted of a 133-foot inclined shaft, 355 feet of drifts, a 300-foot crosscut tunnel and two stopes (Lowell 16:582; Castello 20:117; Laizure 28:89).
	Black Spring	Not determined	18	5S	18E	MD	Upper Agua Fria Creek near Mt. Bullion. One of Las Mariposas Grant mines. Active in 1905, 1906, and 1909. During this period 56.82 tons of ore yielded $620.90 or an average of $11.09 per ton at the prevailing price of gold. (Logan 35:188).
	Blowout						See Bean Creek group.
	Blue Bell (Jumbo)	Not determined; probably Hiram Branson, Midpines	9	4S	20E	MD	On the north slope of the Chowchilla Mountains close to the south Fork of the Merced River 6 miles northeast of Jerseydale. Northwest-striking vein having an average width of 5 feet dips 72° northeast. Porphyry footwall and slate and granite hanging wall. Workings in 1941 consisted of a 145-foot crosscut tunnel and 80 feet of drift. Property consists of three unpatented claims, the Blue Bell, Jumbo, and Blue Bell extension.

LODE GOLD (CONT.)

MAP NO.	CLAIM, MINE, OR GROUP	OWNER NAME, ADDRESS	LOCATION SEC.	T.	R.	B & M	REMARKS
	Blue Bell (Jumbo) (continued)						(Lowell 16:577; Castello 20:107; Laizure 28:80.)
	Blue Jay						See Black Spider.
	Blue Lead						See Garibaldi.
	Boarding House Vein	Not determined	18	5S	18E	MD	Las Mariposa Grant mine. Active 1902-03 producing 133.69 tons of ore yielding $2060.30 or an average of $15.43 per ton (Logan 35:188).
42	Bob McKee (McKee)	Not determined	SE¼ 20	2S	18E	MD	Located 5 airline miles northeast of Coulterville or 1½ miles east of Greeley School. Quartz vein 1 to 3 feet wide strikes N. 20° E. and dips 65° NW. Wall rocks are slate and hornfels. Vein matter consists of milky quartz with pyrite and wall rock inclusions. Workings consist of two shafts, both caved, and a series of open cuts along the vein which may represent caved stopes. Idle in August 1954. Mine unequipped.
43	Bogan and Baitelle (B. B.)	John Wildt, Midpines	W½ 26	4S	18E	MD	Patented property consisting of 5.15 acres in the Colorado district 1 mile east and slightly south of Colorado School. Quartz vein 2½ feet wide (average) strikes northwest and dips steeply northeast. Wall rocks are metasediments of the Paleozoic Calaveras group. In 1914 mine consisted of four adit levels, respectively 60, 100, 250, and 300 feet long. Some specimen ore was produced from pockets. In 1924, fifty tons of ore yielded 47.76 oz. of gold and 1 oz. of silver (U.S. Bur. Mines records). A small tonnage (26) mined in 1937-39 yielded 5 oz. of gold. (Lowell 16:577; Castello 20:107; Laizure 28:81.)

LODE GOLD (CONT.)

MAP NO.	CLAIM, MINE, OR GROUP	OWNER NAME, ADDRESS	LOCATION				REMARKS
			SEC.	T.	R.	B & M	
44	Bonanza I	Lena Giusto Box 66, Angels Camp	29	3S	17E	MD	Located 7 miles southeast of Coulterville. One of Flyaway group of patented claims. Quartz vein in greenstone wall rocks strikes roughly N. 70° W. and dips steeply north. Claim largely undeveloped. Adjoins Southern Cross claim on west and Sweetwater drain on east.
	Bonanza II						See Spread Eagle group.
	Bonanza III						See Quail.
45	Bondurant (Hathaway-Bondurant)	A. E. Adams, et al., 368 Maude Avenue, San Leandro	25	2S	17E	MD	A patented property consisting of 8 acres near the North Fork of the Merced River 12 miles east of Coulterville (Goodyear 88:347; Storms 96:216,225; Lowell 16:577; Castello 20:108; Laizure 24:81,124,35:46; herein). Adjoins the Louisiana mine on the east.
	Booth	Not determined	5	5S	20E	MD	Located 2 miles east of Darrah on Footman Ridge. Quartz vein a few inches to 18 inches wide strikes northwest and dips about 45° NE. Granitic wall rocks. In 1914 there were two shafts 28 and 95 feet deep and a tunnel 120 feet long (Lowell 16:577).
46	Boston	Walter Lautenschlager, 626 South Catalina Street, Los Angeles 5	32,33	2S	16E	MD	Property consists of one patented claim in Mahoney Gulch 1½ miles northwest of Coulterville and three-quarters of a mile north of State Highway 132. There is no record of production from this mine and there are no visible workings. Probably located because veins penetrate the claim at depth.
	Boulder						See Permit.
	Bouvier						See Carson.

LODE GOLD (CONT.)

MAP NO.	CLAIM, MINE, OR GROUP	OWNER NAME, ADDRESS	SEC.	T.	R.	B & M	REMARKS
	Bower Cave						See Horseshoe I mine.
	Bowman	Not determined	26	4S	18E	MD	Mine not distinguishable in field by available description. Claim embraced. Reputed to be a pocket mine. Several shafts and prospect holes in 1914 (Lowell 16:577).
47	Bozeman	Walter J. Lautenschlager, 626 South Catalina Street, Los Angeles	N½ SE¼	3S	16E	MD	Property consists of 1 patented claim located parallel to and about 2000 feet east of the Malvina claim half a mile southwest of Coulterville. There is no record of activity or production from this mine and there are no visible workings. Veins may have been encountered at depth in process of working the adjoining Malvina mine.
	Brooks						See Doss.
48	Brown Bear	Gertrude Chilton and Imogene Flore, 123 Schreiner Place, San Antonio, Texas	7	4S	20E	MD	Patented property consisting of 41 acres near the South Fork of the Merced River. Two claims, the Brown Bear and Brown Bear South Extension.
	Buckeye						See Granite King.
	Buckingham Mountain						See Mount Buckingham.
49	Buena Vista group (Busch, Washington-- Buena Vista)	C. W. and Velma Worley, Mariposa	20,21 28,29	4S	18E	MD	Located in Colorado district 5 airline miles north and slightly west of Mariposa. Property consists of three patented claims, the Washington, Buena Vista, and Phoenix. Formerly included the Talc, Lucky Lindy, and Charles. Close to the Cleveland claim on the north but not adjoining it. (Lowell 16:578; Castello 20:109; Laizure 24:120-121; herein.)

LODE GOLD (CONT.)

MAP NO.	CLAIM, MINE, OR GROUP	OWNER NAME, ADDRESS	SEC.	T.	R.	B & M	REMARKS
50	Buffalo (San Domingo)	Vernon and Clara Tharp, 1850 Central Avenue, Alameda	SW¼ 13	4S	18E	MD	Located 4 miles north of Midpines on Trabucco Creek (Castello 20:136; Boelich 22:365; Laizure 28:115; 35:29, herein).
	Bull Dog						See Permit.
	Bullion Hill						See American Eagle under Zinc.
	Bull Pup						See Permit.
51	Bunce	D. D. Martin, Box 1148, Mariposa	SE¼ 22	3S	18E	MD	On the east side of a tributary to Halls Gulch, half a mile west of the Briceburg-Kingley road. Quartz vein 1 to 3 feet wide strikes N. 35° W. and dips steeply east. Wallrocks are black slate of the Paleozoic Calaveras group. Workings consist of several shafts, one at least 60 feet deep. Worked intermittently by the owner.
52	Bunker Hill (Squirrel)	Not determined	36	3S	19E	MD	On Pinoche Ridge 1½ miles southeast of Hites Cove. Located in 1851. Had one of the first quartz mills built in California. Equipped with wooden stamps. Owner (1920) Harry Taylor. Previous owner Charles Lewis. Rocks in this area are slate and dark hornfelses. Veins strike northwest and dip southwest. Mine was not precisely located during this investigation (Castello 20:109).
	Burkhart						See A-J.
	Burns Creek	Not determined; probably Frank Trabucco, Hornitos	34	4S	16E	MD	

LODE GOLD (CONT.)

MAP NO.	CLAIM, MINE, OR GROUP	OWNER NAME, ADDRESS	LOCATION SEC.	T.	R.	B&M	REMARKS
53	Burr (Miles)	Not determined; mineral rights in this area reserved to T.H. Carlon	E½ NW¼ and SW¼ NW¼ 33	2S	15E	MD	Discovered about 1860 by a prospector named Miles or Miler. Early production, according to a private report by Carl B. Gilbert, was estimated to be $350,000. Shallow workings cover several acres. Blanket vein has a general north strike and a gentle east dip through greenstone wall rocks. A small tonnage of low-grade ore was mined during the period 1937-40.
	Busch						See Buena Vista group.
54	Butte	Walter J. Lautenschlager, 626 South Catalina Street, Los Angeles 5	NW¼ 4	3S	16E	MD	A patented property. Adjoins the Potosi and D. Cook mines. There is no record of production for this mine and it was probably worked as part of the Potosi and D. Cook mines.
55	Butterfly	Not determined	5	4S	15E	MD	See Arcturas.
	B. V. D.						
56	Cabinet	Not determined; probably Ralph E. and Libbie N. Dailey, 1165 West 22d Street, Merced	15,16 22	3S	16E	MD	Located 1½ miles east of Horseshoe Bend on the Merced River. Adjoins the Crystal and Lookout claims. Two veins intersect approximately at right angles. The Cabinet vein, striking northwest and dipping steeply east, is intersected by the Crystal vein striking northeast. The latter is reported to contain cinnabar as well as gold. Wall rocks are greenstone (Turner and Ransome 97:7; Storms 96:216).
	Cader Idra						See Kaderitas.
	Caderitas						See Kaderitas.
	Calderon	Edna M. Bauer, Route 3, Box 252B, Merced	7	5S	16E	MD	Located 1 mile northwest of Hornitos. Unpatented. Quartz vein cuts metavolcanic wall rocks.

LODE GOLD (CONT.)

MAP NO.	CLAIM, MINE, OR GROUP	OWNER NAME, ADDRESS	SEC.	T.	R.	B & M	REMARKS
	Calender and Calendonia						See Mountain King group.
	Calendonia						See Mountain King group.
	Cal-Penn-Tex group						See Black Bart group.
56A	Campo (Campodonica, Italian)	Not determined	NW¼ 16	5S	16E	MD	Located 1 mile east of Hornitos (Storms 96:216; Julihn and Horton 40:125; herein).
	Campodonica						See Campo.
	Canyon Wren	Not determined; last known owner (1937) was T. J. Patrick, Mariposa	21	6S	18E	MD	Located 1 mile northwest of Moore Hill or 7 miles east and slightly south of Cathay. Last active in 1937 when 45 tons of ore yielded 41 oz. of gold, 45 oz. of silver and 2747 pounds of copper (U. S. Bur. Mines records).
	Captain Aiken						See Virginia.
	Caroline						See Bean Creek group.
	Carrie Todd						See Texas Hill.
57	Carson (Bouvier)	Not determined; probably Inez Bouvier, et al., 1431 Moraga Street, San Francisco	3,4	4S	16E	MD	In Hunter Valley 8 airline miles north of Hornitos. Adjoins Iron Duke mine on the south (Castello 20:108; herein).
	Castagnetto I						See Pyramid.
	Castagnetto II						See under Copper.
	Castagnetto III						See Chemisal under Copper.

LODE GOLD (CONT.)

MAP NO.	CLAIM, MINE, OR GROUP	OWNER NAME, ADDRESS	LOCATION SEC.	T.	R.	B & M	REMARKS
	Castagnetto IV	Mrs. Mary Castagnetto, 2739 Ritchie Street Oakland 5	26	2S	16E	MD	Located 2 miles northeast of Coulterville on the Greeley Hill road. Small tonnage of low-grade ore was mined in 1941. Quartz vein in greenstone and granodiorite.
	Cavagnaro						See Orange Blossom.
	Centuary (Century)						See Hasloe.
	Century						See Hasloe.
	Challenge, Pine Crest, and Pine Crest Extension	Not determined; probably P. W. Gallis, Hornitos	30	3S	16E	MD	One of several prospects in the vicinity of the Barrett, Castagnetto II, and Chemisal mines, Hunter Valley district. May be same property once known as the Ferrari mine. Geology and occurrence of ore similar to Barrett.
58	Champion I	Car Da Mining Company, a trust estate (½); Adelaide M. Ray (½), c/o C. P. Rose, 1259 North Fuller Ave, Hollywood	28 33	2S	16E	MD	On Blacks Creek 1-3/4 miles northwest of Coulterville (Lowell 16:578; Castello 20:110; Laizure 28:83; Logan 34:183; herein).
59	Champion II	Belle McCord Roberts, 2625 East 10th Street Long Beach	34	4S	18E	MD	Located 1½ airline miles south of Colorado School and 5 airline miles northeast of Mariposa (Goodyear 90:302-304; Storms 96:216; Castello 21:110; Laizure 28:84, 101; herein).
	Champion III	Not determined	16	5S	17E	MD	Close to Mt. Bullion. Last active in 1903, producing 13.88 tons of ore yielding $332.75 or an average of $23.97 per ton at the prevailing price of gold (Logan 35:188). Quartz vein in slate.

LODE GOLD (CONT.)

MAP NO.	CLAIM, MINE, OR GROUP	OWNER NAME, ADDRESS	LOCATION				REMARKS
			SEC.	T.	R.	B & M	
	Champion IV	Not determined	26 27	3S•	19E	MD	Located 4 airline miles northeast of Kinsley. Probably an old name for one of the Hite group of claims. (Aubury et al. 04:10). Quartz vein in slate.
	Chemisal (Castagnetto III)	Not determined					See under Copper.
	Chenoweth	Not determined	20	4S	17E	MD	Last active in 1901, producing 39.72 tons of ore which yielded $682.25 or $17.19 per ton (Logan 35:188).
	Cherokee I	Maizie Erickson, Le Grande	9,10	4S	15E	MD	Near Akoz zinc mine. Wall rocks are schistose meta-volcanic rocks and massive greenstone. Mine last active in 1950 with a small production of gold and a larger production of silver.
	Cherokee II						See Lovely Rogers.
	Choteau and Sheridan						See Mary Harrison group.
	Clark Mines						See Our Chance.
60	Clearinghouse group (Anderson I, Ferguson, Original, Moonstone, Golden Rule, El Portal)	Frank E. Gallagher, 211 26th Street, Merced	16 21	3S	19E	MD	Six miles west of El Portal at Clearinghouse on the north bank of the Merced River (Storms 96:218; Lowell 16:593; Castello 20:106; Laizure 28:107; Young 29:320; Laizure 35:39; herein).
61	Cleveland	Not determined, probably H. H. Odgers, Midpines	28 29	4S	18E	MD	Lies between the Buena Vista lode and Cannon Gulch placer claims. Not patented. A 5-foot quartz vein strikes N. 70-75° E. and dips 70° N. Schistose greenstone wall rocks. Workings once consisted of an 86-foot shaft and 130-foot drift with several open stopes. Ore reported to occur on the hanging wall side of the vein (Laizure 28:83).

LODE GOLD (CONT.)

MAP NO.	CLAIM, MINE, OR GROUP	OWNER NAME, ADDRESS	SEC.	T.	R.	B & M	REMARKS
62	Colorado	Corinne Kretzer, 555 Thirty Second Street, Richmond (1/6); F. E. and P. W. Judkins, 2817 San Pablo Avenue, San Pablo (1/3); Charles and Irma St. Johns, 619 Humboldt Street, Richmond (1/6); and Ellen N. Weston, 443 Ninth Street, Richmond (1/3).	27	4S	18E	MD	(Lowell 16:579-80; Castello 20:111; Leisure 28:83; herein.)
	Columbia						See Quartzburg group.
	Combination	Geo. W. and Bessie L. Welch, Bagby	SW¼ 31	3S	17E	MD	A pocket mine in serpentine. Property consists of three unpatented claims. Intermittent moderate production, 1933-49, chiefly from pockets of high-grade ore. Inactive in September 1954.
63	Comet	T. A. Clarke, Comet Mine, Mariposa	22	4S	19E	MD	Located 1 mile northeast of Jerseydale on a fork of Skelton Creek (Storms 96:216; Lowell 16:580; Castello 20:111; herein).
	Compromise and Eubanks						See Marble Springs.
64	Confidence and Confidence Extension (Black)	Alma Investment Company, 1257 Shrader Street, San Francisco	SW¼, SW¼, NW¼ 12	4S	19E	MD	Patented property consisting of 41 acres located 7 miles north of Darrah via the Devils Gulch truck trail or 1 mile south of Sims Cove on the South Fork of the Merced River.
	Congo	H. J. Buchenau, Star Route, Box 17, Madera	6 (?)	3S	17E	MD	Unpatented property 4 miles east of Coulterville. In 1894 workings consisted of a 300-foot cross-cut

LODE GOLD (CONT.)

MAP NO.	CLAIM, MINE, OR GROUP	OWNER NAME, ADDRESS	SEC.	T.	R.	B & M	REMARKS
	Congo (continued)						tunnel which intersected the vein 200 feet from the surface. Vein strikes northwest and dips 50° NE in slate and greenstone wall-rocks. Vein was described by Storms as large. Ore minerals are free gold, pyrite, and chalcopyrite. (Storms 94:168; 96:217.)
65	Consolidated Eureka II, Lafayette and Eastern Star	Frank A. Smith, c/o Mrs. Carol B. Shultis, 2716 Stuart Street, Berkeley 5	4,5 8,9	3S	16E	MD	Located 2 airline miles southwest of Coulterville at the northern end of the Horsehoe Bend Mountains, and south of Maxwell Creek. Worked in 1890's by Boglieli and Gotelli. Sold in September 1900 to Columbus Consolidated Gold Mining Company for $6000 (Min. and Sci. Press 1900, vol. 70, no. 13, p. 377). Patented property 59.44 acres. Northwest-trending northeast-dipping quartz vein cuts massive greenstone of the Upper Jurassic Penon Blanco volcanics (Taliaferro 43:283). Principal activity 1900-02 by the Columbus Company. No record of recent activity. Workings consist of several tunnels all in need of cleaning out.
	Consolidated Whitlock and Alabama						See Whitlock group.
	Contact Lode						See Big Break.
	Contention and Black Oak	Not determined	NW¼ 11	3S	18E	MD	See Big Break.
	Cook mines or Cook Estates						See Melvina group and Mary Harrison group.
	Cook, D.						See Melvina group.
	Coronado						See Golden Key group.

LODE GOLD (CONT.)

MAP NO.	CLAIM, MINE, OR GROUP	OWNER NAME, ADDRESS	SEC.	T.	R.	B & M	REMARKS
66	Cotton Creek	Not determined. Last operator was W. H. Hauser, 2469 Mavis Street, Oakland	NW¼ 24	4S	16E	MD	Located a quarter of a mile east of the Hunter Valley road on Cotton Creek, 7 airline miles northeast of Hornitos (Julihn and Horton 40:120; herein).
	Coward						See Hasloe.
	Coyote Hall	Wallace L. Owen, 4443 Belmont Avenue, Fresno 2	11	6S	18E	MD	Located 1 mile southeast Indian Gulch. Northwest-trending quartz vein in slate wall rocks.
	Cranberry						See Rutherford and Cranberry.
	Cripple Creek	Not determined; probably Mariposa Commercial and Mining Company, c/o Eileen Milburn, Mariposa	11	5S	18E	MD	Located a quarter of a mile east of Highway 140. Northwest-trending aplite dike 3 to 4 feet wide dips 45° SW. Serpentine and slate wall rocks. Gold occurs native and in auriferous arsenopyrite. Workings consist of a series of open cuts, an 80-foot inclined shaft, and several short drifts and crosscuts. Ore is very rich but in pockets. (Julihn and Horton 40:157-158.)
67	Crown Lead	Joseph and Emilie Kopf, 512 Seabright Street, Santa Cruz	36	3S	16E	MD	Patented property of 44.72 acres dating back prior to 1864. Operated at various times in conjunction with the Crown Peak and Red Bank or Stevenson group of claims. Has a long record of intermittent activity but only a small record of production. Well-defined quartz vein 3 to 10 feet wide striking N. 45° W. and dipping steeply northeast follows roughly the contact between a serpentine intrusion and slate of the Upper Jurassic Mariposa formation. Claim extends for more than 5500 feet along a branch of the Mother Lode. Vein matter is quartz-mariposite-ankerite rock. Principal workings were a 50-foot inclined shaft and a tunnel of unknown length, now caved. Idle. (Storms 96:217; Lowell 16:581.)

LODE GOLD (CONT.)

MAP NO.	CLAIM, MINE, OR GROUP	OWNER NAME, ADDRESS	SEC.	T.	R.	B & M	REMARKS
68	Crown Peak	Crown Peak Mining Company Ltd., Bear Valley	35 36	3S	16E	MD	Patented fractional claim of 13 acres adjoining Crown Lead mine on southeast. Operated jointly with Crown Lead and Red Bank or Stevenson group of mines at various times. There are two roughly parallel veins on either side of a dike-like intrusion of serpentine which cuts slate of the Mariposa formation. These strike N. 45°W. and dip northeast at an average of about 65°. Workings in 1914 consisted of a 45-foot inclined shaft and a 50 foot open cut. The shaft and adjoining workings are caved. (Storms 96:217; Lowell 14:581; Castello 17:113; Laizure 28:84.)
	Crown Point						See Mount Buckingham.
69	Crystal (Penobscott)	Ralph E. and Libbie N. Daley, 1165 West 22d Street, Merced	15 16	3S	16E	MD	Patented property consisting of 17.19 acres. Two parallel quartz veins, the Cabinet and Lookout, strike N. 40° W. and dip 35° NE. These range from 1 to 3 feet in width and contain chalcopyrite and bornite as well as native gold. These veins are crossed approximately at right angles by the vertical crystal vein which contains small crystals and blebs of cinnabar. Quicksilver was never recovered in commercial amounts. The Crystal vein is also gold-bearing (Turner and Ransome 97:7).
	Cuneo						See Talc.
	Daisy						See Red Bank.
70	Deliah	Walter J. Lautenschlager, 626 South Catalina Street, Los Angeles 5	10 11	3S	16E	MD	Patented property consisting of 9.48 acres. One of numerous Mother Lode claims worked from time to time in conjunction with other claims and for which there is no separate record of operation or production. Claim is astride a small, southwest-trending auxiliary vein, part

LODE GOLD (CONT.)

MAP NO.	CLAIM, MINE, OR GROUP	OWNER NAME, ADDRESS	SEC.	T.	R.	B & M	REMARKS
	Deliah (continued)						of the Mother Lode system. Geology similar to Mary Harrison mine. One of Cook Estate mines.
	Dana						See Diana.
	Democrat						See Quartzburg group.
71	Diana (Dana, North Fork)	H. Barnes and E. Henriksson, c/o Diana Mine, Coulterville	SE¼ 24	2S	17E	MD	Steep-dipping, northeast-trending quartz vein in slate and hornfels wall rocks. Small production in 1901 and 1941. Last ore mined ran 47 oz. of gold per ton.
72	Dillon	Walter J. Lautenschlager, 626 South Catalina Street, Los Angeles 5	4	3S	16E	MD	One of the claims of the Potosi-Malvina group of mines near Coulterville. Lies between the Ninety Four and Bozeman claims and east of the Potosi claim. Apparently never was operated as a separate mine.
73	Diltz (Diltz and Mann, W.Y.O.D.)	Diltz Mines, a trust estate, c/o John P. Pulham, Mariposa	29	4S	18E	MD	On Sherlock Creek 5¼ miles north and slightly west of Mariposa (Lovell 16:581; Castello 21:113; Laisure 28:85; Laisure 35:29; Julihn and Horton 40:146; herein).
	Diltz and Mann						See Diltz.
74	Dolman (Oyler Lode, Oiler Lode, Hickman)	Not determined; probably Ellen T. Simpson et al.	29 33	4S	17E	MD	Former Las Mariposas Grant mine (Logan 35:188; Laisure 35:31-32; Julihn and Horton 40:128-130; herein).
	Dolph						Same as Hayseed. See Golden Key group.
	Dorothy						See Artru.
75	Doss (Doss and Thorne, Gimaca, Brooks)	Frank Trabucco Jr., Hornitos	28	5S	16E	MD	(Storms 96:218; Castello 21:114; Laisure 35:32; herein.)

LODE GOLD (CONT.)

MAP NO.	CLAIM, MINE, OR GROUP	OWNER NAME, ADDRESS	LOCATION				REMARKS
			SEC.	T.	R.	B & M	
	Doss and Thorne						See Doss.
	Douglass Mines						See Malvina group and Mary Harrison group.
76	Duncan	George D. Turner Route 1, Box 1049, Ceres	16 21	5S	16E	MD	Patented property consisting of 44 acres (Storms 96: 218; Castello 21:114; Laizure 28:85; 35:32; herein).
	Dusenberry						See Golden Key group.
77	Early (Louisa, Felix, Revel, George Placer) group.	Robin H. Jackson and Edythe E. Jackson, Box 142, Mariposa	20 21	4S	19E	MD	Patented property of 80 acres (Lowell 16:581, 595; Castello 21:114, 135; Laizure 28:113; herein).
78	East Rutherford	Jessie and Winnie May Black, Route 1, Box 28, Cove, Arkansas	22	3S	19E	MD	Patented property consisting of 13.12 acres. Probably was once a part of the Rutherford and Cranberry mine.
	Eastern Star						See Consolidated Eureka, Lafayette, and Eastern Star.
	Eclipse						See Bandarita.
	El Carmen						See Mexican II.
	El Portal						See Clearinghouse.
79	Elizabeth (Stepping Stones, Nighthawk, Fools Choice)	Frank A. Casaccia, Mariposa	NW¼ 5	5S	17E	MD	Former Las Mariposas Grant mine. Property consists of 4 acres on Corbetts Creek 3 miles south of Bear Valley. Active 1905-10, producing 132.37 tons of ore yielding $518.36 in gold or an average of $19.00 per ton at the old price of gold. Produced 8 tons in 1938 that yielded 7 oz. of gold. Vein averages 6 feet in width, strikes N. 40° W., and dips 50° NE. It is on or close to the contact between meta-andesite greenstone on the west and Mariposa

LODE GOLD (CONT.)

MAP NO.	CLAIM, MINE, OR GROUP	OWNER NAME, ADDRESS	SEC.	T.	R.	B & M	REMARKS
	Elizabeth (continued)						slate on the east. Workings in 1920 consisted of a 100-foot shaft, about 100 feet of drifts and a stope 100 feet high by 5 feet wide. (Storms 94:168; Lowell 16:581; Castello 21:114; Laizure 35:37.)
80	Ely						See Mary Harrison group.
81	Emma I	T. A. Clarke, c/o Comet Mine, Mariposa	36	3S	19E	MD	Located 1½ miles southwest of Brown's Peak and 1½ miles west of the Williams Brothers mine. Formerly part of the Little Wonder group of mines. According to Castello (21:115) and Lowell (16:582) there are two roughly parallel quartz veins striking northwest and dipping steeply northeast. The veins pinch and swell rapidly and ore shoots are small but commonly rich. In 1920 there was a 90-foot tunnel driven in the vein and several caved tunnels and surface workings above the land. One of the veins is reported to carry values chiefly in argentiferous galena.
82	Emma II	Fisher Research Laboratories, Inc., 1961 University Avenue, Palo Alto	29 32	4S	18E	MD	Property consists of two fractional claims, the Emma 1 and 2, totaling 26.92 acres. Patented. Adjoins the Sultana claim on the east and the Mohawk claim on the north. On the south extension of the North-trending Spencer vein. A second vein diverges from the main vein toward the northeast. Developed by a 40-foot shaft and short lateral workings. Greenstone wall rocks. Inactive in 1956.
	Empire						See Spread Eagle group.
83	Enterprise (Barcroft)	H. L. Gorton, Route 4, Box 184, Turlock owns two unpatented claims, the Enter-	34 2,3	4S 5S	16E 16E	MD MD	Mine discovered prior to 1880. R. W. Barcroft of Hornitos was first owner of record. Quartz vein strikes northeast and dips 30° SE. Slate and greenstone wall rocks. Intermittently active 1880-1912, 1935, 1940.

LODE GOLD (CONT.)

MAP NO.	CLAIM, MINE, OR GROUP	OWNER NAME, ADDRESS	SEC.	T.	R.	B & M	REMARKS
	Enterprise (continued)	prise and Enterprise Extension No. 1. Horace Meyer, Cathay, et al., owns one patented claim, the Enterprise Extension No. 2, consisting of 20.66 acres.					Last two shipments of ore ran about 0.3 oz. of gold per ton and 1/2 oz. of silver. Had a 120-foot inclined shaft in 1904. (Wilkinson 04:10.)
	Epperson						See Bean Creek group.
	Ethel May	J. V. Lloyd, 3154 Lincoln Boulevard, Omaha, Nebraska	11	4S	17E	MD	Patented property consisting of 20.27 acres. Northwest-trending vein in slate and schist wall rocks of the Paleozoic Calaveras group.
	Eubanks						See Marble Springs.
84	Eureka I (Eureka 1 and 2)	Russell G. Fournier, Box 21, Mariposa	2	5S	19E	MD	One unpatented claim on the west slope of Mount Buckingham close to the Mount Buckingham mine. Quartz vein averages 2 feet in width, strikes east and is nearly vertical. Slate wallrocks. Workings consist of three adits driven on the vein, the lowermost 400 feet long, the middle 300 feet long and the upper 30 feet long. The lower two tunnels are 50 feet apart and the upper adit is 100 feet above the middle adit. Produced several hundred tons of ore between 1927 and 1940 that averaged 0.845 oz. of gold and 0.32 oz. of silver per ton (Laizure 28:86; 35:32; Julihn and Horton 40:154-155).
	Eureka II						See Horseshoe.
85	Eureka III (Eureka Sky High)	Not determined	23 35	3S	19E	MD	A very old property active prior to 1868. Snyder vein is 3½ feet wide. Slate walls. Vein matter is galena, sphalerite, pyrite, native gold, and tellurides in

LODE GOLD (CONT.)

MAP NO.	CLAIM, MINE, OR GROUP	OWNER NAME, ADDRESS	SEC.	T.	R.	B & M	REMARKS
	Eureka III (continued)		or 36?				quartz. Much of the ore mined between 1868 and 1880 ran between $40 and $100 per ton at the prevailing price of gold. Ore was milled in a water-powered arrastra. (Min. and Sci. Press, 1868, vol. 17, no. 15, p. 230; 1879, vol. 38, no. 13, p. 197; 1879, vol. 39, no. 5, p. 5.)
	Eureka IV						See Consolidated Eureka, Lafayette and Eastern Star.
	Eureka V						See Quartz Mountain.
86	Evans I	Pacific Mining Company, Crocker Building, San Francisco 4	8	4S	17E	MD	Located half a mile east of the Pine Tree mine and Highway 49. North-trending quartz vein dips steeply east. Wall rocks are chiefly massive greenstones of unknown age. Work done mostly prior to 1880. Principal working is a drift adit with short connecting workings totaling about 400 feet. Last worked in 1941 when 2 tons of high-grade ore were taken out. Idle.
87	Evans II	Not determined	14	5S	19E	MD	Located 1 mile northeast of Mariposa adjacent to Grizzly Gulch. An old working long abandoned and almost obliterated. According to Storms (96:169) one vein strikes north and dips 70° E. in greenstone walls. A second vein system consisting of several parallel veins strike northwest and dip 20-30° E. A third, well-defined vein strikes northwest and dips 45° W., wall rocks being slate and a dike rock. Soil and the softer parts of the veins near the surface have been extensively placered. Underground workings probably are not extensive. A pocket mine.
88	Exchequer	Not determined	7	4S	16E	MD	At confluence of Cotton Creek and the Merced River. Partly under waters of Exchequer Lake. A well-defined quartz vein striking northwest crosses the Merced River.

LODE GOLD (CONT.)

MAP NO.	CLAIM, MINE, OR GROUP	OWNER NAME, ADDRESS	SEC.	T.	R.	B & M	REMARKS
	Exchequer (continued)						Worked in 1880's. Property once consisted of seven claims.
	Fanny						See Spread Eagle group.
	Farber group						See Lena Farber and Sunset.
	Farmers Hope (Miners Hope)						See Spread Eagle group.
89	Feliciana	Gold Ledge Mining Company, c/o Walter Gleeson, Merchants Exchange Building, San Francisco 4	12 13	4S	18E	MD	Includes Feliciana and Feliciana Extension Nos. 1-7 (Lowell 16:582; Castello 21:115; Laizure 28:86-88; 35:33; herein).
	Felix						See Early.
	Ferguson						See Clearinghouse.
	Ferrari						See under Copper.
90	Fick and Oxford	Not determined	27	4S	18E	MD	Adjoins Colorado mine on the north, just west of Colorado School and 6 miles northeast of Mariposa. The extension of the Colorado vein averages 18 inches in width and is developed by three open cuts and a 60-foot crosscut adit. Samples from open cuts assayed $4.50 to $7 per ton in gold (Julihn and Horton 40:153).
	Flannigan						See Bob McKee.
91	Floranita	Not determined	6	2S	16E	MD	

LODE GOLD (CONT.)

MAP NO.	CLAIM, MINE, OR GROUP	OWNER NAME, ADDRESS	LOCATION				REMARKS
			SEC.	T.	R.	B & M	
	Fools Choice						See Elizabeth.
	Formation Logging Company	Formation Logging Company, E. R. Peters, Los Angeles	31	3S	17E	MD	A small mine first opened about 50 years ago. Workings consist of a crosscut adit driven east and a short north drift. Wall rocks are greenstone. Under development in the summer of 1956.
92	Fortuna--Esperanza-Cube group	Manuel G. Parra, Route 4, Box 3189-A Sacramento	26 27	3S	17E	MD	Property consists of several claims east of Solomon Gulch and northwest of the Black Bart mine. Northwest-trending northeast-dipping vein series cuts greenstone. Small production of $35 per ton ore in 1952. District is noted for its pockets of high-grade ore.
	Fournier						See B and M.
	Frances						See Francis.
93	Francis (Frances)	Gold Recovery Corp., c/o T. S. O'Brien, Box 223, Ione	SE¼ 19, NE¼ 30	6S	18E	MD	Patented property of 20 acres discovered in 1850's. Intermittently active until 1903. Some development but little or no recorded production in 1930's. In 1900 was equipped with a 10-stamp mill and a 40-ton cyanidation plant. Quartz vein 4 to 5 feet wide strikes northeast and dips 45° SE. Granodiorite and schist wall rocks. In 1903 workings consisted of a 300-foot inclined shaft with levels at 100, 200 and 300 feet. There were 180 feet of drifts on the 100-level, 300 feet on the 200-foot level and 200 feet on the 300-level (Min. and Sci. Press, 1900, vol. 70, no. 3, p. 77). Prior to that time there were two adits, a lower one 140 feet long and an upper one 250 feet long, and a 70-foot crosscut. The upper adit was connected to the surface by a raise (Min. and Sci. Press, 1886, vol. 52, no. 7, p. 76 and vol. 59, no. 11, p. 204). Inactive in 1954.

LODE GOLD (CONT.)

MAP NO.	CLAIM, MINE, OR GROUP	OWNER NAME, ADDRESS	SEC.	T.	R.	B & M	REMARKS
	Franklin and McKinley	Harold H. and Anita Bondshu, Mariposa	19	4S	18E	MD	Unpatented property on the northwest extension of the Milburn-Permit-Geary vein system. Northwest-trending quartz vein in greenstone wall rocks. No recent activity.
94	French	Pacific Mining Company, 1022 Crocker Building, San Francisco	9,16	4S	17E	MD	One mile east of the Pine Tree mine and Highway 49 (Laizure 35:33; herein).
	French Camp	Not determined	4	5S	18E	MD	Northwest-trending vein system in greenstone. Active 1908-12, producing 109 tons of ore which yielded $2177.34 or an average of $19.96 per ton. Former Las Mariposas Grant mine (Logan 35:188).
	Frenchman I						See Mount Gaines.
	Frenchman II						See Mebold I.
	French Pocket						See French.
	Funk						See Hasloe.
	G. Douglass						See Malvina group and Mary Harrison group.
95	Garibaldi (Blue Lead)	Robert O. Greeves, Box 151, Columbia	4,9	3S	18E	MD	Located 1 mile west of Kinsley Guard station on the west side of Skunk Gulch (Julihn and Horton 40:135; herein).
	Geare						See Geary.

LODE GOLD (CONT.)

MAP NO.	CLAIM, MINE, OR GROUP	OWNER NAME, ADDRESS	SEC.	T.	R.	B & M	REMARKS
96	Geary (Geare)	Permit Mining Company, 1063 Howard Street, San Francisco	SE¼ 30	4S	18E	MD	In Whitlock district between Nutmeg and Permit mines (Lovell 16:582; Castello 21:116; Laizure 28:88; Julihn and Horton 40:152; herein).
	Geneva	Irene Van de Carr, 5421 Broadway, Oakland 18	10	4S	16E	MD	On patented agricultural land on west side of Hunter Valley. Quartz vein in greenstone. Principal working is a 60-foot shaft, caved (Castello 21:116). No record of recent activity.
	Gentry Gulch						See Hasloe.
97	Georgia Point	John A. Woods, et al., 522 Hoyt Street,. El Monte	26 27 35	3S	19E	MD	Patented property consisting of the Spring Tunnel, Georgia Point, and South Side claims and the South Side Mill site. Vein is probably an extension of the Hite system. Worked principally in the late 1870's by the South Hite Mining Co.; workings believed to be superficial. Geology similar to Hite mine. Idle.
	Georgiana	Not determined; probably Pacific Mining Company, 1022 Crocker Building, San Francisco	10 15	4S	17E	MD	A very old property long abandoned. In vicinity of the Evans mine. Quartz vein 8 to 18 inches wide strikes East and stands vertical. Vein matter is milky quartz and pyrite with native gold. Former Las Mariposas Grant mine (Storms 96:218).
	Ghirardelli						See Lafayette.
	Gibbs						See Williams Brothers.
	Gilmore vein	Not determined	34 35	5S	18E	MD	In Buckeye Gulch north of the Mormon Bar-Le Grande road. Former Las Mariposas Grant mine. Last active 1900-02 and 1912-13. During these periods 85.12 tons of ore was mined that yielded $1633.05 or an average of

LODE GOLD (CONT.)

MAP NO.	CLAIM, MINE, OR GROUP	OWNER NAME, ADDRESS	LOCATION				REMARKS
			SEC.	T.	R.	B & M	
	Gilmore vein (continued)						$19.20 per ton at the old price of gold (Logan 35:188). May be an extension of the Buckeye vein system.
	Ginaca						See Doss.
98	Gold Bar	Earl A. Robinson, Coulterville	32	2S	18E	MD	In the Bull Creek district three-quarters of an airline mile east of the Marble Spring mine and three-quarters of a mile southwest of Bull Creek School. Opened in 1936, development work done intermittently up to present. Quartz vein in slate, quartz biotite hornfels, and greenstone. Workings consist of a 35-foot shaft and a 138-foot crosscut driven to intersect the vein below bottom of shaft. No recorded preoduction. Equipped with a wooden headframe and ore bin.
	Gold Bring	Not determined	4	5S	17E	MD	Former Las Mariposas Grant mine. One mile east of Elizabeth mine and east of Cow and Calf Creek. Quartz vein in slate walls. Last active 1901-02, producing 24.8 tons of ore which yielded $1724.44 or an average of $70.35 per ton at old price of gold. Presumably a pocket mine (Logan 35:188).
99	Gold Bug	N. D. Madden and K. I. Goulder, c/o Boy Construction Company, Ltd., Shelly Building, Vancouver, British Columbia	28 33	3S	17E	MD	In Solomon Gulch 1 mile west of Black Bart mine (Julihn and Horton 40:142; herein).
100	Gold Coin	Not determined; last known operator (1938) was J. K. Wadley, Texarkana, Arkansas	14 23	3S	16E	MD	Pocket mine on west slope of Whites Gulch a quarter of a mile east of Adelaide mine and 3½ miles by dirt road from State Highway 49 and the Mary Harrison mine. On a branch of the Mother Lode vein system which strikes slightly west of north and dips about 70° E. The main

LODE GOLD (CONT.)

MAP NO.	CLAIM, MINE, OR GROUP	OWNER NAME, ADDRESS	SEC.	T.	R.	B & M	REMARKS
	Gold Coin (continued)						vein, which is well defined, is cut by numerous northwest-trending stringers which contain pockets of high-grade ore. Slate wall rocks. About $15,000 was produced from the mine in 1938. The amount produced from pockets prior to 1938 has not been recorded (Julihn and Horton 40:127).
	Gold Hill	J. E. Rea Jr., Box 243, Madera	NE¼ 12	5S	17E	MD	Former Las Mariposas Grant mine. Active 1900, 1905-07, 1912-15, producing 232.10 tons of ore yielding $3519.60 or an average of $15.30 per ton at the old price of gold. Quartz vein strikes N. 62° W. and dips steeply east following the contact between slate of the Mariposa formation (Upper Jurassic) and massive pyroxene andesite greenstone of unknown age (Logan 35:188; Turner and Ransome 95:econ. map).
101	Gold King (Martin-Walling)	Russell G. Rowe, Route 7, Box 1499, Modesto	11	3S	17E	MD	In the Gentry Gulch district in the vicinity of Hasloe and Lovely Rogers mines. Adjoins the Lovely Rogers on the southeast. Discussed herein.
102	Gold King group	Not determined; last operator (1928) was Gold King Mining Company, Box 1648, Modesto	29 32	2S	18E	MD	Last active in 1926-27. Company drove a 1445-foot adit believed to cut the vein 1185 feet from portal (Laizure 28:89).
103	Gold Star	Mrs. E. B. Easton, El Portal	21,22 27	3S	19E	MD	On State Highway 140 midway between Indian Flat and Clearinghouse. Unpatented property consisting of 13 claims. Adjoins Hite group of claims on west and is on the probable west extension of the Hite vein system. Several parallel veins strike N. 40-45° W. and dip very steeply northeast. Wall rocks are slate and hornfelsic slate cut by narrow dikes of aplite and fine-grained quartz diorite. Workings consist of a main

LODE GOLD (CONT.)

MAP NO.	CLAIM, MINE, OR GROUP	OWNER NAME, ADDRESS	SEC.	T.	R.	B & M	REMARKS
							crosscut-adit 412 feet long driven S. 40° W., open and in good condition in September 1954, and a second adit 56 feet long, caved and inaccessible. Mine is equipped with track, mine cars, and hoisting equipment, but no mill. Had small production from 1936-52.
104	Golden Eagle	Worth A. Brown, Capitola	27	4S	16E	MD	Patented property consisting of 8.3 acres 4½ airline miles northeast of Hornitos via the Hornitos-Bear Valley road. Blanket vein strikes N. 70° W. and dips very gently south. Wall rocks are felsitic volcanics belonging to the Upper Jurassic Amador group. Workings consist of numerous open cuts and pits. Some exploration work was being done in September 1954.
105	Golden Gate I (Schlageter group)	Not determined	18 19	4S	17E	MD	Two airline miles northwest of Bear Valley and half a mile northwest of the Yellowstone mine. Broad quartz-mariposite vein strikes northwest and stands vertical. Wall rocks are schistose greenstone. Workings consist of a 210-foot shaft, a crosscut adit 250 feet long, and a 20-foot raise, presumably from the end of the crosscut (Lowell 16:583; Castello 21:116; Laizure 28:89).
	Golden Gate II						See Golden Key group.
106	Golden Key group (Austin group)	Golden Key Mining Company, Route 1, Box 105D, Gilroy	29 32	4S	18E	MD	Includes Golden Gate, Coronado No. 2, Hayseed, Hayvire, and Regan claims. Once known as the Dolph mine (Preston 90:306, 308; Storms 94:170; 96:218,222; Lowell 16: 576, 583; Laizure 28:79: 35:33; Julihn and Horton 40:151; herein).
	Golden Rule						See Clearinghouse.

LODE GOLD (CONT.)

MAP NO.	CLAIM, MINE, OR GROUP	OWNER NAME, ADDRESS	SEC.	T.	R.	B & M	REMARKS
				LOCATION			
107	Gonzales (Houghton and Gonzales)	Laura I. O'Day, 715 Paseo del Mar, Palos Verdes Estates	SE¼ 34	5S	16E	MD	Patented property consisting of 20 acres on the south slope of Santa Cruz Mountain 1 mile northwest of Indian Gulch. Quartz vein in slate and metavolcanic wall rocks. Principal period of activity was in 1880's.
108	Good Luck and Horseshoe	Minnie Meyer, c/o Kenneth Meyer, Midpines	25	4S	18E	MD	Unpatented claims near Highway 140 west of Bear Creek and north of Timber Lodge. Inactive.
	Grand Prize and Badger						See Oakes and Reese.
109	Granite King and Live Oak (Buckeye, Billings)	Edith McElligott, 1404 Poplar Avenue, Fresno	3,10	6S	18E	MD	On southern border of Las Mariposas Grant in Buckeye Creek district (Logan 35:188; Julihn and Horton 40: 134; herein).
	Gray Eagle						See Black Spider.
110	Greens Gulch and Greens Gulch Extension	Mariposa Commercial and Mining Company, c/o Eileen Milburn, Mariposa	12 13	5S	17E	MD	On south side of the Mount Bullion-Hornitos road 3/4 of a mile wide west of Mount Bullion. Las Mariposas Grant mine (Laizure 28: table facing p. 98; herein).
	Grimshaw						See Ruth Pierce.
	Grizzly Gulch	Frank Cassacia, Mariposa	14	5S	18E	MD	A small mine located in Grizzly Gulch, a tributary to Missouri Gulch, half a mile north of Mariposa. Active in 1909 and 1915, producing 40.5 tons of ore which yielded $428.95 or an average of $10.60 per ton at the old price of gold. Workings have been largely obliterated. Former Las Mariposas Grant mine. Geology similar to Mariposa mine.
	Guadeloupe	Not determined	32	5S	18E	MD	In Guadeloupe Mountains near Agua Fria Creek, 18-inch quartz vein in granodiorite. Active 1904-05, 1913,

LODE GOLD (CONT.)

MAP NO.	CLAIM, MINE, OR GROUP	OWNER NAME, ADDRESS	SEC.	T.	R.	B & M	REMARKS
	Guadeloupe (continued)						1915, producing 83.04 tons of ore which yielded $1932.45 or an average of $23.21 per ton at the old price of gold (Lowell 11:16,583; Logan 35:188).
	Guest	Not determined	35	4S	16E	MD	Narrow quartz vein 2 to 8 inches wide in slate. Had 50-foot shaft in 1914 (Lowell 16:583).
178	Gypsy (North Star and Oro Grande)	Linden E. Foran, Mariposa	32	4S	18E	MD	Southern extension of the South Whitlock Extension claim. Vein of white quartz 1 to 3 feet wide strikes roughly north and dips 30-40° E. Greenstone wall rocks. Workings consist of an inclined shaft, caved, and several pits. Inactive in August 1954. Samples taken in 1927 ran from $4.12 to $50.00 per ton (Laizure 28:104).
	Hacker	Not determined	21	5S	18E	MD	Former Las Mariposas Grant mine located near Arkansas Flat north of Highway 140. Quartz mariposite veins in slate. Active 1906, 1908, 1909, producing 27.22 tons of ore yielding $1338.28 or an average of $49.16 per ton at the old price of gold (Logan 35:188).
	Hard Luck	Mrs. Clara Schilling, Bagby	24	3S	17E	MD	Quartz vein in slate.
	Hardscrabble	Not determined	11	5S	17E	MD	Former Las Mariposas Grant mine active 1903-04, producing 16.84 tons of ore which yielded $301.84 or an average of $17.91 per ton at the old price of gold.
	Hartford						See Quail.
111	Basloe (Gentry Gulch, Funk, Coward, Basloe and Centuary)	Walter D. McLean, Joseph Dupret Jr., Ralph J. Jacobs and R.S. Hudgson, c/o Walter D. McLean, Coulterville	1,2	3S	17E	MD	In Gentry Gulch 5½ airline miles southeast of Greeley Hill and 8 airline miles east of Coulterville (Browne 68:32; Storms 94:169; 96:218; Castello 21:118; Laizure 28:126; herein).

LODE GOLD (CONT.)

MAP NO.	CLAIM, MINE, OR GROUP	OWNER NAME, ADDRESS	LOCATION				REMARKS
			SEC.	T.	R.	B & M	
	Hasloe and Century						See Hasloe.
	Hathaway-Bondurant						See Bondurant.
	Hauser						See Cotton Creek.
	Hayseed						See Golden Key group.
	Haywire						See Golden Key group.
112	Helena	Walter J. Lautenschlager, 626 South Catalina Street, Los Angeles 5	NW¼ ¼	3S	16E	MD	A patented property, one of the claims of the Mahoney group. Probably never operated as a separate mine. See Mahoney.
	Hickman						See Dolman.
113	Hite (Arkell, Hite and Wynant)	Cyrus Bell, 160 South Fairfax, Los Angeles	22 26 27	3S	19E	MD	Includes Priest and Coleman, Hite, Glitner, Summit, McConley, and Old Dominion claims as well as several mill sites. At one time, probably included the Hite Central, Spring Tunnel, Georgia Point, and South Side claims. (Browne 68:34; Goodyear 88:344; Storms 94:170; 96:218; Lowell 16:583-4; Castello 21:118; Laizure 28:90, 125-128; Julihn and Horton 40:145-6; herein.)
114	Hite Central	Nick Flaco and Willard Woodbury, Mariposa (3/4), and Maddox Bros., et al., c/o M. Maddox, Box 483, Austin, Texas (1/4)	27	3S	19E	MD	Patented property consisting of 4 acres. Adjoined Hite mine on the southeast. Geology similar to Hite mine. Superficial workings. Idle.

LODE GOLD (CONT.)

MAP NO.	CLAIM, MINE, OR GROUP	OWNER NAME, ADDRESS	SEC.	T.	R.	B & M	REMARKS
	Hite and Wynant						See Hite.
	Home						See Schroeder group.
	Homestake						See Last Chance I.
115	Horseshoe I (Bower Cave)	Joseph Dupret Jr. and Irene Dupret, Star Route, Box 21B Cazadero	NE¼ 23	2S	17E	MD	Patented property consisting of 20 acres on Jordan Creek 3/4 of a mile northwest of Bower Cave and 8 airline miles northeast of Coulterville (herein).
116	Horseshoe II (Eureka)	Not determined	23	4S	16E	MD	At the south end of Hunter Valley close to and on the west side of the paved Hunter Valley road. An old property active mainly in 1895-97. Quartz vein 2 to 5 feet wide strikes N..60°W. and dips 65° SW. In 1895 the shaft was 100 feet deep, had a north drift 100 feet long and a south drift 40 feet long. Ore averaged about $11 per ton at the old price of gold. Ore contains a large percentage of pyrite (Min. and Sci. Press, 1895-97, vols. 73, 75; Eng. and Min. Jour. no. 24, p. 568, 1895; Castello 21:119).
	Houghton and Gonzalez						See Gonzales.
117	Independence I (Twin Springs)	Charles Jordan 633 Green Avenue San Bruno	NE¼ 23	3S	18E	MD	On the Jenkins Hill road half a mile (airline) east of the Briceburg-Kinsley road but about 3½ miles from the Briceburg road via the Jenkins Hill road. Described herein.
	Independence II or Independencio						See Juniper.
	Indian Gulch	Louis and Dorothy E. Erickson, Horn-itos	SW¼ 34	5S	16E	MD	Patented property consisting of 20 acres. Inactive over 50 years.

LODE GOLD (CONT.)

MAP NO.	CLAIM, MINE, OR GROUP	OWNER NAME, ADDRESS	SEC.	T.	R.	B & M	REMARKS
	Iowa and New York	Charles H. Carleton, 1544 Jonquil Terrace, Chicago 26, Illinois	29	4S	19E	MD	Unpatented property in Jerseydale district near Sweetwater mine. Quartz vein in granitic wall rocks. No recent production.
118	Iron Duke	Mrs. Louise M. Broad, 3469-20th Street, San Francisco, 10	NW¼ 4 / SW¼ 23	4S / 3S	16E	MD	In Hunter Valley 8 airline miles north of Hornitos. (Laizure 35:34; Julihn and Horton 40:119; herein.)
119	Isola	Not determined	4	3S	16E	MD	See Malvina group.
	Italian						See Campo.
120	Jackson	Not determined	1	3S	15E	MD	Located 1½ miles east of Granite Spring School on the Coulterville-Hayward-La Grange road. Northwest-trending quartz vein in greenstone and green schist. Idle.
121	Jenkins Hill		18	3S	19E	MD	On west side of Miller Gulch half a mile north of the Merced River and state Highway 140 adjacent to former property of Yosemite Portland Cement Company. Discovered in 1850's. Mill swept away in flood of 1862. Worked in 1919 and 1927. Owned in 1927 by W. J. Schofield of Oakland and A. B. Smith of San Francisco. Vein of sugary quartz 4 inches to 3 feet wide contains auriferous sulfides but little or no free gold. Wall rocks are slate and granite. Workings consist of two drift adits 90 feet apart, one above the other, connected by a raise. Lower adit is 432 feet long (Laizure 28:90-91).
	Jenny Lind						See Washington.
	Josephine						See Pine Tree and Josephine.
	Josie						See Number Nine.

LODE GOLD (CONT.)

MAP NO.	CLAIM, MINE, OR GROUP	OWNER NAME, ADDRESS	LOCATION				REMARKS
			SEC.	T.	R.	B & M	
	Jubilee						See Red Bank.
	Jumbo						See Blue Bell.
122	Juniper	Not determined	30	4S	17E	MD	On Bear Valley Mountain half a mile south of the Juniper mine (Laizure 28:91; 35:34; herein).
123	Juniper (Juniper and Patricia)	Pereno O. Zirker, Pearl Crowell, Ella B. Edwards, and Hallett B. Hammatt, 832 Franklin Street, Monterey	19 30	4S	17E	MD	Property consists of one patented claim of 20 acres and several unpatented claims on Bear Valley Mountain 2¼ miles west of Bear Valley (Storms 94:171; Laizure 35:42; herein).
	Juniper and Patricia						See Juniper.
124	Kaderitas (Caderitas, Cader Idra, Little Wonder)	T. A. Clarke, c/o Comet Mine, Mariposa	31	3S	20E	MD	On the south slope of Brown's Peak 3½ airline miles southwest of El Portal. Adjoins the Mexican (El Carmen) mine on the south. Once called Mina de la Libertad. Quartz vein strikes N. 40° W. and dips 70° NE. Vein thickens and thins rapidly, ranging from a few inches to 4 feet wide. Vein matter is gold, pyrite, arseno-pyrite, galena, and sphalerite in a gangue of milky quartz and fractured wall rock. Wall rocks are graphit-ic slate and schist with intrusions of biotite grano-diorite. Mine workings are an adit 100 feet long and several superficial workings above this level. Last worked in 1953-54 by Pacific Placers Engineering Company of El Portal. Idle in 1955. Worked for a time in 1920-26 as part of the Little Wonder group of mines. (Wilkinson 04:10; Laizure 28:94.)
125	Kane	Mariposa Mining and Commercial Company,	13	5S	18E	MD	On southwest slope of Kanes Hill at junction of Stockton Creek and Rocky Gulch, 1 mile east of Mariposa. Lam-

LODE GOLD (CONT.)

MAP NO.	CLAIM, MINE, OR GROUP	OWNER NAME, ADDRESS	LOCATION				REMARKS
			SEC.	T.	R.	B & M	
	Kane (continued)	c/o Eileen Milburn, Mariposa					inated quartz vein carrying gold, galena and pyrite strikes N. 20° W. and dips steeply SW. Vein averages 3 to 4 feet wide and is exposed for 1300 feet. Principal working is a 170-foot crosscut-adit driven into the south-west slope of the hill which taps the vein 100 feet below the surface. There is also a 50-foot shaft directly a-bove these workings and numerous open cuts along the sur-face outcrop of the vein. Wall rocks are pyroxene ande-site greenstone, granite dike rocks, and serpentine. Last worked 1938-1940 by G. C. and L. G. Kane of Mariposa. Idle. (Julihn and Horton 40:157.)
	Kane Quartz						See Barney Kane.
	Kangaroo	Not determined. Probably Mariposa Mining Company, c/o Eileen Milburn, Mariposa	4	5S	17E	MD	Quartz vein in slate walls. Last active in 1903, pro-ducing 5.52 tons of ore from which $162.01 in gold was extracted, or an average of $29.38 per ton (Logan 35:188).
	Kate Kearney						See Red Cloud.
	Key Extension						See Number Nine.
	Key Lode						See Number Nine.
	Kid Saxon						See Mebold 1 group.
126	King Midas	Not determined; former Las Mariposas Grant mine.	8	5S	18E	MD	Discovered in 1925. Under lease in 1927-28 to J. N. Knight and associates of Mt. Bullion. Close to State Highway 49 one mile east of Mt. Bullion. Gold occurs chiefly in serpentine in a series of 5 parallel shear zones striking northwest and dipping about 70° NE. Gold has been deposited in thin seams of gouge along joint planes and in fractures in talc-serpentine schist. Ex-cept for several surface cuts the main workings are an

LODE GOLD (CONT.)

MAP NO.	CLAIM, MINE, OR GROUP	OWNER NAME, ADDRESS	SEC.	T.	R.	B & M	REMARKS
	King Midas (continued)						Inclined shaft 50 feet deep and a vertical shaft 100 feet deep located 200 feet to the southeast. Mill runs on vein matter taken out during development work returned $10 per ton with mill tailings running $4.29 per ton and sulfide concentrates running as high as $190 per ton (Laizure 28:93).
	King Saxon						See Mebold I group.
	King Saxon and Queen Saxon						See Mebold I group.
127	King Solomon	John M. Graham, 1090 Carolyn Avenue, San Jose 25	SW¼ 14	4S	18E	MD	Patented claim and mill site totaling 25.16 acres in the Briceburg district 2 miles south and slightly east of Briceburg. Near Highway 140, three-fourths of a mile north of its intersection with Trabucco Creek. Adjacent to Roma and Lena Farber group of claims. Last worked in 1920, last production in 1911. Quartz vein 1½ to 2 feet wide strikes northeast, and dips 70° SE. Pyroxene andesite greenstone wall rocks. Workings in 1920 consisted of a 70-foot tunnel connected with a 60-foot raise and some drifts. Idle. (Castello 21:137; Laizure 28:117.) Formerly, operated as part of the Sierra Rica mine.
	Klondike and Last Chance						See under Copper.
	Kockel						See Permit.
	Kohler						See Bean Creek group.
	Lacy						See Mockingbird I.

LODE GOLD (CONT.)

MAP NO.	CLAIM, MINE, OR GROUP	OWNER NAME, ADDRESS	SEC.	T.	R.	B & M	REMARKS
	LaFayette and Eastern Star						See Consolidated LaFayette, Eureka, and Eastern Star.
	LaFayette and LaFayette Extension (Ghirardelli)	Louis and Gabriel Queriolo, 861 San Ramon Way, Sacramento	15	4S	16E	MD	On patented agricultural land astride the Hunter Valley Road, 6½ airline miles northeast of Hornitos. (Castello 21:119; Laizure 28:93; herein.)
	Lakeview						See Last Chance II.
128	Landrum (Simeon Landrum, Barley Field)	George Matlock, Mariposa	27 34	4S	18E	MD	Located 1 mile south of Colorado School and 5½ airline miles northeast of Mariposa (Storms 96:216; Lowell 16; 585; Castello 21:120; Laizure 28:94; herein).
129	Last Chance I (Homestake)	E. D. Foster and Hazel Hansen, c/o E. D. Foster, 1279 Eagle Vista Drive, Los Angeles 41	15	3S	17E	MD	Unpatented property of four claims, the Pine Flat Nos. 1, 2, 3, and the Homestake. An old property once equipped with a five-stamp mill. Quartz vein in slate walls. Reported to contain high-grade ore. Present owner cleaned out and retimbered the inclined shaft to a depth of 125 feet. Total depth of shaft believed to be 300 feet. A hoist and pump are on the property. Idle at present.
130	Last Chance II (Lakeview, Mispah)	Mrs. Mary Attleweed, 2114 Mission Street, and Frank Davis, 729 Shotwell Street, San Francisco; C. W. Jessen, Hornitos	NE¼ 31	3S	16E	MD	In the northwest of Hunter Valley, about 9 airline miles north of Hornitos. Three unpatented claims. Country rock is greenstone. Property was surveyed in 1936, map shows 118-foot shaft with three levels and 187 feet of drifts driven along a quartz vein which strikes northwest. This shaft is now caved and a new 25-foot shaft has been opened. Produced a small tonnage of ore of good grade 1931-38.
	Last Chance III						See Klondike and Last Chance.

LODE GOLD (CONT.)

MAP NO.	CLAIM, MINE, OR GROUP	OWNER NAME, ADDRESS	SEC.	T.	R.	B & M	REMARKS
131	Last Hope	Alice S. Owsley, Midpines	19 24 30	4S	18E	MD	Series of stringer veins in slate strike N. 35° W., dip 55° to 60° northeast. Workings consist of an open cut, several adit drifts on the vein system and several partly caved stopes. Idle in August 1954. Property consists of Last Hope 1 and 2 lode claims, Home placer claim, and Last Hope mill site, all unpatented.
132	Leach	M. Van de Carr, Briceburg	SE¼ 33	3S	18E	MD	Located 1½ miles northwest of Briceburg and about 1 mile by unimproved dirt road from the graded Briceburg-Kinsley road. First worked about 1900-05 by Bell, Leach, and Charles Hartley. Quartz vein 1 to 3 feet wide strikes N. 25° W. and dips 75-80° NE. Vein is accompanied over much of its strike length by a fine-grained, light-green granitic dike. Ore minerals occur partly in ribboned quartz, partly in altered dike material and partly in altered wall rock next to the vein. They consist chiefly of pyrite, chalcopyrite, and native gold. Chief operating working is an inclined shaft about 50 feet deep with drifts and stopes aggregating 300 to 350 feet. An older crosscut tunnel about 60 feet long has extensive drifts at the end which apparently connect with the workings from the shaft. Present owner works the mine intermittently.
133	Lena Farber and Sunset IV (Farber group)	William O. Bolden, Midpines	23 14	4S	18E	MD	Patented property consisting of 32.4 acres in the Briceburg district just east of Highway 140 about midway between Midpines and Briceburg. Operated in the 1920's by the Feliciana Gold Mining Company. Owned in 1928 by United Building and Development Company of San Francisco. Adjoins Roma, King Solomon, and Sierra Rica mines and geology is similar. Workings thickly brushed over but believed to be superficial.

LODE GOLD (CONT.)

MAP NO.	CLAIM, MINE, OR GROUP	OWNER NAME, ADDRESS	LOCATION SEC.	LOCATION T.	LOCATION R.	LOCATION B & M	REMARKS
134	Lewis Brothers (North-west Extension of Princeton)	Mariposa Commercial and Mining Company, c/o Eileen Milburn, Mariposa	13	5S	18E	MD	Described by storms (96:219) as a small, rich vein opened to a depth of 60 feet by several shallow shafts and connecting drifts. Active in 1921 producing 8.60 tons of ore yielding $105.86 or an average of $12.30 per ton (Logan 35:188). Typical Mother Lode vein in slate walls. A minor vein of the Mother Lode system.
135	Little Bear and Little Bear Extension (Pete Gordon)	Paul Miller, Box 1058, Merced	19	4S	19E	MD	On little Bear Creek 2 miles north of Midpines. There are three narrow veins on the property, the main one striking northwest and dipping 45° NE. Wall rocks are slate and various granitic rocks. Workings in 1920 consisted of a 175-foot adit, a 30-foot adit on the opposite side of the creek, a 12-foot winze, presumably from the main adit and a shallow shaft. (Lowell 16:584; Castello 21:120.)
	Little Charlie						See Spread Eagle group.
	Little Judge						See Oro Rico.
	Little Wonder						See Kaderitas.
	Live Oak						See Granite King and Live Oak.
	Live Oak group						See Live Oak and Governor.
136	Live Oak and Governor (Stud Horse Flat group, Live Oak group, White Oak group)	North American Gold Mines, Inc., Ada Stewart, 25 Delmar Avenue, San Jose 10	35	3S	16E	MD	South of the Merced River 2 miles northwest of Bagby (Lowell 16:598; Castello 21:139; Laizure 28:94; 35:34; Julihn and Horton 40:127; herein).
	London						See Banner group.

LODE GOLD (CONT.)

MAP NO.	CLAIM, MINE, OR GROUP	OWNER NAME, ADDRESS	LOCATION				REMARKS
			SEC.	T.	R.	B & M	
137	Long Mary	Not determined	17	5S	17E	MD	Close to the Mount Bullion-Hornitos road 6 miles west of Mount Bullion (Lowell 16:586; Castello 21:121; herein).
138	Lookout	Not determined	15 22	3S	16E	MD	Located 3 airline miles south of Coulterville and 1 mile east of Horseshoe Bend on Exchequer Reservoir (Lake McClure). On the southwest slope of Horseshoe Bend Mountain adjacent to the Crystal and Cabinet mines. Claim is astride a northwest-trending quartz vein cutting pyroxene andesite greenstone of Upper Jurassic age. Idle over 50 years.
139	Los Parker	Los Parker Mining Company, c/o Bill Parker, Midpines (1954)	13	4S	18E	MD	Feliciana Mountain district, about 2½ airline miles southeast of Briceburg, on east side of Trabucco Creek. Six unpatented claims, the Bonanza nos. 1 and 2, Buckskin, Chilenian Nos. 1 and 2. Country rocks are metavolcanic and granitic. Mine opened in the 1930's. Quartz vein explored by a tunnel, now caved. In mid-1954 a new tunnel was begun. Equipped with air compressor and five-stamp mill.
140	Louis and Louis Extension	Mariposa Commercial and Mining Company, c/o Eileen Milburn, Mariposa	11	5S	17E	MD	Las Mariposas Grant mine, part of a parcel of land containing 240 acres (Lowell 16:586; Castello 21:121; Laizure 28:96; herein).
141	Louise I (Louise, Louise Point)	Walter J. Lautenschlager, 626 South Catalina Street, Los Angeles 5	3	3S	16E	MD	Astride Highway 49 a quarter of a mile south of Coulterville (Storms 96:219; Lowell 14:586; Laizure 28:96; Logan 35:183; herein).
	Louise II						See Early.

LODE GOLD (CONT.)

| MAP NO. | CLAIM, MINE, OR GROUP | OWNER NAME, ADDRESS | LOCATION | | | | REMARKS |
			SEC.	T.	R.	B & M	
142	Louisiana	Sara W. Treat, c/o Mrs. Roger Sherman, 975 Roble Ridge, Palo Alto	25 26	2S	17E	MD	Located 12 miles by road east of Coulterville via the Greeley Hill, McDiermid station and Dutch Creek road (Min. and Sci. Press, 1866, vol. 12, no. 3, p. 40; Storms 96:219; Castello 21:96; Julihn and Horton 40:136-138; herein).
	Louise						See Louisa I.
	Louise Point						See Louisa I.
143	Lovely Rogers (Cherokee II, Shimer)	Mrs. Nettie Hauck, et al., Yosemite National Park	2,11	3S	17E	MD	Half a mile south of Gentry Gulch and 1 mile southwest of the Hasloe mine (Browne 68:32; Storms 96:217; Lowell 16:587; Castello 21:122; Laizure 28:96; herein).
144	Lucky Boy	R. B. Sharp, Bear Valley	10	4S	16E	MD	Adjoins the French mine on the east and is on the east extension of the French vein. Principal working is a 5- by 7-foot crosscut adit more than 100 feet long connecting with a long drift and air shafts between the mill building and the cabin. The main adit is equipped with track, ore car, ore bins, grizzly, jaw crusher, arrastra and ball mill. There is also a secondary crosscut adit several hundred feet long and also equipped with track and cars. Last recorded production 1950-52, a small tonnage of ore being mined which yielded about 0.6 ounces of gold per ton. Intermittently worked by the owner.
	Lucky Strike						See Manzanita.

LODE GOLD (CONT.)

MAP NO.	CLAIM, MINE, OR GROUP	OWNER NAME, ADDRESS	LOCATION				REMARKS
			SEC.	T.	R.	B & M	
	Ludwig vein	Not determined; probably Mariposa Commercial and Mining Company, c/o Eileen Milburn, Mariposa	18	5S	18E	MD	In vicinity of the Princeton mine to the northwest. According to Storms (96:219) the vein is in slate with a dike on the hanging-wall side. In 1896 workings consisted of three shafts, 180, 100 and 25 feet deep, respectively, and several hundred feet of drifts. Production has been recorded in 1900, 1902-05 and 1910. During this period 736.59 tons of ore yielded $9836.86 or an average of $13.37 per ton at the old price of gold. There has been no recent activity and the workings are hard to identify.
	Mahoney						See Malvina group.
145	Malone (Bear Creek)	Not determined	5	5S	19E	MD	Half a mile east of Bear Creek School and State Highway 140 (Preston 90:300-302; Lowell 16:587; Castello 21:122; Laizure 28:97; 35:36; herein).
146	Malvina group (Potosi, Mahoney, Douglass, D. Cook, Merced Co.,)	Walter J. Lautenschlager, 626 South Catalina Street, Los Angeles	4,9 10 32 33	3S 2S	16E 16E	MD MD	Patented property of over 300 acres includes the Boston, Mahoney, Glendive, G. Douglass, Miles, D. Cook, Helena, Potosi, Ninety Four, Dillon, Malvina No. 1, Malvina No. 2, Bozeman and Regina claims (Goodyear 88:347; Fairbanks 90:39; Storms 94:172; 96:220, 222; Lowell 16: 587, 594; Castello 21:123; Laizure 35:36; Logan 35: 183-84; Julihn and Horton 40:107; herein).
	Mammoth						See Quail.
	Mammoth group						See under Copper.
	Manzanita I	James A. Low and Edward A. Zeus, Box 29, Incline	31?	2S	18E	MD	About 13 miles east of Coulterville, probably near the Marble Springs mine. Worked in a small way before 1900, the quartz being crushed in an arrastra. A small production of gold and silver has been recorded from this mine for the period 1940-46. (Storms 96:220.)

LODE GOLD (CONT.)

MAP NO.	CLAIM, MINE, OR GROUP	OWNER NAME, ADDRESS	SEC.	T.	R.	B & M	REMARKS
	Manzanita II (Lucky Strike)	C. W. Jessen, Hornitos	31	3S	16E	MD	In the northwest end of Hunter Valley, about 9 airline miles north of Hornitos. Two unpatented claims. County rock is greenstone. Idle.
147	Marble Springs (Compromise and Eubanks)	George G. Glenn, 3134 East 10th Street, Oakland 1	30 31	2S	18E	MD	In the Bull Creek district 1½ miles west of Bull Creek School and 2 miles south of Bower Cave. (Browne 68:32; Wilkinson 104:112; Lowell 16:580; Castello 21:123; Laizure 28:97; 35:46; herein.)
148	Margaret	Walter J. Lautenschlager, 626 South Catalina Street, Los Angeles 5	33 34	2S	16E	MD	Patented property containing about 20 acres just west of Highway 49 in Coulterville. Adjoins the Louisa mine on the northwest. Worked intermittently prior to 1900 as part of the Mary Harrison group of mines, and earlier as an independent property. A typical Mother Lode vein occurrence similar to others on the east branch of the lode. Workings largely caved and inaccessible but probably not extensive. Details of production not recorded. Idle many years.
	Margaretta						See Silver Lead.
149	Marguerite	Louis Ferretti, c/o Magaret Tiscornio, 1633 Argonne Drive, Stockton 4	NE¼ 6	3S	17E	MD	Pocket mine in northwest-striking shear zone near the contact between granitic rock and slate and hornfels metasediments of the Paleozoic Calaveras group. Intermittently operated by the owner.
150	Mariposa	Mrs. Frank E. Gallagher, 211 Twenty-Sixth Street, Merced	23 24	5S	18E	MD	On the southeast outskirts of Mariposa close to the Mormon Bar road (Browne 68:28-29; Storms 94:173, 220; 00:142; Lowell 14:587; Castello 21:124; Laizure 28:99; Logan 35:9, 184; Laizure 35:38; Julihn and Horton 40:155-157; herein).
	Martin-Walling						See Gold King.

LODE GOLD (CONT.)

MAP NO.	CLAIM, MINE, OR GROUP	OWNER NAME, ADDRESS	LOCATION SEC.	T.	R.	B & M	REMARKS
151	Martinez	Mrs. E. J. Hannah, 909 East Yosemite Avenue, Merced	21	5S	16E	MD	Patented property. Adjoins Duncan mine on southeast and is on the extension of the Duncan vein. This strikes N. 50° W., dips 45° NE and is 2 to 5 feet wide. Vein matter is milky quartz with pyrite and native gold. Wall rocks are schistose and hornfelsic volcanics and metasediments of the Upper Jurassic Amador group. Probably discovered prior to 1870. Produced 1000 ounces of gold between 1897 and 1899 from an unknown tonnage of ore. In 1917 lessees extracted $6,000 from a pocket (Eng. and Min. Jour., 1917, vol. 103, no. 2, p. 126). Between 1933 and 1939 a total of 835 tons of ore yielded 317.4 ounces of gold and 49 ounces of silver. (U. S. Bur. Mines records.) Idle since 1939.
152	Mary Harrison group (Ely, Choteau and Sheridan)	Walter J. Lautenschlager, 626 South Catalina Street, Los Angeles 5	3,10 11	3S	16E	MD	Patented property totaling nearly 80 acres near Coulterville on Highway 49. Includes the Balance, Choteau and Sheridan, Dahlia, Ely, and Venture claims and several mill sites. (Browne 68:34; Goodyear 88:346; Fairbanks 90:11, 124; Storms 94:173; 96:220; 00:146; Castello 21:124-25; Laizure 28:100; Logan 35:118; herein.)
	Massa Garden	Not determined; former Las Mariposas Grant mine	12	5S	17E	MD	In vicinity of Mount Ophir and Greens Gulch mines. Quartz vein in slate. Active 1900, 1905-07, 1910-15 producing 232.1 tons of ore yielding $3519.60 or an average of $15.30 per ton at the old price of gold. (Logan 35:188.)
	Mayflower						See Barney Kane.
	McCall						See Number Nine.
	McKee						See Bob McKee.

LODE GOLD (CONT.)

MAP NO.	CLAIM, MINE, OR GROUP	OWNER NAME, ADDRESS	LOCATION SEC.	LOCATION T.	LOCATION R.	LOCATION B & M	REMARKS
153	Mebold I (King Saxon, Queen Saxon, Kid Saxon, Frenchman)	Thomas V. Bell Jr. et al., 317 East 26th Street, Merced owns the Kid Saxon claim. Others not determined.	21	4S	18E	MD	Unpatented group of claims on the east side of Saxon Creek 2 miles northwest of Colorado School and half a mile south of the Schroeder group of mines. Quartz vein 1 to 5 feet wide strikes N. 25° W. and dips southwest at about 45°. Wall rocks are chiefly black slate of the Paleozoic Calaveras group. Slaty cleavage strikes roughly parallel with the vein and dips steeply northeast. Workings consist of two drift adits 450 and 300 feet in length and several shallow shafts and short crosscut adits. Inactive in September 1955. Workings are strung out for about half a mile in a line roughly parallel to Saxon Creek. Ore reported to run from $4.12 to $22.40 per ton, from assay samples (Laizure 28:93).
154	Mebold II	Not determined	17	4S	19E	MD	On Sweetwater Creek northwest of Sweetwater mine. Quartz vein 2 feet wide in granitic rock. In 1916 had a 50-foot crosscut tunnel and short drifts. (Lowell 16:588.)
155	Menlo Consolidated (Aden)	Catherine Aden, c/o Dorman R. Mangels, 508½ Sacramento Street, Vallejo	22 23 26 27	4S	18E	MD	Patented property of 57 acres half a mile northeast of Colorado School. Includes the Ernestine, Menlo, and Mountain Queen claims. Northwest-trending vein system cutting metasediments and metavolcanics of the Paleozoic Calaveras group. Workings in 1916 consisted of a 300-foot crosscut, 180 feet of drifts and two shafts 105 and 40 feet deep respectively. (Lowell 16:575.)
	Merced Gold Mining Company						See Malvina group and Mary Harrison group.

LODE GOLD (CONT.)

MAP NO.	CLAIM, MINE, OR GROUP	OWNER NAME, ADDRESS	SEC.	T.	R.	B & M	REMARKS
156	Merced River I	Percy L. Pettigrew, Box 639, Palo Alto, and Horace Meyer, Cathey	31 6	3S 4S	17E 17E	MD MD	Half a mile northwest of Bagby between the Merced River and Highway 49. Patented property of 19.58 acres. Northwest-trending quartz vein averaging 4 feet in width on or near the contact between serpentine on the east and slate on the west. Vein dips 60° NE. In 1904 workings consisted of a shaft 160 feet deep; three adits, the longest 165 feet; and at least 100 feet of drifts. Worked in 1879 with a water-powered arrastra, decomposed vein matter running $18 per ton (Wilkinson 04:11-12; Min. and Sci. Press, 1879, vol. 38, no. 13, p. 197).
157	Merced River II	Not determined	15 16 22	3S	16E	MD	In Horseshoe Bend district in the vicinity of the Crystal and Cabinet mines. May once have included these properties. In 1916, the R. A. Keller and James Burns Company owned 13 unpatented claims. At that time workings consisted of a main shaft 100 feet deep with two levels and 210 feet of drifts. There was a second shaft 90 feet deep and a 230-foot crosscut. Once equipped with a 10-stamp mill (Lowell 16:588).
157	Mexican I	Not determined; probably Ellen T. Simpson et al.	28	4S	17E	MD	Located 1½ miles south of Bear Valley (Laizure 35:31; Julihn and Horton 40:130; herein).
158	Mexican II (El Carmen)	Pacific Placers Engineering Company, El Portal	31	3S	19E	MD	On the south slope of Brown Peak 3¼ airline miles southwest of El Portal. Adjoins the Kaderitas mine on the north. Reached via the Williams Brothers and Kaderitas mine roads. Discovered about the same time as the Hite mine by Mexicans. Ore commonly high-grade but pockety. Two roughly parallel quartz veins 2 to 6 feet wide cross the property, striking N. 40-50° W. and dipping 70-80° NE. Developed by an old inclined shaft 150 feet deep

LODE GOLD (CONT.)

MAP NO.	CLAIM, MINE, OR GROUP	OWNER NAME, ADDRESS	LOCATION				REMARKS
			SEC.	T.	R.	B & M	
	Mexican II (continued)						with 625 feet of drifts, and two tunnels 240 and 380 feet long, respectively. The shaft connects with one of the tunnel levels. Vein matter is ribbon quartz. Wall rocks are slate and massive quartz-biotite rock (Laizure 28:101). Inactive except for assessment work.
159	Mexican Diggins	Gold Ledge Mining Corporation, Merchants Exchange Building, San Francisco	7,8	4S	19E	MD	Located 1½ miles northeast of Feliciana Mountain. Sampled and explored 1949-51 by Gold Ledge Mine Corporation. Produced a small tonnage of high-grade ore from an old dump. Currently idle.
160	Midas	C. C. Kellogg and Adolph Bulla, 50 Twenty-Sixth Street, Merced	15	3S	16E	MD	On the west branch of the Mother Lode 1½ miles south of the Malvina group at the confluence of Buckhorn and Maxwell Creeks. Patented. Discovered in late 1860's or early 1870's, probably by Fred MacCrellish. Vein is 6 feet wide, strikes N. 32° W. and dips 65° NE. Slate wall rocks. Principal working is a shaft 150 feet deep with 625 feet of drifts. Shaft is watered and partly caved. Idle. (Wilkinson 04:11-12.)
161	Milburn	Beatrice Peregoy, 711 North Crescent Avenue, Lodi	31	4S	18E	MD	Adjoins the Permit mine on the south and is under lease to the Permit Mining Company. Quartz vein strikes N. 20° W. and dips very steeply east. Wall rocks are sheared greenstone of unknown age. Shaft is caved. Hoist house, compressor shed, and headframe remained on property in September 1954. In addition to the shaft there are several open cuts. Idle. Produced a small tonnage of ore between 1937 and 1948 which averaged about 0.25 ounces of gold and 0.1 ounces of silver per ton.
	Mildred and Abbie	Rudolph Vlasich, Box 69, El Portal	30	3S	19E	MD	Unpatented property consisting of one lode and one placer claim.

LODE GOLD (CONT.)

MAP NO.	CLAIM, MINE, OR GROUP	OWNER NAME, ADDRESS	SEC.	T.	R.	B & M	REMARKS
	Miles						See Burr.
	Miners Hope						See Spread Eagle group.
	Mispah						See Last Chance II.
	Missouri Gulch	Not determined; former Las Mariposas Grant mine.	14	5S	18E	MD	A small working last active in 1903 when 7.33 tons of ore were mined that yielded $54.88 or an average of $7.48 per ton (Laizure 28:98).
162	Mockingbird I (Talc, Lacy)	David W. and Maxie S. Dukes, 612 Blackburn, Watsonville	27	4S	18E	MD	Located 1 mile south of Colorado School and 5½ miles, airline, northeast of Mariposa. Patented property consisting of 20 acres (Preston 90:304; Storms 96:219; Lowell 16:588; Castello 21:126; Laizure 28:101; herein).
	Mocking Bird II	Not determined; former Las Mariposas Grant mine.	30	5S	18E	MD	At the foot of Mocking Bird Ridge in Guadeloupe Valley. Under the lease to W. J. Howard and William Boldt. Last active between 1901 and 1908 producing 87.47 tons of ore yielding $2135 or an average of $24.42 per ton. Quartz vein in granitic rock (Logan 35:188; Eng. and Min. Jour. 1902, vol. 74, no. 5, p. 160).
163	Mohawk (North Whitlock Extension)	Alma Clark and Clyde Smith, c/o T. C. Smith, Route 1, Box 209, Bakersfield	31	4S	18E	MD	On the north extension of the same vein system that traverses the Whitlock and North Whitlock properties. Vein strikes north to N. 10° E.; dip ranges from 75° E to vertical. Wall rocks are greenstone. Vein has two branches, each ranging from a few inches to 2 feet in width. Worked prior to 1916 by G. L. Kennedy. Later by Nick Mullins (Castello 21:126-127). Produced a 1-ton pocket of ore in 1933 that yielded more than 9 oz. of gold.
	Monarch						See Spread Eagle group.

LODE GOLD (CONT.)

MAP NO.	CLAIM, MINE, OR GROUP	OWNER NAME, ADDRESS	SEC.	LOCATION T.	R.	B & M	REMARKS
164	Montara	Margaret I. Fisher and O. H. Knowlton, Sonora	30 31	2S	18E	MD	Adjoins the Marble Springs mine and was formerly worked by the Marble Springs Mining Company. See Marble Springs mine. Produced 62 tons of ore during the period 1931-33 which averaged about 0.6 ounces of gold and 1 ounce of silver per ton.
	Monte Christo group	Not determined	20	4S	20E	MD	In a tributary to Devils Gulch 4 miles east of the Cornett mine. According to Lowell (16:588-589) the vein is 6 feet wide and is developed by a 65-foot inclined shaft and several open cuts. Ore carries arsenopyrite and tetrahedrite as well as native gold.
165	Moonlight	R. W. Feals, A. W. Jackson, and C. Musson, 3700 Cutting Boulevard, Richmond	30 31	2S	18E	MD	On the North Fork of the Merced River just west of the Marble Springs mine. Accessible by jeep trail from Greeley Hill and the McDiermid Guard Station via the Bondurant mine road. Geology similar to Marble Springs mine. Undergoing development work in 1954 and 1955.
	Moonstone						See Clearinghouse.
	Moore Hill						See Arcturas.
166	Mormon Bar	Not determined; formerly part of Las Mariposas Grant	SE¼ 25	5S	18E	MD	Quartz vein a few inches to 4 feet wide strikes east and dips very gently north. Vein is accompanied by a dike of dark granitic rock. Vein matter is quartz with very finely divided gold and few or no sulfide minerals. Developed by an inclined shaft 120 feet deep (measured along the incline) with 200 feet of drifts on the 80-foot level, two underhand stopes and one raise. Last period of activity was 1933-34 when 160 tons were mined which ran 0.28 ounces of gold per ton (Laizure 35:38).

LODE GOLD (CONT.)

MAP NO.	CLAIM, MINE, OR GROUP	OWNER NAME, ADDRESS	SEC.	LOCATION			REMARKS
				T.	R.	B & M	
167	Morning Star	Edith Trabucco, et al., Mariposa	11 12	4S	16E	MD	Patented property discovered prior to 1882. Owner in 1882 was Phil Gossar. Equipped with a 10-stamp mill. Sold for $16,000 in 1884 to a Mr. Rogers of San Francisco. In 1909, produced 100 tons of ore which yielded 62.98 ounces of gold and 10 ounces of silver (U. S. Bur. Mines records). Some development work was done in 1935 by J. H. Kelm of Bagby but there apparently was no production. Workings were caved and inaccessible in September 1954. Quartz vein strikes northeast, dip not observed. Wall rocks are massive pyroxene andesite agglomerate greenstones belonging to the Upper Jurassic Amador group. Vein matter is brecciated quartz carrying conspicuous pyrite. Workings consist of a crosscut adit at least 100 feet long driven N. $50°$ W. and with at least 100 feet of drifts. According to Wilkinson (04: 12) there was once a 200-foot inclined shaft. (See also Laizure 35:38.)
	Mount Belle group		25	5S	17E	MD	A group of claims which included the Artru or Dorothy, White Oak, Blue Quartz and Champion mines. These were controlled by Belle McCord Roberts and the Consolidated Gold Fields of Mariposa in 1927, but since have been broken up into separate properties under multiple ownerships. (See Artru and Champion.)
168	Mount Buckingham (Vanderbilt, Sunset III)	Helen M. Ketler et al., 3325 Kempton Avenue, Oakland	1,2 11 12	5S	19E	MD	Near southeast end of Mt. Buckingham above Snow Creek 1 mile west of Darrah (Storms 94:176; 96:223; Wilkinson 04:13; Lowell 16:589; Castello 21:127; Laizure 28: 102; herein).
169	Mount Gaines	J. W. Radil, 444 California Street, San Francisco	35 36	4S	16E	MD	On a tributary to Burns Creek 4½ airline miles northeast of Hornitos (Storms 96:221; Wilkinson 04:12; Lowell 16:589; Castello 21:128; Laizure 28:103; 35:39; Julihn and Horton 40:121-125; herein).

LODE GOLD (CONT.)

MAP NO.	CLAIM, MINE, OR GROUP	OWNER NAME, ADDRESS	SEC.	LOCATION T.	R.	B & M	REMARKS
170	Mount Ophir	Mariposa Commercial and Mining Company, c/o Eileen Milburn, Mariposa	12	5S	18E	MD	One mile northwest of Mount Bullion close to the south side of Highway 49 (Browne 68:29,30; Storms 94:173; 96:221; Wilkinson 04:12; Lowell 16:590; Castello 21: 128; Laizure 28:98, 103; 35:37; Julihn and Horton 40: 111-112; herein).
171	Mountain Belle	Not determined	NE¼ 25	5S	17E	MD	Located 7½ miles west of Mariposa and a quarter of a mile south of State Highway 140. Either the same as or adjacent to the Sorrel (Sarle) mine. Last worked in 1908. Quartz vein averages 3 feet in width between quartz porphyry walls. Workings originally consisted of a 174-foot inclined shaft and 100 feet of drifts on two levels (Lowell 16:589).
171	Mountain King	J. W. Radil, 444 California Street, San Francisco 4.	31 6	3S 4S	18E 18E	MD MD	On the north side of Merced River Canyon 1 mile east of Quartz Mountain and 6 miles east of Bagley. Patented (Wilkinson 04:12; Van Norden 17:698-703; Castello 21: 129; 28:102; herein).
172	Mountain Queen, Mount Queen)	Pete Mulas and Gus Schwing, Mariposa	29	4S	18E	MD	Close to Sherlock Creek east of the Sultana and Spread Eagle groups of mines. Adjoins and partially overlaps the W.Y.O.D. claim of the Diltz mine. Quartz vein 2 feet wide in greenstone. Workings in 1916 consisted of three shafts 85, 60, and 45 feet deep respectively. Idle except for assessment work (Lowell 16:591).
173	Mountain View I	Mariposa Commercial and Mining Company,	11.	5S	17E	MD	Located 2 miles northwest of Mount Bullion. Adjoins Louis mine on the northwest (Lowell 16:591; Laizure

LODE GOLD (CONT.)

MAP NO.	CLAIM, MINE, OR GROUP	OWNER NAME, ADDRESS	SEC.	T.	R.	B & M	REMARKS
	Mountain View I (continued)	c/o Eileen Milburn, Mariposa					28:102; Logan 35:188; herein).
174	Mountain View II	Mrs. Adeline Giusto, Box 25, Altaville	20 or 29	3S	17E	MD	One of Flyaway group of mines at the head of Flyaway Gulch near the Schilling Ranch, 6½ airline miles southeast of Coulterville. Discovery and assessment work only--no extensive workings. Geology similar to Big Lode and other mines of Flyaway district. See Big Lode.
	Mountain View III						See Early.
175	Nellie Kaho	Harold Hanson, Mariposa; under lease to Harmon and Gault (1956)	4 33	5S 4S	17E 17E	MD MD	Three miles south of Bear Valley just west of Cow and Calf Road (Julihn and Horton 40:125-126; herein).
176	New Years I	Not determined	29	3S	15E	MD	Located 8½ miles north of Merced Falls on Webb Station-Hayward road. Quartz vein 2 to 4 feet wide strikes northwest and dips very steeply east. Drift adit 150 feet long has been driven on footwall side in hydrothermally altered, massive greenstone. No data on production available, but little or no stoping has been done. Idle in 1955.
	New Years II						See Barrett under Copper.
	Nighthawk						See Elizabeth.
177	Ninety Four						See Malvina group.
	North Fork						See Diana.
178	North Star and Oro Grande						See Gypsy.

LODE GOLD (CONT.)

MAP NO.	CLAIM, MINE, OR GROUP	OWNER NAME, ADDRESS	LOCATION				REMARKS
			SEC.	T.	R.	B & M	
179	North Whitlock	Nettie M. Miller, 106 Lexington Avenue, Redwood City	SW¼ 32	4S	18E	MD	Patented property consisting of 20.51 acres. Whitlock vein strikes north to N. 10° E. Dip is 75° E. to vertical. Vein matter chiefly milky quartz with native gold and auriferous pyrite. Greenstone wall rocks include flow lava, agglomerate, and intrusive pyroxene andesite porphyry. An old tunnel, open for 400 feet, starts on North Whitlock property, ends at boundary of adjoining Whitlock claim. Above tunnel is second adit, 40 feet long, and 100-foot nearly vertical shaft, the principal active workings. Mill includes a jaw-crusher, ball-race mill, two flotation cells and amalgamation barrel. Under lease to Tom Grunewald of Mariposa (August 1954).
180	Number One (Martin Quartz)	Not determined	34, 35	4S	16E	MD	Property consists of one unpatented claim located in 1875 Operated intermittently 1896-1917. Quartz vein averages about 3 feet in width, strikes northwest and dips 43° SW. Greenstone and schistose metavolcanic wall rocks. Principal working in 1914 was a 360-foot shaft with three levels, 1150 feet of drifts, and a 130-foot crosscut. Once equipped with a 10-stamp mill (Lowell 16:591-92; Castello 21:130; Laizure 28:105).
	North Whitlock Extension						See Mohawk.
181	Number Five (Monte Carlo)	Not determined	2,11	5S	16E	MD	Located 5 miles northeast of Hornitos and 1 mile northeast of the Number Nine mine (Lowell 16:592; Castello 21:130; Laizure 28:104; herein).
182	Number Eight	Not determined	10	5S	16E	MD	Half a mile west of the Number Nine mine. Quartz vein 3 to 6 feet wide strikes N. 25-30° W. and dips 65-70° SW. Wall rocks are quartz-mica schist and schistose granitic rock, probably altered granodiorite. Main vein has two shafts about 300 feet apart. All workings flooded. Inactive in September 1954.
183	Number Nine (Yosemite, Ginaca, Bill Jones, and McCall)	W.R. Plunkett and U.M. Peyrellade, 1278-26th Avenue, Oakland	10	5S	16E	MD	Located 2 airline miles northeast of Hornitos. Patented property of more than 50 acres (Storms 94:173; 96:221; Castello 21:130; Laizure 28:130; herein).

LODE GOLD (CONT.)

MAP NO.	CLAIM, MINE, OR GROUP	OWNER NAME, ADDRESS	LOCATION				REMARKS
			SEC.	T.	R.	B & M	
184	Nutmeg	Sam Le Barry, c/o Permit Mining Company, Midpines	30	4S	18E	MD	On Whitlock Creek 4 airlines miles northeast of Mount Bullion and 1 mile north of the Permit mine (Julihn and Horton 40:152; herein).
185	Oakes and Reese (Grand Prize and Badger)	W. A. Hayes, Martin S. Heller, and J. P. Warren, 1900 Leimert Boulevard, Oakland 2	4 32 33	4S 3S	16E 16E	MD MD	Patented property of 26.15 acres in the northwestern part of Hunter Valley 2 miles east of Exchequer reservoir on Temperance Creek. (Browne 68:30-31; Raymond 10:23-24; Castello 21:131; Laizure 28:106; Julihn and Horton 40:117-119; herein.)
	Oak Springs	Not determined; former Las Mariposas Grant mine	9	5S	18E	MD	Located 2½ miles east of Mount Bullion, close to the south side of Highway 49. Quartz vein strikes N. 60-65° W. and dips steeply east on or close to the contact between schists of the Paleozoic Calaveras group and serpentine. Last active 1900-06, producing 74.81 tons of ore yielding $1689.80 or $22.60 per ton (Laizure 28: 98).
186	Odell group	Not determined	14	4S	17E	MD	Property consists of a group of six claims located 1500 feet above the Merced River near the confluence of Sherlock Creek and the Merced River. Discovered in 1924. Two crosscut adits were driven between 1924 and 1928, the upper 80 feet long and the lower 175 feet long. Upper adit cuts two veins 35 and 70 feet from the portal, respectively. The lower adit intersects the veins 55 feet and 135 feet from the protal. Some specimen ore was found in the 18-inch front vein and lower-grade ore in the 3-foot rear vein (Laizure 28:106). Idle.
	Oiler						See Dolman.

LODE GOLD (CONT.)

MAP NO.	CLAIM, MINE, OR GROUP	OWNER NAME, ADDRESS	SEC.	T.	R.	B & M	REMARKS
187	Old Timer	Herman L. Bullard and Thomas M. Fagan, Incline	NE¼ 16	3S	19E	MD	Adjoins the Clearinghouse mine on the northwest, half a mile north of the Merced River and Highway 140 at Clearinghouse. Old Timer vein strikes N. 10-20° W. and dips 45-50° NE, 2 to 3 feet wide. Vein matter is milky quartz with pyrite, chalcopyrite and native gold. Wall rocks are hornblende quartz diorite, black quartz-biotite hornfels and mixed, injected contact rocks. Vein at surface is near the contact between metasediments and a granitic intrusion. Ore taken out during development was averaging $10 per ton with some assays running as high as $70 per ton. Only working is a short crosscut tunnel striking N. 55° E. with short northwest and southeast drifts. Strike of metamorphic rocks near the mine is N. 20-25° W. A relatively new working active in 1954-55.
188	Old Timers Club	Ray R. and Mary L. Scott, Mariposa	S½ 15	4S	19E	MD	Unpatented property consisting of 10 acres, 1½ miles north of Jerseydale.
	Old Wilcox						See White Oak and Wilcox.
	Onyx	Not determined	10	6S	17E	MD	Close to Highway 140 near Cathay. Quartz vein 3 feet wide strikes northwest and dips very steeply southwest between walls of granitic rock. Vein matter is milky quartz, pyrite, and native gold. Vein is traceable for 1½ miles. Workings consist of a 60-foot shaft and open cuts. Idle in 1954. Last operators were L. H. Rowland and S. B. Givens of Cathay (1927).
189	Orange Blossom (Cavagnaro)	Mrs. Charles B. Cavagnaro, Hornitos	NE¼ 21	4S	16E	MD	Located 5¼ airline miles northeast of Hornitos. Accessible from the Hornitos-Bear Valley road via the Silver Lead mine. A pocket mine located near the contact between contact-altered slate of the Mariposa formation and massive greenstone. Produced $18,000 in 1933-34.

LODE GOLD (CONT.)

MAP NO.	CLAIM, MINE, OR GROUP	OWNER NAME, ADDRESS	SEC.	T.	R.	B & M	REMARKS
	Orange Blossom (continued)						including 2 pockets containing $4000 each (Laizure 35:39). Last operated in 1935 by H. A. McKenzie of Hornitos. Idle in 1954.
	Original						See Clearinghouse.
	Oro Grande						See Gypsy.
190	Oro Rico (Penon Blanco)	Oro Rico Mines Company, c/o A. D. Vencile, Room 14, 1584 West Washington Boulevard, Los Angeles	19 20 29	2S	16E	MD	Located 2 miles northwest of Coulterville on Highway 49. Property includes the Old Judge, South Judge, Little Judge, Penon Blanco, North Penon Blanco, and Star lode claims, a total of more than 100 acres. (Browne 68:35; Fairbanks 90:43; Storms 96:221; Lowell 16:493; Castello 21:132; Laizure 28:109; 35:40; Logan 35:135-136; herein.)
191	Ortega	Frank A. Cassacia, Mariposa	24 19	5S 5S	18E 17E	MD MD	Located 8 miles west of Mariposa astride Highway 140. Adjoins the Sorrel (Sarle) mine (Laizure 35:37; Julihn and Horton 40:131-133; herein).
192	Oso	Not determined; former Las Mariposas Grant mine	20	4S	17E	MD	In Oso Gulch half a mile west and slightly south of Bear Valley. Northwest-striking shear zone in slate. Ore occurs in a thin sheet of talcose material probably derived from serpentine. According to Browne(68:29) about $400,000 was taken from a shaft 50 feet deep and 7 feet wide. Last active in 1906, producing 7.85 tons of ore yielding $164.02 or an average of $20.91 per ton (Laizure 28:98).
193	Our Chance (Clark Mines, Albert Austin group)	Not determined	29	4S	18E	MD	On Sherlock Creek 4 airline miles northeast of Mount Bullion. Adjoins the Diltz mine on the west. (Lowell 16:576; Castello 21:106; Laizure 28:79; Julihn and Horton 40:147-148; herein.)

LODE GOLD (CONT.)

MAP NO.	CLAIM, MINE, OR GROUP	OWNER NAME, ADDRESS	SEC.	T.	R.	B & M	REMARKS
	Oyler Lode						See Dolman.
	P and L	Ray W. Cokeley, 415 Hillside Crest, Piedmont 11	21 or 28	3S	20E	MD	Located 2½ miles south of El Portal.
	Paines shaft	Not determined; former Las Mariposas Grant mine	24	5S	18E	MD	Near the east border of the town of Mariposa in the vicinity of the Mariposa mine. Probably a minor working on the Powell vein. Active 1910-14 producing 1324.75 tons of ore yielding $7551.52 or an average of $5.71 per ton at the old price of gold (Laizure 28: table facing p. 98).
	Patricia and Charles						See Juniper.
194	Pendola Garden	Not determined; formerly Las Mariposas Grant mine	10	5S	17E	MD	On Cow and Calf Creek 3½ airline miles west and slightly north of Mount Bullion. Quartz vein strikes northwest. Slate wall rocks. Active in 1908 producing 17 tons of ore yielding $278.58 or an average of $16.40 per ton at the old price of gold. Small production in 1949-50 (Laizure 28:98).
	Penobscott						See Crystal.
	Penon Blanco						See Oro Rico.
195	Permit (Boulder, Kockel, Bulldog, Bullpup)	Permit Mining Company, c/o H. H. Odgers, Midpines	23 31	4S	18E	MD	On Whitlock Creek 6 miles north of Mariposa or 3 airline miles northeast of Mount Bullion (Wilkinson 04:10; Lowell 16:578; Castello 21:109; Laizure 28:110; 35:40; herein).
	Perrin and Craigue						See Rutherford and Cranberry.
	Pete Gordon						See Little Bear and Little Bear Extension.

LODE GOLD (CONT.)

MAP NO.	CLAIM, MINE, OR GROUP	OWNER NAME, ADDRESS	SEC.	T.	R.	B & M	REMARKS
	Phoebe	Not determined; former Las Mariposas Grant mine.	18	5S	18E	MD	A small working probably on the Childs vein on the northeast quarter of the section. Last active in 1908 producing 130 tons of ore from which $272.58 in gold was extracted, or an average of $2.10 per ton (Laizure 28: table facing p. 28).
	Piedra de Goza						See Roma.
	Pine Cone (Bald Mountain)	Not determined		4S	16E	MD	Small production from pockets in 1930 by F. A. Cavagnaro, Hornitos. Multiple quartz vein 2½ feet wide (Laizure 28:110).
	Pine Crest						See Challenge, Pine Crest, and Pine Crest Extension.
	Pine Crest Extension						See Challenge, Pine Crest, and Pine Crest Extension.
196 196A	Pine Tree and Josephine	Pacific Mining Company, 1022 Crocker, Building, San Francisco	8, 9 16 17	4S	17E	MD	Located 2 miles north of Bear Valley on Highway 49. (Browne 68:29; Raymond 70:23, 24; Fairbanks 90:35-36; Storms 94:173-175; OO:143; Wilkinson 04:11, 12; Lowell 16:594; Castello 21:133; Laizure 28:110; Knopf 29:83-84; Logan 35:186-189; Laizure 35:40-41; Julihn and Horton 40:107-110; Bradley 54:21, 32; herein.)
	Pioneer						See Argo.
197	Pittsburg Landing	Frank B. Ewing, Mariposa	8	5S	18E	MD	A patented property containing 33 acres located 1½ miles east of Mount Bullion in vicinity of King Midas mine. Pocket mine in talcose serpentine. Former Las Mariposas Grant mine.
	Platinum King	C. F. Kingery, 847 Nineteenth Street, Merced	31	3S	17E	MD	Unpatented property consisting of five claims, the Platinum King, Platinum King Extension, Platinum Queen, and Inspiration claims. Pocket gold mine in greenstone. No platinum has ever been recovered.

LODE GOLD (CONT.)

MAP NO.	CLAIM, MINE, OR GROUP	OWNER NAME, ADDRESS	LOCATION SEC.	LOCATION T.	LOCATION R.	LOCATION B & M	REMARKS
	Poole (Poole Track)	Not determined	10	5S	16E	MD	Discovered about 1872 by former state assemblyman D. M. Poole who pounded out several thousand dollars worth of gold with a hand mortar at the discovery site. Worked by Poole during the 1870's and by Live Oak Mining and Milling Company during the 1880's. Last work known to authors was done in 1909. Once equipped with an eight-stamp mill. Vein 3 to 6 feet wide strikes approximately north and is nearly vertical. Vein matter milky quartz with abundant pyrite with free gold chiefly in the oxidized zone. Much of the ore ran $4 to $50 per ton at the old price of gold. The principal working is a 150-foot shaft with considerable drifting on several levels. (Min. and Sci. Press, numerous entries 1872 to 1889.)
	Poole Track						See Poole.
	Porphyry Hill	Not determined; former Las Mariposas Grant mine.	20	5S	18E	MD	Close to Agua Fria Creek, 2 miles southeast of Mount Bullion. Workings not identified in 1954-55. Active 1901, 1905, 1907-10, producing 37.97 tons of ore yielding $2164.48 or an average of $57.09 per ton (Laizure 28: table facing p. 98).
198	Potosi						See Malvina group.
	Prescott						See Badger.
	Priest and Coleman						See Hite.
199	Princeton, New Princeton, and Princeton Extension	Mariposa Commercial and Mining Company, c/o Eileen Milburn, Mariposa	13 18	5S	18E	MD	On the southern outskirts of Mount Bullion. Formerly part of Las Mariposas Grant (Browne 67:41-42; 68:29-31; Raymond 71:30-31; Fairbanks 90:33; Storms 00:143; Lowell 16:594; Laizure 28: table facing p. 98; Knopf 29:84-85; Logan 35:189; Julihn and Horton 40:113-116; herein).

LODE GOLD (CONT.)

MAP NO.	CLAIM, MINE, OR GROUP	OWNER NAME, ADDRESS	SEC.	T.	R.	B & M	REMARKS
	Pumkin						See Black Hill.
200	Pyramid (Castagnetto I)	Lloyd A. Mason, Hornitos	14,15 23	4S	16E	MD	In Hunter Valley close to the west bank of Cotton Creek (Castello 20:110; Laizure 28:83; Julihn and Horton 40:120-121; herein).
201	Quail (Alvina, Hartford)	H. E. Moerlien, San Martin	15 16	3S	17E	MD	In Indian Gulch 6 miles southeast of Greeley Hill or 3 miles southeast of Dogtown. Includes four patented and seven unpatented claims, about 380 acres (Storms 94:175; Lowell 16:595; Castello 21:134; Laizure 28:112; 35: 42, 46; Julihn and Horton 40:141; herein).
202	Quartzburg group (Argonaut, Columbia, Democrat, Spring)	Quartzburg Mining Company, c/o Peter H. Thurbon, 2554 - 21st Street, San Pablo	32	3S	16E	MD	In the northwest part of Hunter Valley, about 8½ airline miles north of Hornitos. Four unpatented claims. Country rock is greenstone. Quartz vein strikes northwest and dips east, developed by incline shaft now caved. Equipment has been removed but headframe and several buildings remain. Evidently mine has been closed for many years. (This property may be confused with another Quartzburg group in sec. 32, T. 4 S., R. 16 E., about 2½ miles north of Hornitos.)
	Quartz Mountain	Not determined	21	5S	16E	MD	Approximately 2 miles southeast of Hornitos and about a quarter of a mile southeast of the Duncan and Martinez mines. Discovered in September 1867 by Mexicans who took out a $3000 pocket. Worked in 1880's by Webber and Rogers, and by Branson. Purchased in 1896 by S. W. Parker of New York (Hornitos Gold Mining Company). Last development work was done about 1917. Ore was high in sulfides and much of it was low grade. Quartz vein strikes northwest and dips steeply northeast. Principal working is a southwest-trending adit with

LODE GOLD (CONT.)

MAP NO.	CLAIM, MINE, OR GROUP	OWNER NAME, ADDRESS	SEC.	T.	R.	B & M	REMARKS
	Quartz Mountain (continued)						connecting drifts and stopes. Wall rocks are slate, spotted slate, and schist and hornfels of Upper Jurassic age (Storms 94:175; 96:222. Also numerous entries in Min. and Sci. Press).
	Queen Saxon						See Mebold I group.
203	Queen Specimen	Pacific Mining Company, 1022 Crocker Building, San Francisco	8	4S	17E	MD	On Highway 49 one fourth of a mile northwest of the Pine Tree mine mill or 3 miles north of Bear Valley (Laizure 28:112, 188; Knopf 29:84; Logan 35:188; Julihn and Horton 40:fig. 44 facing p. 108; herein).
	Recorder						See Rex.
204	Red Bank (Stevenson group)	Percy L. Pettigrew, Box 639, Palo Alto, and Horace Meyer, Cathay	36	3S	16E	MD	On the north bank of the Merced River 1½ miles northwest of Bagby (Storms 96:223; Castello 21:135; Laizure 28: 113; 35:43; herein).
205	Red Cloud I (Kate Kearney)	Carl Harper, Coulterville	22 27	2S	17E	MD	Located 3½ miles east of Greeley Hill (Goodyear 88:345; Storms 94:175; Castello 21:135; Laizure 28:113; 35: 43; herein).
	Red Cloud II						See Blue Moon, under Zinc and Washington, under Gold.
	Red Oak						See Live Oak and Governor.
	Reed	Not determined	6	3S	18E	MD	On the North Fork of the Merced River near the head of the Bandarita ditch. Quartz vein 6 to 14 inches thick cuts slate and massive hornfelsic claystone and siltstone. Equipped with a Huntington mill. Has produced between $10,000 and $15,000 (McLean, Walter, personal communication, 1956).

LODE GOLD (CONT.)

MAP NO.	CLAIM, MINE, OR GROUP	OWNER NAME, ADDRESS	LOCATION				REMARKS
			SEC.	T.	R.	B & M	
206	Reeds Flat	Not determined; former Las Mariposas Grant mine	21	4S	17E	MD	Located half a mile east of Bear Valley. Pockets and small shoots of ore occur in serpentine. Between 1909 and 1913 some 243 tons of ore was mined by lessees which yielded $4,191 or an average of $17.22 per ton at the old price of gold (Laizure 28: table facing p. 98). Last active 1938-39 producing 620 tons of ore and 50 tons of old tailings yielding 99 oz. of gold and 26 oz. of silver (U. S. Bur. Mines records).
	Regan						See Golden Key group.
	Revel						See Early.
207	Rex (Recorder, Rumley)	Charles M. Schroeder, et al., Box 169, Mariposa	20	4S	18E	MD	West of Saxon Creek near the Mebold mine. Pocket mine in greenstone. Rumley vein varies from 4 inches to 18 inches in width, strikes north and dips 30° E. Another north-striking, west-dipping vein 2 inches to 2 feet wide was encountered in running a crosscut that does not show at the surface. This vein produced high-grade pockets in 1928 when Robert MacLean of Mariposa recovered as much as $200 per day from ore broken in a hand mortar (Laizure 28:113). Between 1924 and 1938 the mine produced 63 tons of high grade ore yielding 233.95 ounces of gold and 34 ounces of silver (U. S. Bur. Mines records).
208	Reynolds	Not determined; probably Mariposa Commercial and Mining Company, c/o Eileen Milburn, Mariposa	11	5S	17E	MD	Vein parallels the Mother Lode 500 feet northeast of the Mountain View mine. Active 1900, 1904, 1906, producing 168.6 tons of ore yielding $2217.95, or an average of $12.56 per ton at the old price of gold (Laizure 28: table facing p. 98). Wall rocks are Mariposa slate. Vein is probably part of the Mother Lode system, which is multiple in this vicinity.

LODE GOLD (CONT.)

MAP NO.	CLAIM, MINE, OR GROUP	OWNER NAME, ADDRESS	SEC.	T.	R.	B & M	REMARKS
	River Tunnel (continued)						average of 83 cents per ton (Laizure 28: chart facing p. 98; Julihn and Horton 40: fig. 44, facing p. 109).
	Robinson and Orr group	Not determined	21	4S	18E	MD	Between 1932 and 1936, Frank D. Robinson of Merced and N. P. Orr of Mariposa operated group of seven claims. A tunnel 700 feet long was driven which reached a depth of 200 feet below the surface and a short crosscut was run. Gold occurs in quartz stringers in a quartz porphyry dike associated with graphitic and talcose schist. The dike strikes northwest and is almost vertical. (Laizure 35: 43.)
211	Roma (Piedra de Goza)	Frank Harris and Nellie Harris, 595 Tunnel Avenue, San Francisco 24	14	4S	18E	MD	A patented property of 25 acres close to the Yosemite All-Year Highway, 1½ miles south and slightly east of Briceburg (Storms 00:146; Lowell 16:596; Castello 21: 136; Laizure 28:113; herein).
	Roosevelt	Not determined	9	5S	18E	MD	Workings were not identifiable by the authors. Section 9 is underlain by greenstone and serpentine. Mine was last know to be active in 1905-06, producing 14.3 tons of ore yielding $159.60 or an average of $11.17 per ton. (Laizure 28: chart facing p. 98).
212	Rose	Not determined; former Las Mariposas Grant mine	8	5S	18E	MD	East of Highway 49 one mile east of Mount Bullion. Adjoins King Midas mine on the southeast and geology is similar. See King Midas.
	Royal group						See Sultana group.
	Rumley						See Rex.
213	Ruth Pierce (Amelia, Grimshaw)	Tennessee Gold Mining Company, c/o Pete Jericoff, 1219 Rhode	13	5S	16E	MD	On the Mount Bullion-Hornitos road 4 miles east of Hornitos (Lowell 16:596; Castello 21:136; Boalich 22: 365; Laizure 28:114; 35:43; Julihn and Horton

LODE GOLD (CONT.)

MAP NO.	CLAIM, MINE, OR GROUP	OWNER NAME, ADDRESS	SEC.	T.	R.	B B M	REMARKS
214	Ruth Pierce (continued)	Island Street, San Francisco 10					40:126-127; herein).
214A	Rutherford and Cranberry (Perrin and Craigue)	L. J. Goodrich, 1522 State Street, Santa Barbara, owns one partial claim of 10.34 acres. Ownership of other claims not determined.	15 22	3S	19E	MD	On the Merced River close to Highway 140 about 4 miles west of El Portal (Goodyear 88:348; Storms 96:217; Lowell 16:580, 596; Castello 21:112, 136; Laizure 28:114; 35:43; herein).
	St. Gabriel and Honeycomb						See Silver Lead.
	St. Marys	Not determined	SE¼ 29	5S	16E	MD	Located 1 mile southwest of Doss mine on the Hornitos-Indian Gulch road. Northwest-trending, northeast-dipping quartz vein in dark slate and schist of Upper Jurassic age. Principal working is a shaft with connecting short laterals. Idle.
	San Domingo						See Buffalo.
215	Santa Clara	Myrnie Koepp, Midpines	25	4S	18E	MD	A patented property containing 21 acres west of Highway 140 and half a mile south of Bear Creek Lodge. Quartz vein strikes N. 55° W. in greenstone wall rocks. In 1896 workings consisted of a 60-foot inclined shaft and a 100-foot tunnel (Storms 96:222).
216	Santa Rosa	Not determined	25	4S	18E	MD	Apparently adjoins the Santa Clara mine and is on the same ore zone. An old mine with extensive workings prior to 1895. Worked intermittently from 1895 to 1911 by R. B. Stockton of Madera. Small production from pockets 1911, 1926, 1929-34. Had 10-stamp mill built in 1895. Stockton and Eckter of Madera milled considerable

LODE GOLD (CONT.)

MAP NO.	CLAIM, MINE, OR GROUP	OWNER NAME, ADDRESS	LOCATION SEC.	T.	R.	B & M	REMARKS
	Santa Rosa (continued)						ore between 1934 and 1936 that ran $7.50 per ton exclusive of pockets of high-grade ore. Ore occurs in a nearly vertical stringer zone striking N. 55° W. Zone is cut by numerous cross veins or cutters some of which carry ore. Workings in 1935 consisted of a 200-foot adit, a 100-foot adit and three shafts, 35, 50, and 50 feet deep, respectively. Ore zone is 7 to 20 feet wide (Laizure 35:43).
	Sarle						See Sorrel.
	Saxon King and Saxon Queen						See Mebold I.
	Schlageter						See Golden Gate group.
217	Schoolhouse	Frank A. Maschio and John Maschio, Merced, and Joe Maschio, Snelling	10	4S	16E	MD	A patented property in central Hunter Valley 7 airline miles northeast of Hornitos (Castello 21:737; Laizure 28:115; herein).
218	Schroeder group	C. M. and R. F. Schroeder, Box 169, Mariposa	16	4S	18E	MD	On high ground west of Saxon Creek Canyon 3 airline miles northwest of Colorado School (Storms 96:223; Lowell 16: 596, 600; Castello 21:137; Laizure 28:115, 127; 31: 44; herein).
	Scorpion	Not determined; former Las Mariposas Grant mine.	20	4S	17E	MD	Close to Bear Valley on the west. Active under lease from Las Mariposas Grant in 1902, producing 46.74 tons of ore yielding $341.21 or an average of $7.30 per ton. No further information was available and the workings could not be identified (Laizure 28; chart facing p. 98).

LODE GOLD (CONT.)

| MAP NO. | CLAIM, MINE, OR GROUP | OWNER NAME, ADDRESS | LOCATION | | | | REMARKS |
			SEC.	T.	R.	B & M	
	Sebastopol	Not determined	32	5S	19E	MD	Located 1½ miles southwest of Bootjack. According to Laizure (28:116) there is a well-developed quartz vein prospected by surface cuts and several adits. A report by Techow and Davis of Sacramento listed sample assays ranging from $1.65 to $5.60 per ton. U. S. Bureau of Mines records show a small production in 1926 and 1927 and for the period 1936-41. The property was not visited during this investigation. Wilkinson (04:16) describes it as a pocket line on a 4-foot quartz vein striking north and dipping 40° S.Gneiss walls. He indicates presence of a 100-foot shaft. Owner 1904-27 was L. L. Hart of Mariposa. Belle McCord Roberts of Long Beach purchased the mine in 1927 (Laizure 28:116).
	Seneca						See Doss.
	Shimer						See Lovely Rodgers.
219	Sierra Rica	J. W. Graham and Amelia Wilson, 1090 Carolyn Avenue, San Jose	14	4S	18E	MD	Just east of Highway 140 three airline miles northwest of Midpines in the vicinity of the King Solomon, Roma, and Lena Farber mines. Accessible by half a mile of trail from Highway 140. A patented fractional claim of 2.88 acres. Operated through most of its history in conjunction with the King Solomon or King Solomon and Roma mines. Early production reported to be $300,000. Had a very small recorded production in 1910 and some work was done on the mine in 1919 by lessees. According to Laizure (28:116) the vein averages 4 feet in width, strikes northeast and dips 70° SE between greenstone and slate walls. Workings in 1928 consisted of a 450-foot crosscut tunnel and a 125-foot raise with some drifts from both these workings. Not visited because of dense brush. Apparently idle many years (Wilkinson 04:13; Lowell 16:597; Castello 21:137; Laizure 28:116).

LODE GOLD (CONT.)

MAP NO.	CLAIM, MINE, OR GROUP	OWNER NAME, ADDRESS	SEC.	T.	R.	B & M	REMARKS
	Silver Bar						See under Silver.
	Silver Lane						See under Silver.
220	Silver Lead (St. Gabriel and Honeycomb, Silver Lead and Margaretta)	E. G. Branson et al., c/o Horace Meyer, Cathay	28	4S	16E	MD	Located 4½ miles north and slightly east of Hornitos (Lowell 16:597; Castello 21:138; Boalich 23:365; Laizure 28:116-117; 35:45; herein).
221	Silver Right (Silver Cloud)	Annie L. Spence, 2242 Tasso Street, Palo Alto	19, 30	3S	17E	MD	Located 7 miles southeast of Coulterville. One of Flyaway group of mines. Patented property consisting of 18 acres. Lies between Alta and Sweetwater claims. Explored by several short tunnels and drifts. Quartz vein strikes N. 50° W. and dips 55°-60° NE. Vein is closely associated and parallel to a system of albitite and diorite dikes. Dike system generally divides greenstone and serpentine wall rocks. Property is not equipped. Idle.
	Simeon Landrum						See Landrum.
222	Simpson vein	Not determined; former Las Mariposas Grant mine	30	5S	18E	MD	Located 1½ miles south and slightly east of the Ortega and Sorrel mines on Highway 140, or 1-3/4 miles west of Guadeloupe Valley. In rugged country and adjacent to La Mineta Gulch. North-trending quartz vein in granitic wall rocks. Active 1902-03, producing 12.03 tons of ore yielding $317.42, or an average of $26.37 per ton (Laizure 28: chart facing p. 28).
	Sirocco						See Yellowstone.
223	Sorrel (Sarle)	Not determined; probably Frank A. Casaccia, Mariposa	NE cor. 25	5S	18E	MD	On Highway 140 just south of the Ortega mine and on the same vein system Active in 1909 under F. L. Wallingford who was operating a new mill (Eng. and Min. Jour., 1909, vol. 87, no. 15, p. 773). North-trending quartz vein in granitic rocks. Adjacent to the Ortega mine--see Ortega.
224	South Cranberry and South Cranberry Extension	Not determined	22	3S	19E	MD	Near the Merced River and Highway 140 about 4 miles west of El Portal. Adjoins Cranberry mine. See Rutherford and Cranberry. Not active recently. Owner (1927) was Helen Bass, 280 Union Street, San Francisco.

LODE GOLD (CONT.)

MAP NO.	CLAIM, MINE, OR GROUP	OWNER NAME, ADDRESS	LOCATION				REMARKS
			SEC.	T.	R.	B & M	
	South Side						See Georgia Point.
225	Southern Cross	Lena Giusto, Box 66, Angels Camp	29	3S	17E	MD	Patented property 20.38 acres, 8 miles southeast of Coulterville. One of Flyaway group of claims. Adjoins Bonanza claim on the east. Vein strikes N. 70° W. and dips northeast between greenstone walls. Gold occurs in pockets. Idle.
226	Specimen, Pershing, and Specimen Extension	Helmuth H. Lane, 128 North La Jolla Street, Los Angeles	35, 36?	3S	16E	MD	Located 1¼ miles northwest of Bagby. Pocket mine in greenstone. Active in 1952 under E. J. Mariah of Mt. Bullion and J. E. Costa of Bear Valley. Produced a small tonnage of 3-oz. per ton ore. Temporarily idle in September 1955. Workings consist of a 60-foot shaft with 40 feet of drifts. Total production 1905-52 is about $12,000 realized from approximately 894 tons of ore or an average of about $13.42 per ton (U. S. Bur. Mines records).
227	Spencer	Leroy H. Dart et al., 28 Domingo Avenue, Berkeley	SW¼ 29	4S	18E	MD	Whitlock district 4 miles north and slightly west of Mariposa. Patented property of 20 acres located parallel and close to the Sultana and Bald Eagle claims. Quartz vein averaging 3 feet in width strikes slightly west of north, and dips about 70° E. Wall rocks are massive to schistose, platy greenstone of unknown age. Discovered prior to 1883. In 1896 had a 480-foot drift adit, cutting pay shoots 45 and 55 feet long, respectively, and a 100-foot shaft, unusable (Storms 96:223). Lowell(16:597) identified workings consisting of a 150-foot working shaft, 150-foot airshaft, 500 feet of drifts, 220 feet of crosscuts, a 50-foot winze and a stope 350 feet long and 150 feet high. In 1927 the active workings were a 150-foot crosscut connecting with a 50-foot winze and a large stope (Laizure 28:117). The property was inactive and the workings were inaccessible

LODE GOLD (CONT.)

MAP NO.	CLAIM, MINE, OR GROUP	OWNER NAME, ADDRESS	SEC.	T.	R.	B & M	REMARKS
	Spencer (continued)						and difficult to identify in November 1955. A small production was recorded for the mine in 1898, 1900, and 1932.
228	Spread Eagle group (Empire, Farmers Hope, Miners Hope, Mohawk, Monarch, Tollgate, Bonanza II)	Mack C. Lake, c/o Carlo S. Morbio, 58 Sutter Street, San Francisco	29 31 32	4S	18E	MD	In the Whitlock district 4 miles north of Mariposa. A patented property including the Empire, Fanny, Miners Hope, Little Charlie, Mohawk, Monarch, Spread Eagle, and Tollgate claims, and the Empire millsite (Preston 90: 308; Lowell 16:582, 597; Castello 21:126; Laizure 28: 117; 35:38; Julihn and Horton 40:148-150; herein).
	Spring						See Quartzburg group.
	Spring Hill and Spring Extension						See Mountain King.
	Spring Tunnel						See Georgia Point.
	Squirrel						See Bunker Hill.
229	Standard	Dorothy Clough, 101 Salinas Road, Watsonville	33	4S	18E	MD	A patented property of 21 acres on Sherlock Creek southeast of the Diltz mine and near the Weston placer ground. Long inactive; workings were not identifiable to the authors.
	Star	Not determined	7	3S	18E	MD	Adjoins the Texas Hill mine and is on the west extension of the Carrie Todd vein. Main working is an adit level 150 to 200 feet long with several connecting winzes, but minor stoping only. Equipped with a 10-stamp mill 1890-1902 and there had been one mill previous to 1890. See Texas Hill mine.
	Stepping Stones						See Elizabeth.

LODE GOLD (CONT.)

MAP NO.	CLAIM, MINE, OR GROUP	OWNER NAME, ADDRESS	SEC.	T.	R.	B & M	REMARKS
	Stevenson group						See Red Bank.
230	Stockton Creek and Stockton Creek Tunnel	Not determined; former Las Mariposas Grant mine.	24	5S	18E	MD	Half a mile east and slightly south of Mariposa. Adjoins the Mariposa mine on the southeast. Discovered in the 1850's. Worked intermittently up to 1938. Principal recorded production was between 1900 and 1915 during which time 632.98 tons of ore was mined which yielded $11,996.94 or an average of $18.95 per ton. (Laizure 28: chart facing p. 98.) There are two veins, each about 2½ feet wide. Vein most extensively worked, believed to be an extension of the Mariposa vein, strikes about N. 65° W. and dips 45° SW. Workings in 1928 (Laizure p. 118) consisted of a 135-foot inclined shaft with two levels. Upper level has 250 feet of drifts and the lower level 300 feet of drifts. There is also a stope 3 feet wide, 100 feet long and 70 feet high, presumably driven from the first level (Lowell 14:598; Castello 21:139; Laizure 28:98).
	Stud Horse Flat group						See Live Oak and Governor.
	Succeedo						See River Tunnel and Queen Specimen.
	Sultan						See Sultana group.
231	Sultana group (Royal group, Bald Eagle, Sultan, Sultana, and Sultana Extension)	Thomas E. Faxon and Thomas M. Bains III, Mariposa	29 32	4S	18E	MD	A patented property containing 45.86 acres east of Whitlock Creek and 4 airline miles north and slightly west of Mariposa. Includes Sultan, Sultana Extension, Sultana, and Bald Eagle fractional claims. Operated prior to 1914 by Tresidder, Bains, and Tresidder. Sultana vein trends north and dips 45-60° E. In the Sultana Extension claim several east-trending cutter veins which dip moderately south cross the Sultana vein. The Sultana vein ranges from a few inches to about 18 inches

LODE GOLD (CONT.)

MAP NO.	CLAIM, MINE, OR GROUP	OWNER NAME, ADDRESS	LOCATION SEC.	LOCATION T.	LOCATION R.	LOCATION B & M	REMARKS
	Sultana group (continued)						in width. Wall rocks are massive pyroxene andesite greenstone of unknown age. Developed by a shaft 75 feet deep serving 200 feet of drifts and connected by a drift to a 30-foot deep winze. A small production was recorded in 1936 but the property has been largely idle for several years except for assessment work. The Sultan claim was recently patented, the last claim in the group to achieve this status. Mr. Bains lives on the property (Lowell 16:596). The mine adjoins the Spread Eagle vein on the east.
232	Sunset I (Trujillo) group	Robert B. Trujillo, Route 3, Box 249, Merced	25	4S	16E	MD	At the headwaters of Burns Creek 5¾ airline miles northeast of Hornitos. Accessible via the paved Bear Valley--Hornitos road and 1 mile of unimproved dirt road. Property consists of the unpatented Sunset, South Sunset Extension, La Venue, and Nacimiento del Oro. A pocket mine in greenstone of the Upper Jurassic Amador group. Produced high-grade ore in 1910-12, 1918, and 1933-39-- a total of 757 tons from which 984 oz. of gold and 216 oz. of silver were extracted (U. S. Bur. Mines records). Ore has averaged nearly $27.00 per ton, mostly at the old price of gold.
	Sunset II						See Quail.
	Sunset III						See Mount Buckingham.
	Sunset IV						See Lena Farber and Sunset IV.
	Sunshine						See Martin-Walling.

LODE GOLD (CONT.)

MAP NO.	CLAIM, MINE, OR GROUP	OWNER NAME, ADDRESS	LOCATION				REMARKS
			SEC.	T.	R.	B & M	
233	Sunshine group	Herbert D. Merrill, Mariposa	19 20	4S	18E	MD	On the northeast bank of Sherlock Creek across the canyon from the Golden Key mine. Accessible by Highway 49 and the graded Sherlock Creek road, 11 miles from Mariposa. Adjoins the claims of the Golden Key group. Most of the work on the property was done prior to 1914 when workings consisted of a 100-foot inclined shaft with 280 feet of drifts, a 100-foot raise, and a stope 140 feet long and 100 feet deep (Lowell 16:598). There is also a tunnel of unknown length driven west from the level of Sherlock Creek. There are several veins 1 to 3 feet wide on the property striking generally north and dipping moderately to steeply east. The mine has not been worked recently. Property consists of the Sunshine 1, 2 and 3 lode claims and the Yosemite placer.
234	Sweetwater I	Lena Guisto, Box 66, Angels Camp	29 30	3S	17E	MD	Part of a patented property consisting of 20.38 acres which includes the Southern Cross, Bonanza, and Sweet-water claims. These are all part of the Flyaway group of pocket claims located prior to the 1890's. Quartz vein is parallel to and closely associated with a series of albitite and diorite dikes striking N. 50° W. and dipping 55°·60° NE. Workings consist of several short tunnels and drifts. The dike series generally divides greenstone wall rocks to the east from serpentine wall rocks to the west. Mine is unequipped and idle (Sept. 1954).
235	Sweetwater II	Hudson River Gold Mines, Ltd., c/o H. Vanel, P.O. Box 76, San Rafael; under lease to Clyde Foster of Jerseydale.	17 20	4S	19E	MD	On the west side of Sweetwater Creek 2 airline miles northwest of Jerseydale (Castello 21:140; Laizure 28: 118; 35:45; herein).

LODE GOLD (CONT.)

MAP NO.	CLAIM, MINE, OR GROUP	OWNER NAME, ADDRESS	LOCATION SEC.	T.	R.	B & M	REMARKS
236	Talc (Lacy Cuneo)	Estate of Catharine Cuneo, 1011-22d Street, Merced	19	3S	17E	MD	Located 7 miles southeast of Coulterville at head of Flyaway Gulch. One of the Flyaway group of claims. Patented property consists of 20 acres. Adjoins Alta claim on the west. Workings consist of several short tunnels and drifts and a few shallow open cuts. Vein strikes N. 60° W. and dips 50° NE. Occurs at contact of greenstone, on the northeast, and serpentine on the southwest. Unequipped and idle (Sept. 1954).
	Teats (Teets)						Same as Spread Eagle; see Spread Eagle group.
	Texas						See Texas Hill.
	Texas Gulch (Texas)	Not determined; formerly part of Las Mariposas Grant. Operated by Noel Coutts of Mount Bullion up to 1952.	19	5S	18E	MD	In Texas Gulch close to Highway 140 between Nigger Hill and the Ortega mine. Northwest-trending quartz vein in slate. Active 1902-03 producing 185.54 tons of ore yielding $2689.73 or an average of $14.51 per ton at the old price of gold. Produced about 1100 tons of low-grade ore between 1941 and 1952. Inactive in August 1955.
237	Texas Hill (Texas, Carrie Todd)	G. Ross Frank and Mary E. Shell, c/o W. J. Beatty, Coulterville	7,8	3S	18E	MD	In the Kinsley district 2 miles southwest of Kinsley guard station of U. S. Forest Service. Property includes one patented claim of 20.66 acres (Wilkinson 04: 10; Castello 21:109, Laizure 28:82,89; 35:45; herein).
	Thomas and Bousson	Not determined; former Las Mariposas Grant mine.	17	5S	18E	MD	Located 1 mile east of Mount Bullion, apparently one of several unidentified shallow workings in serpentine. Last active in 1903 producing 22.37 tons of ore yielding $240.65 or an average of $10.77 per ton at the old price of gold (Laizure 28: chart facing page 98).
	Thorn Extension						See Doss.

LODE GOLD (CONT.)

MAP NO.	CLAIM, MINE, OR GROUP	OWNER NAME, ADDRESS	SEC.	T.	R.	B & M	REMARKS
	Tollgate						See Spread Eagle group.
238	Triumph	Mrs. G. C. Griffith, 427 Twenty-Fourth Street, Merced	29	4S	18E	MD	On Whitlock Creek west of the Spread Eagle group, 4 airline miles north of Mariposa. A north-trending, steeply east-dipping quartz vein cuts greenstone wall rocks. Developed by a 150-foot shaft. No recent activity.
	Trujillo group						See Sunset I.
239	Tulito	Arthur Giles, Hornitos	E½ SE¼ 29	5S	16E	MD	A patented property of 20 acres located 3 miles south of Hornitos near the head of Toledo Gulch. An old property long idle.
	Turner	Not determined; former Las Mariposas Grant Mine	24 29 30	5S	17E 18E	MD MD	Active 1902-03, 1906, 1911, 1914-15 producing 309.56 tons of ore yielding $2904.51 or an average of 9.40 per ton. (Laizure 28: chart facing p. 98.) One of several workings in the vicinity of the Ortega and Sorrel mines astride Highway 140. See Ortega mine.
	Twin Springs						See Independence.
240	Tyro	Louise A. Ward and Laurence Eaton, P.O. Box 223, Ross, Marin County	9 10	3S	16E	MD	A patented property consisting of the Rittershoffen claims of 20.15 acres. Adjoins Malvina group on the south and is 1½ miles southwest of Coulterville (Fairbanks 90:39; Storms 94:176; 96:223; Lowell 16:598-599; Laizure 28:119; Logan 35:190; herein).
241	Uncle Jim, Uncle Jim No. 2, and Bear	Robert W. and Emma V. Landon, 2709 North Ontario Street, Burbank	20 29	3S	19E	MD	A patented property of 60 acres on Highway 140 at the confluence of the Merced River and its South Fork. Placered and ground sluiced in gold rush days. Still worked occasionally for both lode and placer gold. Quartz vein system trending northwest crosses the property. There is a short adit toward the south end of

LODE GOLD (CONT.)

MAP NO.	CLAIM, MINE, OR GROUP	OWNER NAME, ADDRESS	SEC.	T.	R.	B & M	REMARKS
	Uncle Jim, Uncle Jim No. 2, and Bear (continued)						the property and other shallow workings. Last recorded production was in 1936 (Laizure 28:119).
	Union and Summit	S.H.F. McGouran, Box 859, Auburn	17 or 21	4S	19E	MD	Adjoins or is in the vicinity of the Early mine.
242	Union Leader group	Jack G. Bell, Coulterville	24 25 19 30	2S 2S	18E 19E	MD MD	Two miles north and slightly west of Little Grizzly Mountain on Deer Flat. Accessible from the Kinsley-Briceburg road by way of the Bull Creek road. Last active in 1938-40 producing a moderate tonnage of ore averaging $20.60 per ton. Quartz vein 1 to 3 feet wide, strikes slightly west of north and dips 20-45° E.
	Vanderbilt						See Mount Buckingham.
243	Venture						See Mary Harrison group.
244	Violet	C. A. Rinde, 1020 Miller Avenue, Berkeley 8	Center 35	2S	17E	MD	A patented property of 40 acres 4½ miles southeast of Greeley Hill and 2 miles east of the McDiermid guard station of the U. S. Forest Service. Quartz vein 1 to 3 ft. wide strikes northeast and dips 45-50° NW. Wall rocks are slate and hornfelsic slate.
	W.Y.O.D.						See Diltz.
245	Virginia (Virginia-Belmont)	Sam E. Wells, 7533 Cecelia Avenue, Downey	13 14	3S	16E	MD	Just west of Highway 49 about 4 miles south of Coulterville. A patented property of 92 acres (Goodyear 90:41; Storms 96:224; Lowell 16:599; Castello 21:141; Laizure 28:119; 35:46; Logan 35:190; herein).
	Virginia-Belmont						See Virginia.

LODE GOLD (CONT.)

MAP NO	CLAIM, MINE, OR GROUP	OWNER NAME, ADDRESS	SEC.	T.	R.	B & M	REMARKS
	Walsh						See Bean Creek group.
246	Washington I (Jenny Lind, Red Cloud)	Mariposa Land Company, 1022 Crocker Building, 620 Market Street, San Francisco	4,5	5S	16E	MD	A patented property just north of the Hornitos-Bear Valley road 1½ airline miles northeast of Hornitos or 2½ miles by road (Storms 96:224; Castello 20:141-142; Laizure 26:138; herein).
	Washington II						See Buena Vista group.
	Washington-Buena Vista						See Buena Vista group.
	Watson vein	Not determined; former Las Mariposas Grant mine	4	5S	18E	MD	On the Whitlock Creek road near the French Camp mine 3 airline miles north of Mariposa. Active in 1903, 1905, and 1910 producing in that period 142 tons of ore yielding $1235.25 or an average of $8.69 per ton. Workings were not identifiable by the authors but there are several narrow, north-trending quartz veins in greenstone in that vicinity.
247	West Rutherford and W. R. Extension	Not determined	22	3S	19E	MD	Adjoins the Rutherford mine. On the north side of the Merced River and Highway 140 four miles west of El Portal. See Rutherford and Cranberry.
	Westward						See Whitlock group.
	Wet Branch	Not determined; former Las Mariposas Grant mine	17	5S	18E	MD	Located 1 mile east of Mount Bullion. One of several small workings is serpentine near Highway 49. Not distinguishable from other workings. Active in 1901 and 1908 producing 69.6 tons of ore yielding $432.50 or an average of $6.22 per ton (Laizure 26: chart facing p. 98).
	White Gulch						See Virginia.

LODE GOLD (CONT.)

MAP NO.	CLAIM, MINE, OR GROUP	OWNER NAME, ADDRESS	SEC.	T.	R.	B & M	REMARKS
248	White Oak and White Oak Extension	Edward Y. Myers, Box 305, Mariposa	27 34	4S	18E	MD	In the Colorado district four-tenths of a mile south of Colorado School. Lies between the Landrum and Mocking Bird mines. Quartz vein 2 to 4 feet wide strikes N. 25-30° W. and dips 70-80° NE. Vein matter is bluish quartz showing abundant chalcopyrite and some pyrite, with occasional blebs of native gold. Vein is accompanied by thoroughly altered, clayey dike-rock. Veins crop out on an old erosion surface and the granitic country rock is deeply weathered. Mining has been badly hampered by caving ground. Workings consist of a 30-foot shaft, surmounted by a headframe and ore bin, and a second shaft 60 feet deep, partly caved. The extent of the lateral workings could not be determined. Ore is commonly high-grade but pockety and generally occurs where cutters strike the main vein.
249	White Oak group						See Live Oak and Governor.
	White Oak and Wilcox	Not determined	18	3S	19E	MD	Six miles northeast of Briceburg near the mouth of Miles Gulch. East of the old site of Yosemite Portland Cement Company's rock crusher and close to the Merced River and Highway 140. Quartz vein about 2 feet wide strikes north and dip 30° E. Wall rocks are slate and slaty limestone intruded by granitic rocks. Developed by a 450-foot adit near river level and several shallow shafts and short tunnels. (Lowell 16:593; Castello 21:142.)
250	White Porphyry group	Not determined; several of the northernmost workings are held by F. E. Ryberg of Coulterville	34	3S	17E	MD	In the Cat Town district 3 miles southeast of Buckhorn Peak or 7½ airline miles southeast of Coulterville. Discussed herein.

LODE GOLD (CONT.)

MAP NO.	CLAIM, MINE, OR GROUP	OWNER NAME, ADDRESS	LOCATION				REMARKS
			SEC.	T.	R.	B & M	
251	White Quartz	Agnes McIver, et al., c/o E. F. Ducommun, 960 Avondale Road, San Marino	15	5S	16E	MD	Located 1½ airline miles east and slightly north of Hornitos and half a mile southwest of the Number Nine mine. A series of parallel quartz veins strike N. 20-50° E. and dip 45-60° SE. Best developed veins have an average width of 2 feet and locally reach 4-5 feet. Wall rocks are quartz-mica schist and massive quartz-mica rock. There has been considerable stoping on three of the veins from pits, tunnels, and shallow shafts. Vein matter is milky quartz containing pyrite, chalcopyrite, bornite, and finely divided native gold. An inclined shaft of unknown depth is surmounted by an old wooden head frame. There was about 75 tons of ore on the dump in August 1954. This property was probably once part of the Number Nine holdings. There had been no recent activity in the summer of 1954. The patented property consists of 11.17 acres.
252	Whitlock group (Consolidated Whitlock and Alabama)	Dr. Frank E. Gallison, P.O. Box 491, Ventura, E. J. Freethy, 1432 Kearney Street, El Cerrito	32	4S	18E	MD	East of the Whitlock Creek road 3½ airline miles northwest of Mariposa and 4½ miles by graded dirt road from Mount Bullion on Highway 49 (Storms 94:176; 96:224; Lowell 16:599; Castello 21:142; Laizure 28:121; herein).
	Wildcat	Not determined; former Las Mariposas Grant mine	16	5S	17E	MD	A small working in the vicinity of the Princeton mine, long idle. Last active in 1908 producing 12.2 tons of ore yielding $132.86 or an average of $6.22 per ton at the old price of gold (Laizure 28: chart facing p. 98).
253	Williams Brothers (Gibbs)	Earl, Ray, Fred and G. H. Williams, El Portal, or Route 1, Box 1061 E, Modesto	31 32	3S	20E	MD	On the southeast slope of Browns Peak 3 miles south and 1 mile west of El Portal. Described herein.

LODE GOLD (CONT.)

MAP NO.	CLAIM, MINE, OR GROUP	OWNER NAME, ADDRESS	LOCATION				REMARKS
			SEC.	T.	R.	B & M	
254	Yellow Metal group	Berniece Jeffery Moss, Box T, Mariposa	16	4S	18E	MD	An unpatented property consisting of three claims, the North Yellow Metal, South Yellow Metal, and Dos Pinos, on Saxton Creek half a mile south of the Merced River and a fourth of a mile northeast of the Schroeder mine. Workings consist of an east-driven crosscut adit and several lesser drift adits. These were driven to strike a steeply dipping north-trending quartz vein cutting slaty metasediments of the Paleozoic Calaveras group. A modest production was recorded in the 1930's under the Yellow Metal Mining Company. This company also operated the White Oak mine near Colorado School for a time in the 1920's and early 1930's.
255	Yellowstone (Sirocco, Yellowstone-Sirocco)	Not determined	19	4S	17E	MD	Located 1½ miles west of Bear Valley just north of the Bear Valley-Hornitos road (Storms 94:195; 96:224; herein).
	Yellowstone Sirocco						See Yellowstone.
	Yosemite						See Number Nine.

PLACER GOLD

MAP NO.	CLAIM, MINE, OR GROUP	OWNER NAME, ADDRESS	SEC.	T.	R.	B & M	REMARKS
256	Agatha	Elsie R. Dovidio, 450 Jones Street, San Francisco	10	4S	18E	MD	On the Merced River at Briceburg. A patented property of 19 acres. Stream bottom and bench gravels were worked in the early days.
257	Ah Wai Drift	Not determined	29, 32	4S	18E	MD	On the southwest side of Sherlock Creek half a mile southeast of the Diltz mine. Boulder-bearing bench gravel, probably Pleistocene. Developed by 2000 feet of ditches. Workings largely inaccessible in 1954. (Lowell 16:600.) Worked by John Hand and W. Weston during World War I period and there were extensive workings prior to that time.
258	Barker Corporation dredge grounds	Not determined	11, 14	5S	16E	MD	Located 2½ miles northeast of Hornitos. Company also operated dredges on properties in Hunter and Bear Valleys during the period 1940-42. Deposits yielded between 17 and 27 cents per yard and grossed over $1,000,000 during the above period (Averill 46:261).
259	Bennett-Pate and Baker dredge ground	Not determined	36, 1,2, 31	7S, 8S, 7S	16E, 16E, 17E	MD, MD, MD	On Mariposa Creek near the point of its debouchment onto the valley floor east of Planada. Close to the K. R. Nutting Company dredge ground. Drilled and sampled in 1934 by Walter C. Vaughn of San Francisco and found to be too low-grade to work by the available equipment (Laizure 35:46). Worked in 1936 by Placer Properties, Inc. of Oregon House, California. Several hundred thousand yards of material were processed that yielded 13 cents per cubic yard. The deposit was about 600 feet wide and 14 feet deep, mainly gravel and cobbles. Bedrock is greenstone, granite, and slate (for a description of the processing plant see Julihn and Horton 40:160).
260	Big Chick	Mrs. Mabel E. Escobar, Coulterville (1950)	1	3S	16E	MD	On Maxwell Creek 2 miles east and slightly south of Coulterville. Produced a small tonnage of material yielding

PLACER GOLD (CONT.)

MAP NO.	CLAIM, MINE, OR GROUP	OWNER NAME, ADDRESS	SEC.	T.	R.	B & M	REMARKS
	Big Chick (continued)						about 1.75 per ton in 1949-50 (U. S. Bur. Mines records).
261	Bull Creek	Not determined	4	3S	18E	MD	Close to the Briceburg-Kinsley-Coulterville road 1 mile north of Kinsley guard station of the U. S. Forest Service. Dredged in 1947.
262	Costa Placer	Ralph and Libbie Dailey, 1165 West 22nd Street, Merced	9 10	4S	16E	MD	In Hunter Valley half a mile west of the Schoolhouse mine. A patented property of 20 acres.
263	Crocker-Ruffman Land and Water Company property	Not determined; probably Thurman Gold Dredging Company, 625 Market Street, San Francisco	8 (?)	5S	16E	MD	On Burns Creek near Hornitos. Operated 1941-42 by Thurman and Wright of San Francisco. Also active 1948-51. Between 1941 and 1951 about 4,843,109 yards of gravel were processed which yielded about $650,000 or an average of about 13.5 cents per cubic yard.
264	Fremont Placer	Pacific Mining Company, 1022 Crocker Building, San Francisco	6	4S	17E	MD	In Hell Hollow near its confluence with the Merced River at Bagby. A patented property of 8.15 acres worked prior to 1900.
265	Fremont Drifting and Mining Company property	George W. and Bessie Welch, Bagby	6	4S	17E	MD	On the Merced River at Bagby. A patented property of 43.68 acres. Worked prior to 1900.
266	Garden (Weston)	John Val and L. P. Pucinelli, Box 316, Angels Camp	33	4S	18E	MD	On Sherlock Creek 1½ miles southeast of the Diltz mine. A patented property of 20 acres. A contemporary stream deposit of gravel and sand without boulders. Considerable coarse gold was recovered. Worked during the early 1900's by W. D. Weston (Lovell 16:600).
	George Placer						See Early under Lode gold deposits.

PLACER GOLD (CONT.)

MAP NO.	CLAIM, MINE, OR GROUP	OWNER NAME, ADDRESS	SEC.	T.	R.	B & M	REMARKS
267	Golden Eagle	Gladys Telfer Brashears, P.O. Box 283, Clovis	3	5S	19E	MD	On Jones Creek 3 miles west of Darrah. A patented property of 60 acres. Gold probably derived from weathered upper parts of quartz veins in the vicinity. No recent activity.
268	Hanas Placer ground	Not determined	9	7S	17E	MD	On Mariposa Creek 2 airline miles northwest of White Rock silica mine. The ground extends 6000 feet along Mariposa Creek and is 9 to 25 feet deep, but the v-shaped canyon is very narrow and only a small area of the 40-acre tract is pay gravel. Bedrock is slate. Much of the gravel averages 50 cents per yard. So far as the authors have been able to determine only part of the gravel was worked by T.D. Hanas in 1938. See Julihn and Horton (40:162) for details of the washing equipment.
269	Huelsdonk Placer	Not determined	5,6	4S	15E	MD	On Browns Creek near the Butterfly mine. Some gold and a little platinum were produced by W. A. Huelsdonk in 1926.
270	Jones Flat Placer	Gladys Telfer Brashears, P.O. Box 283, Clovis	24	4S	15E	MD	On Jones Flat half a mile northwest of the Blue Moon mine or 5½ miles, airline, northwest of Hornitos. Pleistocene bench gravels were worked in early days by hand methods and later by dragline dredge.
271	Jordan dredge ground	R. C. Jordan, Box 277, Ahwahnee (1950)	33 34	5S	19E	MD	On the West Fork of Chowchilla Creek half a mile south of Bootjack. Last worked in 1950 by Jordan.
272	Kockel Placer	Not determined	24	4S	17E	MD	On a tributary to Sherlock Creek 1½ airline miles northwest of the Nutmeg mine. A few pounds of coarse gold nuggets were recovered in 1938 by Traugott Kockel. The largest nuggets weighed 1½ oz. They were taken from the edge of a slide which had buried rich placer ground. (Julihn and Horton 40:162).

PLACER GOLD (CONT.)

MAP NO.	CLAIM, MINE, OR GROUP	OWNER NAME, ADDRESS	SEC.	T.	R.	B & M	REMARKS
273	Mariposa Sand and Gravel Company ground	Mariposa Sand and Gravel Company, Mariposa	25 36	5S	18E	MD	On Mariposa Creek opposite the fair grounds, 1½ miles south of Mariposa. Small production of by-product gold and silver saved in the washing plant of the sand-gravel aggregate operation.
274	New Year Diggins (Prouty Ranch)	Not determined	5,6,7 31,32	4S 3S	15E 15E	MD MD	On Browns Creek and its tributaries 7 airline miles north of Merced Falls. Worked in 1927-28 by George W. Tobin and associates of La Grange (Tacoma Mining Company). Deposit estimated to contain 350,000 to 400,000 yards of gravel ranging from a few feet to 47 feet in depth. Values were in fine gold and a little platinum. Some of the gravel yielded 75 cents to a dollar a yard. Gold is believed to have come in part from the numerous pockety veins in the vicinity (Laizure 28:122).
275	K. R. Nutting Company dredge ground	Not determined	1	8S	16E	MD	Adjoins the Bennett-Pate dredge ground on the west and south. Much of the land is on the level floor of the Great Valley just inside the Mariposa County line. Pay gravel is 700 feet wide and 13 feet deep, on the average. Gold content averaged about 15 cents per yard and recovery was 13 cents per yard. In 1938 about 180,000 cubic yards of overburden (soil) was stripped from the deposit and several hundred thousand yards of gravel processed. (Julihn and Horton 40:167.)
276	C. C. Pierce property	Not determined	19(?)	6S	17E	MD	On Corbet Creek 2 miles east of the Ruth Pierce mine. Trebor Corporation processed 100,000 yards of gravel in 1943 (3½ months) which yielded approximately $10,170 or an average of 10.17 cents per cubic yard (Averill 46:262).
	Prouty Ranch						See New Year Diggins.

PLACER GOLD (CONT.)

MAP NO.	CLAIM, MINE, OR GROUP	OWNER NAME, ADDRESS	SEC.	T.	R.	B & M	REMARKS
277	Schroeder Placer	Chas. M. and C. F. Schroeder, Mariposa	16	4S	18E	MD	On high ground west of Saxon Creek 1½ miles north of the Buena Vista mine. Ground was described by Lovell (16:600) as bench gravel containing clay and boulders. Washed material remaining on the ground in 1954 was bedrock remnants and soil mantle and it is doubtful if this was an old channel deposit. About 5 acres were ground sluiced by the Schroeder family during the period of World War I. At that time there were 5½ miles of ditches and 1500 feet of pipe lines. Water was supplied by Saxon Creek. Much of the ground was very rich.
278	Success Nugget	Ulysses S. Martin, 156 Second Avenue, San Francisco 18	25	2S	17E	MD	Located 5 miles east and slightly south of Greely Hill. Adjoins the Bondurant mine on the north. Ground consisted of soil mantle and the weathered parts of several veins of the Louisiana vein system. Most of the placering was done prior to 1900.
279	Thoms Ranch dredge ground	Not determined	35, 2	7S, 8S	17E, 17E	MD, MD	On border between Mariposa and Merced Counties on Mariposa Creek; 27,000 yards processed in 1950 by Midstate Dredging Company of the Le Grand yielded less than 10 cents per yard.
280	Trebor Corporation dredge grounds	Not determined; formerly part of Mariposa Grant	25, 36	5S	18E	MD	On Mariposa Creek at Mormon Bar. Dredged principally during the summer and fall of 1938 by dragline and floating washing plant. Average depth of gravel was only 6 feet and bedrock pinnacles and large boulders hampered operations; 150,000 to 200,000 yards of material was processed that averaged 22 cents per cubic yard (Julihn and Horton 40:161-162). In 1940-42 the Trebor Corporation operated its dredge on the Fretz, Gaskill, Machado, Trabucco, Turner, and Waltz properties in Hunter Valley and on Agua Fria Creek.

PLACER GOLD (CONT.)

MAP NO.	CLAIM, MINE, OR GROUP	OWNER NAME, ADDRESS	LOCATION SEC.	LOCATION T.	LOCATION R.	LOCATION B & M	REMARKS
281	Upper Bear Creek	Not determined	29 32	4S	17E	MD	Between Bear Valley and May Rock. Thin gravels on slate bedrock. Last worked in 1947.
282	Waltz property	Ed P. Waltz, 19 North Sierra Street, Reno, Nevada	12 7	6S 6S	15E 16E	MD MD	On Burns Creek 3 miles south of the Merced-Hornitos road just east of the Mariposa-Merced County line. Worked by dragline dredge 1940-42 by Trebor and Barker Corporations. Also in 1953 by the Adobe Mining Company of Madera. Produced over 355,000 cubic yards of pay ground averaging about 15 cents per cubic yard in 1940 and 1941.

LEAD

MAP NO.	CLAIM, MINE, OR GROUP	OWNER NAME, ADDRESS	LOCATION SEC.	LOCATION T.	LOCATION R.	LOCATION B & M	REMARKS
	American Eagle						See under Zinc.
	Blue Moon						See under Zinc.

MANGANESE

MAP NO.	CLAIM, MINE, OR GROUP	OWNER NAME, ADDRESS	SEC.	T.	R.	B & M	REMARKS
283	Gale prospect	Not determined	1	3S	15E	MD	On the south side of Highway 132 two hundred yards southeast of the Gale Ranch house and 7.3 miles west of Coulterville. Chert beds striking north and dipping 70° E. contain thin partings and stains of black manganese oxide. The chert, which is white, yellow, or pale red, is part of the Upper Jurassic Hunter Valley formation. This formation extends south-east through the Caldwell mine at Jasper Point. (Wilson, I. F., in Trask, et al. 50:109.) There is one shaft 6 feet square and 20 feet deep, but no systematic development work has been done.
284	Caldwell (Daly)	Not determined	NE¼ 14	3S	15E	MD	Located 1½ miles south of Granite Springs School and 7 miles west of Coulterville via Highway 132. On patented agriculture land (Wilson, I. F. in Trask, et al. 50:106-108).
	Carrier						See Donnelly.
	Daly						See Caldwell.
285	Donnelly (Carrier)	Not determined	17	4S	19E	MD	On the west bank of Sweetwater Creek one-tenth of a mile south of the Donnelly Ranch house, 1 mile north of the Sweetwater gold mine, and 15 miles northeast of Mariposa. Pockets and stringers of black oxides of manganese assaying about 30 percent Mn occur in fine-grained quartzite (metachert) of the Paleozoic Calaveras group of rocks. The metachert is enclosed in mica schist and is cut by quartz veins. The chert belt strikes N. 15° W., dips 70° W., reaches a width of 23 feet and a length of 300 feet. Developed by one open

MANGANESE (CONT.)

MAP NO.	CLAIM, MINE, OR GROUP	OWNER NAME, ADDRESS	LOCATION SEC.	T.	R.	B & M	REMARKS
	Donnelly (continued)						cut. No record of production and no recent activity. (Wilson, I. F., in Trask, et al. 50:108.)
286	Granite Springs	Dr. Robert D. Dunn, 300 Homer Avenue, Palo Alto	11	3S	15E	MD	Located 1 mile south of Granite Springs School near Willow Creek. Geology and mineral occurrence similar to the Caldwell or Daly mine, which lies farther south. A small tonnage of ore was produced in 1954.
287	Kelm prospect	Not determined	19	3S	16E	MD	At Jasper Point 100 yards south of the old Yosemite Railroad bed. Inaccessible by motor vehicle. Close to the Jasper Point rock quarry on the west. Lenticular masses of black manganese oxide are found in a zone averaging about 6 feet in width. Individual lenses average about 1.5 feet in thickness and contain ore between 45 and 50 percent manganese. One ore body is 35 feet long. The ore zone is in thick-bedded red and green chert of the Upper Jurassic Hunter Valley formation. This formation is several hundred feet thick and is underlain by pillow basalt greenstone. The chert strikes N. 0° to 20° W. and dips 35° to 40° E. The ore occurs near the base of the chert section. Ore is exposed in an open cut 8 by 10 feet in plan and 15 feet deep at the highest face. The ore zone is cut by faults of small displacement and the base of the ore zone is sheared. Fifty to 100 feet above the quarry is a second ore zone at about the same horizon as the other but probably not connected with it. A 30-foot tunnel in jasper encountered no manganese ore, failing to reach the ore horizon. There is no record of production and the property is inactive. (Wilson, I. F., in Trask, et al. 50:109-110.)
	M-Q						See Surprise.
	Mebold and Camin						See Strickland.

MANGANESE (CONT.)

MAP NO.	CLAIM, MINE, OR GROUP	OWNER NAME, ADDRESS	LOCATION SEC.	T.	R.	B & M	REMARKS
288	Robie	Not determined	2	3S	15E	MD	On the Robie Ranch 0.8 of a mile west of Granite Springs School and 7 miles west of Coulterville via Highway 132. Thin bands and stringers of black oxide occur in thick-bedded white chert of the Hunter Valley formation. One shallow pit was the only working in 1942 (Wilson I. F., in Trask et al. 50:110).
	Strickland (Mebold and Camin) prospect	Not determined	8	4S	19E	MD	On the west bank of Sweetwater Creek 4.2 miles north of the Sweetwater gold mine, 3.2 miles by road north of the Donnelly manganese prospect, and 16 miles northeast of Mariposa. On patented land. Manganese oxides occur disseminated in gray metachert 13 feet in maximum thickness, but averaging about 5 feet. The chert, which is enclosed in mica schist, strikes N. 60° W. and dips 75° SW. These metasediments are part of the Paleozoic Calaveras group of rocks. In addition to disseminated manganese oxides there are scattered pockets and small lenses of high-grade ore containing 45 percent manganese. A little relict pink rhodonite indicates the character of the primary ore (Wilson, I.F., in Trask et al. 50:111).
289	Surprise (M-Q) prospect	Not determined	E½ 23	3S	17E	MD	On Indian Creek 0.2 of a mile northwest of its confluence with the North Fork of the Merced River, 2 miles southeast of the Quail gold mine and 8 miles northeast of Bagby. Accessible only by trail. Unoxidized rhodonite-rhodochrosite ore occurs in lenticular masses in quartz schist and quartzite. The sheets of quartz schist in turn are enclosed by blue-gray phyllite. These metasediments belong to the Paleozoic Calaveras group of rocks. They strike N. 60°-70° W., dip very steeply east or are vertical. Pink, brown, and buff rhodochrosite-rhodonite occurs with some spessartite garnet. The small amount of black oxide ore present contains 41.3% manganese, 2-8% iron, 8.1% silica, 0.04% sulfur and 0.13%

MANGANESE (CONT.)

MAP NO.	CLAIM, MINE, OR GROUP	OWNER NAME, ADDRESS	LOCATION				REMARKS
			SEC.	T.	R.	B & M	
	Surprise (M-Q) prospect (continued)						phosphorus pentoxide (Taliaferro, N. L., in Trask et al. 50:111-112).

QUICKSILVER

| MAP NO. | CLAIM, MINE, OR GROUP | OWNER NAME, ADDRESS | LOCATION | | | | REMARKS |
			SEC.	T.	R.	B & M	
69	Crystal	Ralph E. and Libbie N. Daley, 1165 West 22d Street, Merced	15 16	3S	16E	MD	A patented property of 17.19 acres. Cinnabar occurs as drusy crystals and blebs in calcite vein matter and in hydrothermally altered greenstone immediately adjacent to the vein. The principal occurrence noted by Turner and Ransome (97:7) is shown on the folio map to be about midway between the intersection points of the Lookout and Cabinet veins. See tabulated list under Gold, the Crystal mine.

SILVER

MAP NO.	CLAIM, MINE, OR GROUP	OWNER NAME, ADDRESS	SEC.	T.	R.	B & M	REMARKS
	Bryan						See Silver Bar.
	Goat Camp prospect	Not determined	14	2S	18E	MD	Located 1½ miles west of Pilot Peak. Quartz vein containing galena-pyrite ore occurs between a 100-foot wide belt of talcose rock, forming the hanging wall, and a foot-wall of chert and schist. Some of the ore ran $30 to $40 per ton, mainly in lead and silver (Walter McLean, personal communication, 1956).
	Goldenrod						See Silver Lane.
	Johnny Green						See under Copper.
	Silver						See Silver Bar.
290	Silver Bar (Bryan, Silver)	Jeffery Investment Company, 7041 Thornhill Drive, Oakland	8,9	6S	19E	MD	Three airline miles south and slightly west of Bootjack and 1 mile north of the Silver Lane mine (Laizure 28:123; 35:44; Julihn and Horton 40:163; herein).
291	Silver Lane	Not determined	8,9 16 17	6S	19E	MD	Four airline miles south and slightly west of the Bootjack and 1/2 mile south of the Silver Bar mine (Laizure 35:44; Julihn and Horton 40:164; herein)
	Silver Lead						See under Gold.
	Wellman Ranch						See Silver Lane.

TUNGSTEN

MAP NO.	CLAIM, MINE, OR GROUP	OWNER NAME, ADDRESS	SEC.	T.	R.	B & M	REMARKS
292	Big Grizzly group	J. E. Diffenbaugh, L. E. Atwood, et al., 2110 Grape Street, Abilene, Texas	17	3S	19E	MD	On the east side of Ned Gulch 1 mile northwest of Clearinghouse on the Merced River. Gold-bearing quartz veins also contain scheelite. Six roughly parallel veins 1 to 3 feet wide cross the property which strike approximately N. 45° W. and dip northeast at angles ranging from 30° to 70°. The veins have been developed by several short adits and open cuts. (Tucker and Sampson 41:480.)
293	Blue Dipper group	L. W. and W. J. Barnett, Box 33, El Portal; John Milanovitch, Box A, El Portal; and Otto Mayer, Box 33, El Portal (1954)	2 11	3S	20E	MD	On steep terrain on the east side of Dry Gulch 3½ airline miles northwest of El Portal. Approached by about 12 miles of dirt road from Crane Flat on Highway 120. Discussed herein.
	Blue Moon group	Otto Mayer et al., Box 357, Mariposa	8?	3S	20E	MD	
294	Blue Spot group	John Milanovitch, Box A, El Portal; Barnett Brothers, Box 33, El Portal	7	3S	20E	MD	On steep terrain 1¼ airline miles northwest of El Portal. Approached by 2¼ miles of steep dirt road from Rancheria Flat on Highway 140. Extensive road construction work to make this group of claims accessible for exploration was never completed and so far as the authors know there has been no production thus far. Scheelite occurs in limy beds near the contact between granitic rocks and a metasedimentary series made up chiefly of massive quartz-biotite rock.
295	Blue Star group	L. W. and W. J. Barnett, Box 33, El Portal	23	3S	19E	MD	Half a mile east of Indian Flat on State Highway 140. Approached by half a mile of dirt road leading south from Highway 140 about half a mile east of Indian Flat guard station of the U. S. Forest Service. Tungsten

TUNGSTEN (CONT.)

MAP NO.	CLAIM, MINE, OR GROUP	OWNER NAME, ADDRESS	LOCATION				REMARKS
			SEC.	T.	R.	B & M	
	Blue Star group (continued)						occurs in tactite along the border of a small limestone pendant in granitic rocks. About 200 tons were mined in 1954 and milled at the Incline Mining Company mill in 1955.
	Donna group	Not determined	NW¼	3S	20E	MD	On Crane Creek 4½ airline miles south and slightly east of Crane Flat on State Highway 140. Approached from Crane Flat by 6½ miles of road and 1 mile of trail. A series of prospects with no known production.
	Early						See under Gold.
296	Garnet Queen group	Sierraposa Mining Company, 508 Fourth Street, San Francisco 7 (Hudson, Law, Toye, and Odgers)	8	3S	20E	MD	Half a mile west of El Portal in the canyon running south from Eagle Peak.
	Lucky Spot group	Clyde Gann, El Portal	8 ?	3S	20E	MD	
297	Maybe	John Milanovitch, Box A, El Portal	8	3S	20E	MD	On the south slope of Eagle Peak half a mile north of El Portal.
	Victory Tungsten	Jacob H. Manzey and Lloyd E. Ashley, Box 93, Groveland (1954)	8	3S	20E	MD	Small production from scheelite in tactite (Tucker and Sampson 42:337).
	War Baby	Not determined	9	6S	21E	MD	Small production of scheelite in tactite. Operated in 1939-40 by Bissett and Rea. May now be held under another name (Tucker and Sampson 42:337).

TUNGSTEN (CONT.)

| MAP NO. | CLAIM, MINE, OR GROUP | OWNER NAME, ADDRESS | LOCATION | | | | REMARKS |
			SEC.	T.	R.	B & M	
298	Washington Tungsten	Miller and Associates, Nelson Cove, Mariposa County	20	6S	20E	MD	Near the East Fork of the Chowchilla River half a mile east of Indian Peak close to the Madera County line. A mill run on the ore has been made and the associations plants increase the mill capacity (California Min. Jour. November 1955).

ZINC

MAP NO.	CLAIM, MINE, OR GROUP	OWNER NAME, ADDRESS	SEC.	T.	R.	B & M	REMARKS
299	Akoz (B.A.B., Radium, Asposozien)	Carlin Estate, c/o Mrs. P. Erickson, La Grange	9,10	4S	15E	MD	Near Webb Station 5½ airline miles north of Merced Falls and 7.8 miles by road from Merced Falls (Wiebelt 47: 1-6; herein).
300	American Eagle (Bullion Hill, Blue Bell, and Bonanza)	Jack I. and Irene Kapp (1/3), 1650 Liberty Street, Santa Clara; Edward C. Morrison (1/3), Karbel, and Vallerde Brothers, Box 568, Portola	30	4S	16E	MD	A patented property of 20 acres 3 airline miles northwest of Hornitos. Adjoins the Blue Moon mine on the south (Castello 21:109; Eric and Cox 47:143-144; herein).
	Asposozien						See Akoz.
	B. A. B.						See Akoz.
	Blue Cloud						See Blue Moon.
301	Blue Moon (Blue Cloud, Red Cloud, Porcupine)	Not determined; last operated by Hecla Mining Company, Wallace, Idaho	19, 30	4S	16E	MD	Located 4 airline miles northwest of Hornitos. Adjoins American Eagle mine on the north (Eric and Cox 48:144-147; herein).
	Porcupine						See Blue Moon.
	Radium						See Akoz.
	Red Cloud						See Blue Moon.

ANDALUSITE

| MAP NO. | CLAIM, MINE, OR GROUP | OWNER NAME, ADDRESS | LOCATION | | | | REMARKS |
			SEC.	T.	R.	B & M	
302	Southwest of Three Buttes deposit	Not determined	SW¼ 17	6S	16E	MD	Astride the Indian Gulch-Planada road 1¼ miles southwest of Indian Gulch. Described herein.

BARITE

MAP NO.	CLAIM, MINE, OR GROUP	OWNER NAME, ADDRESS	SEC.	T.	R.	B & M	REMARKS
				LOCATION			
303	Egenhoff (Devils Gulch)	California Barite Corporation, c/o Edwin Earl, 1102 Rowan Building, Los Angeles	17 21	4S	20E	MD	Located 2 miles northwest of Chowchilla Mountain, near the top of the ridge between Granite Creek Canyon and Devils Gulch 3 airline miles east of Jerseydale (Laizure 28:144; 35:46; Bradley 30:52; Julihn and Horton 40:170; herein).
304 304A	El Portal Barite	El Portal Mining Company, c/o R. J. Groves, Trust Department, 464 California Street, San Francisco	18 19	3S	20E	MD	Astride the Merced River and Highway 140, one and one half miles west of El Portal (Lowell 16:571; Castello 21:102; Laizure 28:144; Bradley 30:51; Young 30: 70-71; Julihn and Horton 40:168-170; Braun 50:130-132; herein).

LIMESTONE & DOLOMITE

MAP NO.	CLAIM, MINE, OR GROUP	OWNER NAME, ADDRESS	LOCATION				REMARKS
			SEC.	T.	R.	B & M	
	Bondshu deposit						See Marble Point deposit.
305	Bower Cave deposits	Not determined	19,20 29,30	2S	18E	MD	Astride the Briceburg-Kinsley-Coulterville road 19 miles from Briceburg (Logan 47:252; herein).
306	Cotton Creek deposit	Not determined	18	4S	16E	MD	Just north of the Cotton Creek arm of Exchequer Lake 3 miles southeast of Granite Springs School (Logan 47:252; herein).
	Emory deposit						See Jenkins Hill deposit.
307	Jenkins Hill deposit (Emory, Yosemite Cement, Timertone, Richardson)	John Richardson, 1124 South Fifth Avenue, Arcadia	7 18	3S	19E	MD	One mile north of the Merced River and 8 miles northeast of Briceburg (Lowell 16:601; Laizure 28:148; Stevenson 27:79; herein).
308	Kinsley deposit	Not determined	4,9	3S	18E	MD	Half a mile west of Kinsley guard station of the U. S. Forest Service near the Briceburg-Kinsley-Coulterville road; discussed herein.
309	Marble Point deposit	Not determined	2	4S	19E	MD	On the south side of the South Fork of the Merced River 1¼ miles south and slightly east of Hite Cove (Lowell 16:501; Logan 47:252; herein).
310	O'Brien deposit	Miss Ethel T. O'Brien et al., 1534 Clay Street, San Francisco	N½ NE¼ 11	4S	18E	MD	One mile southeast of Briceburg (Logan 47:253; herein).
	Richardson deposit						See Jenkins Hill deposit.
	Timertone deposit						See Jenkins Hill deposit.
	Yosemite Cement deposit						See Jenkins Hill deposit.

MAGNESITE

| MAP NO. | CLAIM, MINE, OR GROUP | OWNER NAME, ADDRESS | LOCATION | | | | REMARKS |
			SEC.	T.	R.	B & M	
	Ashworth Ranch deposit (Big Spring Hill)	Not determined	30	5S	19E	MD	Half a mile east of Mormon Bar on Big Spring Hill are several magnesite veins of variable width up to 2 feet. These occur in weathered serpentine adjacent to a spring deposit of siliceous sinter. The occurrence is undeveloped and probably is too small to be of economic importance (Laizure 28:148).
	Big Spring Hill						See Ashworth Ranch.

MICA

| MAP NO. | CLAIM, MINE, OR GROUP | OWNER NAME, ADDRESS | LOCATION | | | | REMARKS |
			SEC.	T.	R.	B & M	
311	Brushy Canyon deposits	Not determined	1, 2	7S	17E	MD	In the White Rock district 2 miles north and slightly west of White Rock School. Discussed herein.

ROCK, SAND, & GRAVEL

| MAP NO. | CLAIM, MINE, OR GROUP | OWNER NAME, ADDRESS | LOCATION | | | | REMARKS |
			SEC.	T.	R.	B & M	
312	Humbug Creek deposits	W. J. Saye, Mariposa	10	6S	19E	MD	On Humbug Creek 2¼ airline miles south of Bootjack. Granitic debris in Humbug Creek. Intermittently operated.
313	Jasper Point	Not determined	19	3S	16E	MD	On the south side of the Merced River at Jasper Point on the old road bed of the Yosemite Valley Railroad. Rock is chiefly yellow and red jasperoid chert of the Upper Jurassic Hunter Valley formation. Beds are over 100 feet wide and stand almost vertical. Operated about time of World War I by Stone Company of San Francisco and by various lessees. Idle because of remoteness from markets. (Lowell 16:602.)
314	Mariposa Sand and Gravel Company	George P., J. G., and E. C. Greenamayer, Mariposa	25 36	5S	18E	MD	On Mariposa Creek 1½ miles south of Mariposa. Described herein.
	Yosemite Stone quarry	Not determined	14	4S	15E	MD	Five miles northeast of Merced Falls on the old roadbed of the Yosemite Valley Railroad. Rock is a massive greenstone about 500 feet wide and standing nearly vertical. Bordered by slaty metasediments. Operated during World War I by Ransome-Crummey Company of Oakland (Lowell 16:603).

ROOFING GRANULES & TERRAZZO CHIPS

MAP NO.	CLAIM, MINE, OR GROUP	OWNER NAME, ADDRESS	LOCATION				REMARKS
			SEC.	T.	R.	B & M	
315	Sonora Marble Aggregate deposit	Sonora Marble Aggregates Company, Sonora	NW¼ 19	3S	17E	MD	On the Bagby grade of Highway 49 about 3 miles northwest of Bagby. Described herein.

SILICA

MAP NO.	CLAIM, MINE, OR GROUP	OWNER NAME, ADDRESS	LOCATION				REMARKS
			SEC.	T.	R.	B & M	
	Le Grande						See White Rock.
316	White Rock deposit	Kaiser Aluminum and Chemical Company, 1924 Broadway, Oakland	14,15	7S	17E	MD	Half a mile north of the Mormon Bar-Le Grande road and 1½ miles west of White Rock School. Operated in the 1920's by Kelm and Carman of Le Grande and from 1942-52 by the present owners or the predecessors in the Kaiser organization (Laizure 28:150; herein).

SLATE

MAP NO.	CLAIM, MINE, OR GROUP	OWNER NAME, ADDRESS	SEC.	T.	R.	B & M	REMARKS
	Cunningham quarries	Not determined	6,7,8 17	7S	17E	MD	Located 7 airline miles south of Catbay astride Mariposa Creek. Some sheets of roofing slate were produced prior to 1914. Idle.
317	Miles Creek deposits	Not determined	2,3 35	7S 6S	16E 16E	MD MD	On the northwest side of Highway 140 about 7 miles from Planada. A series of quarries in the Mariposa formation. Idle for more than 10 years.
318	Nigger Hill deposit	Lee I. Rowland, Mariposa	21	5S	18E	MD	On the south side of Agua Fria Creek and Highway 140 at the north end of Nigger Hill. Discussed herein.
319	Pacific Slate Company quarries	Not determined	6	6S	16E	MD	Located 3 miles southwest of Indian Gulch. Produced sheets of roofing slate prior to 1914. Idle.

TALC

MAP NO.	CLAIM, MINE, OR GROUP	OWNER NAME, ADDRESS	SEC.	T.	R.	B & M	REMARKS
	Cuneo						See Talc.
	Lacy						See Talc.
	Gordon deposit	Not determined	24?	2S	16E	MD	On Greeley Hill 1 mile west of Greeley Hill School. Supplied soapstone building blocks for many of the early buildings in Coulterville (Heizer and Fenenga 48:103).
	Prouty Ranch deposit	Not determined	11	4S	15E	MD	Located 1 mile southeast of Webb Station and 4 miles north of Merced Falls. An undeveloped occurrence (Laizure 28:152) on the Robert Prouty Ranch.
	Shaw Talc deposit	Not determined	32	3S	17E	MD	Located 1½ miles northeast of Bagby in rugged country east of Flyaway Gulch. Claims were held in the late 1920's by Robert Shaw of Sonora. Property is relatively inaccessible and is believed to be undeveloped (Laizure 28:152).
	Talc (Cuneo, Lacy)	Catherine Cuneo estate, 1011 Twenty-Second Street, Merced	19	3S	17E	MD	Located 7 miles southeast of Coulterville at the head of Flyaway Gulch. A patented gold property of 20 acres. Soapstone occurs along the southwest contact of the vein-dike system in the workings of the gold mine. There has been no production and the amount of soapstone present is probably small. The talc grades into serpentine. See under Gold.
	Yosemite Talc	Not determined	31	3S	17E	MD	At the lower end of Flyaway Gulch near Highway 49. Operated in the 1920's by T. B. Elliott of Angels Camp. According to Laizure (28:152) the deposit is 75 feet wide and the talc is white. Workings consist of several open cuts and a 75-foot shaft. The talc contains 61% SiO_2, 32.5% MgO, 2.5% R_2O_3 (R = Al + Fe) and 2.3% water.

o

www.ingramcontent.com/pod-product-compliance
Lightning Source LLC
Chambersburg PA
CBHW060327200326
41519CB00011BA/1864